Milan Konecny · Sisi Zlatanova · Temenoujka L. Bandrova (Eds.)

Geographic Information and Cartography for Risk and Crisis Management

Towards Better Solutions

Springer

Editors
Prof. Milan Konecny
Masaryk University
Dept. Geography
Kotlarska 2
611 37 Brno
Czech Republic
konecny@geogr.muni.cz

Dr. Sisi Zlatanova
Delft University of Technology
OTB, GIS Technology
Jaffalaan 9
2628 BX Delft
The Netherlands
s.zlatanova@tudelft.nl

Dr. Temenoujka L. Bandrova
University of Architecture Civil
Engineering & Geodesy
Dept. Descriptive Geometry
Hristo Smirnenski Blvd. 1
1046 Sofia
Bulgaria
bandrova_fgs@uacg.bg

ISSN 1863-2246 e-ISSN 1863-2351
ISBN 978-3-642-26293-7 e-ISBN 978-3-642-03442-8
DOI 10.1007/978-3-642-03442-8
Springer Heidelberg Dordrecht London New York

Cover Design: deblik, Berlin

Printed on acid-free paper

Springer is part of Springer Science+Business Media (www.springer.com)

Lecture Notes in Geoinformation and Cartography

Series Editors: William Cartwright, Georg Gartner, Liqiu Meng, Michael P. Peterson

Table of Contents

Contributors

Orhan Altan Civil Engineering Faculty, Photogrammetry Division, Istanbul Technical University (ITÜ), Istanbul, Turkey, e-mail: oaltan@itu.edu.tr

Alessandro Annoni European Commission, DG Joint Research Centre, Institute for Environment and Sustainability, Spatial Data Infrastructures Unit, Ispra, VA, Italy, e-mail: alessandro.annoni@jrc.ec.europa.eu

Alessandro Annunziato IPSC JRC, Ispra, VA, Italy, e-mail: alessandro.annunziato@jrc.it

Temenoujka Bandrova Department of Photogrammetry and Cartography, University of Architecture, Civil Engineering and Geodesy, Sofia, Bulgaria, e-mail: bandrova_fgs@uacg.bg

Lars Bernard FR Geowissenschaften, Professur für Geoinformationssysteme, Dresden, TU Dresden, Germany, e-mail: Lars.Bernard@tu-dresden.de

Martin Breunig Institute for Geoinformatics and Remote Sensing, University of Osnabrück, Osnabrück, Germany, e-mail: mbreunig@uni-osnabrueck.de

Monica Brezzi Regional Statistics OECD – Public Governance and Territorial Development Directorate, Organisation for Economic Co-operation and Development, Paris, France, e-mail: monica.brezzi@oecd.org

Manfred Buchroithner Institute for Cartography, Dresden University of Technology, Dresden, Germany, e-mail: manfred.buchroithner@tu-dresden.de

Edgar Butwilowski Institute for Geoinformatics and Remote Sensing, University of Osnabrück, Osnabrück, Germany, e-mail: ebutwilo@uni-osnabrueck.de

Valentina Casola Università di Napoli "Federico II", Dipartimento di Informatica e Sistemistica, Napoli, Italy, e-mail: casolav@unina.it

Mathilde Caspard SERTIT, Illkirch Graffenstaden, France, e-mail: mathilde.caspard@sertit.u-strasbg.fr

Stephen Clandillon SERTIT, Illkirch Graffenstaden, France, e-mail: stephen.clandillon@sertit.u-strasbg.fr

Antonio Condorelli Civil and Environmental Engineering Department, University of Catania, Catania Italy, e-mail: acondor@dica.unict.it

Massimo Craglia European Commission, DG Joint Research Centre, Institute for Environment and Sustainability, Spatial Data Infrastructures Unit, Ispra, VA, Italy, e-mail: massimo.craglia@jrc.ec.europa.eu

Luca D'Onofrio Dipartimento di Informatica e Sistemistica, Università di Napoli "Federico II", Napoli, Italy, e-mail: donofrio.luca@libero.it

Ad de Roo European Commission, DG Joint Research Centre, Institute for Environment and Sustainability, Land Management and Natural Hazards Unit, Ispra, VA, Italy, e-mail: ad.de-roo@jrc.ec.europa.eu

Giusy Di Lorenzo Dipartimento di Informatica e Sistemistica, Università di Napoli "Federico II", Napoli, Italy, e-mail: giusydilor@gmail.com

H. Sebnem Duzgun Department of Geodetic and Geographic Information Technologies (GGIT), Middle East Technical University (METU), Ankara, Turkey, e-mail: duzgun@metu.edu.tr

Sandra Eckert European Commission, Joint Research Centre (JRC), Institute for the Protection and Security of the Citizen (IPSC), Global Security and Crisis Management Unit, TP 267, ISFEREA, Ispra, VA, Italy, e-mail: sandra.eckert@jrc.it

David Fraser School of Mathematical and Geospatial Sciences, RMIT University, Melbourne, VIC, Australia, e-mail: david.fraser@rmit.edu.au

Sérgio Freire e-GEO – Center for Geographical and Regional Planning Studies, Faculty of Social and Human Sciences, New University of Lisbon, Lisboa, Portugal, e-mail: sfreire@fcsh.unl.pt

Lucie Friedmannová Faculty of Science, Department of Geography, Laboratory on Geoinformatics and Cartography (LGC), Masaryk University, Brno, Czech Republic, e-mail: lucie@geogr.muni.cz

Simone Gadenz IPSC JRC, Ispra, VA, Italy, e-mail: simone.gadenz@jrc.it

Daniele A. Galliano IPSC JRC, Ispra, VA, Italy, e-mail: daniele.galliano@jrc.it

Roberto Guana IPSC JRC, Ispra, VA, Italy, e-mail: roberto.guana@jrc.it

Thomas Hahmann German Aerospace Center (DLR), German Remote Sensing Data Center (DFD), Wessling, Germany, e-mail: thomas.hahmann@dlr.de

Manuel Hunold German Remote Sensing Data Center (DFD), German Aerospace Center, (DLR), Wessling, Germany, e-mail: Manuel.Hunold@DLR.de

Lorenz Hurni Institute of Cartography, ETH Zurich, Zurich, Switzerland, e-mail: hurni@karto.baug.ethz.ch

Francisco Igualada GIS Centre at UNLB (CS/DFS) United Nations, Brindisi, Italy, e-mail: FranciscoIgualada@unlb.org

Markus Jahn Institute for Geoinformatics and Remote Sensing, University of Osnabrück, Osnabrück, Germany, e-mail: mjahn@uni-osnabrueck.de

Mikael Jern NCVA - National Center for Visual Analytics, ITN/VITA Linköping University, Norrköping, Sweden, e-mail: mikael.jern@liu.se

Simon Jirka 52°North GmbH, Münster, Germany, e-mail: jirka@52north.org

Serkan Kemec Department of Geodetic and Geographic Information Technologies (GGIT), Middle East Technical University (METU), Ankara, Turkey, e-mail: skemec@metu.edu.tr

Gerhard Kemper GGS GmbH, Kämmererstr, Speyer, Germany, e-mail: kemper@ggs-speyer.de

Thomas Kemper Institute for the Protection and Security of the Citizen, Support to External Security Unit, JRC, Ispra, VA, Italy, e-mail: thomas.kemper@jrc.ec.europa.eu

Norman Kerle Earth System Science, International Insitute for Geo-Information Science and Earth Observation (ITC), Enschede, The Netherlands, e-mail: kerle@itc.nl

Milan Konecny Faculty of Science, Institut of Geography, Laboratory on Geoinformatics and Cartography, Masaryk University, Brno, Czech Republic, e-mail: **konecny**@geogr.muni.cz

Jiří Kozel Faculty of Science, Department of Geography, Laboratory on Geoinformatics and Cartography, Masaryk University, Brno, Czech Republic, e-mail: jirikozel@centrum.cz

Olaf Kranz German Aerospace Center (DLR), Wessling, Germany, e-mail: olaf.kranz@dlr.de

Jukka M. Krisp Faculty of Civil Engineering and Geodesy, Cartography (LFK), Technische Universität München (TUM), München, Germany,
e-mail: jukka.krisp@bv.tum.de

Petr Kubíček Faculty of Science, Institut of Geography, Laboratory on Geoinformatics and Cartography, Masaryk University, Brno, Czech Republic,
e-mail: kubicek@geogr.muni.cz

Melanie Kunz Institute of Cartography, ETH Zurich, Zurich, Switzerland,
e-mail: hurni@karto.baug.ethz.ch

Paul Vincent Kuper Institute for Geoinformatics and Remote Sensing, University of Osnabrück, Osnabrück, Germany, e-mail: pkuper@uni-osnabrueck.de

Barend Köbben Department: Geo-information Processing, International Institute for Geo-Information Science and Earth Observation (ITC), Enschede, The Netherlands, e-mail: kobben@itc.nl

Stefan Lang Z_GIS – Centre for Geoinformatics, Salzburg, Austria,
e-mail: Stefan.Lang@sbg.ac.at

Christophe Lienert Institute of Cartography, ETH Zurich, Zurich, Switzerland,
e-mail: lienert@karto.baug.ethz.ch

Tomáš Ludík Masaryk University, Brno, Czech Republic, e-mail: xludik2@fi.muni.cz

Patrik Lundblad Swedish Meteorological and Hydrologic Institute SMHI, Norrköping, Sweden, e-mail: patrik.lundblad@smhi.se

Laban Maiyo Nairobi, Kenya, International Insitute for Geo-Information Science and Earth Observation (ITC), Enschede, The Netherlands, e-mail: maiyo16561@itc.nl

Sandro Martinis German Aerospace Center (DLR), German Remote Sensing Data Center (DFD), Wessling, Germany, e-mail: sandro.martinis@dlr.de

Masashi Matsuoka GEO Grid Research Group, Information Technology Research Institute, National Institute of Advanced Industrial Science and Technology (AIST), Tsukuba, Japan, e-mail: m.matsuoka@aist.go.jp

Michele Maugeri Faculty of Engineering, Department of Civil and Environmental Engineering, University of Catania, Catania, Italy, e-mail: mmaugeri@dica.unict.it

Nicola Mazzocca Dipartimento di Informatica e Sistemistica, Università di Napoli "Federico II", Napoli, Italy, e-mail: nicola.mazzocca@unina.it

Oliver Meyer Con Terra GmbH, Münster, Germany, e-mail: o.meyer@conterra.de

Ernesto Motta Faculty of Engineering, Department of Civil and Environmental Engineering, University of Catania, Catania, Italy, e-mail: emotta@dica.unict.it

Matthias Müller FR Geowissenschaften, Professur für Geoinformationssysteme, TU Dresden, Dresden, Germany, e-mail: Matthias_Mueller@tu-dresden.de

Eva Mulíčková Masaryk University, Brno, Czech Republic, e-mail: eva_mulickova@yahoo.com

Giuseppe Mussumeci Civil and Environmental Engineering Department, University of Catania, Italy, e-mail: gmussume@dica.unict.it

Eva Ortlieb Geoinformatics Research Group (AGIS), University of the Bundeswehr Munich, Neubiberg, Germany, e-mail: eva.ortlieb@unibw.de

Ulrich Raape German Remote Sensing Data Center (DFD), German Aerospace Center, (DLR), Wessling, Germany, e-mail: Ulrich.Raape@DLR.de

Jaroslav Ráček Masaryk University, Brno, Czech Republic, e-mail: racek@fi.muni.cz

Erminia Raciti Faculty of Engineering, Department of Civil and Environmental Engineering, University of Catania, Faculty Catania, Italy, e-mail: eraciti@dica.unict.it

Carl Reed Open Geospatial Consortium, Inc, USA, e-mail: creed@opengeospatial.org

Mark Reichardt Open Geospatial Consortium, Inc, USA, e-mail: mreinhardt@opengeospatial.org

Wolfgang Reinhardt Geoinformatics Research Group (AGIS), University of the Bundeswehr Munich, Neubiberg, Germany, e-mail: wolfgang.reinhardt@unibw.de

Tomáš Řezník Department of Geography, Laboratory on Geoinformatics and Cartography, Masaryk University, Brno, Czech Republic, e-mail: tom@mail.muni.cz

Monika Rusnakova Faculty of Science, Institut of Geography, Laboratory on Geoinformatics and Cartography, Masaryk University, Brno, Czech Republic, e-mail: rusnakova.monika@email.cz

Gustav Šafr Masaryk University, Brno, Czech Republic, e-mail: safr_gustav@volny.cz

Jesús San-Miguel European Commission, DG Joint Research Centre, Institute for Environment and Sustainability, Land Management and Natural Hazards Unit, Ispra, VA, Italy, e-mail: jesus.san-miguel@jrc.ec.europa.eu

Čeněk Šašinka Faculty of Science, Laboratory on Geoinformatics and Cartography, Institute of Geography, Masaryk University, Brno, Czech Republic, e-mail: cesnekar@seznam.cz

Björn Schilberg Institute for Geoinformatics and Remote Sensing, University of Osnabrück, Osnabrück, Germany, e-mail: bjoern.schilberg@uni-osnabrueck.de

Olga Špatenková Helsinki University of Technology, Geoinformation and Cartography Research Group, Faculty of Engineering and Architecture, Finland, e-mail: olga.spatenkova@tkk.fi

Zdeněk Stachoň Faculty of Science, Laboratory on Geoinformatics and Cartography, Institute of Geography, Masaryk University, Brno, Czech Republic, e-mail: zstachon@geogr.muni.cz

Radim Štampach Faculty of Science, Department of Geography, Laboratory on Geoinformatics and Cartography, Masaryk University, Brno, Czech Republic, e-mail: r.stampach@centrum.cz

Christoph Stasch Institute for Geoinformatics (IfGI), University of Münster, Münster, Germany, e-mail: staschc@uni-muenster.de

Tilmann Steinmetz University of Würzburg, Würzburg, Germany, e-mail: Tilmann.Steinmetz@DLR.de

Christian Strobl German Remote Sensing Data Center (DFD), German Aerospace Center, (DLR), Wessling, Germany, e-mail: Christian.Strobl@DLR.de

Václav Talhofer Department of Military Geography and Meteorology, University of Defense, Brno, Czech Republic, e-mail: vaclav.talhofer@unob.cz

Sven Teßmann German Remote Sensing Data Center (DFD), German Aerospace Center, (DLR), Wessling, Germany, e-mail: Sven.Tessman@DLR.de

Andreas Thomsen Institute for Geoinformatics and Remote Sensing, University of Osnabrück, Osnabrück, Germany, e-mail: athomsen@uni-osnabrueck.de

Dirk Tiede Z_GIS – Centre for Geoinformatics, Salzburg, Austria, e-mail: dirk.tiede@sbg.ac.at

Franz-Xaver Trauner Institute for Soil Mechanics and Geotechnical Engineering, University of the Bundeswehr Munich, Neubiberg, Germany, e-mail: franz-xaver.trauner@unibw.de

André Twele German Aerospace Center (DLR), German Remote Sensing Data Center (DFD), Wessling, Germany, e-mail: andre.twele@dlr.de

Rico Vogel FR Geowissenschaften, Professur für Geoinformationssysteme, TU Dresden, Dresden, Germany, e-mail: Rico.Vogel@tu-dresden.de

Vit Voženílek Department of Geoinformatics, Palacky University Olomouc, Olomouc, Czech Republic, e-mail: vit.vozenilek@upol.cz

Tuong-Thuy Vu Geoinformatics, Royal Institute of Technology, Stockholm, Sweden, e-mail: thuyvu@infra.kth.se

Alexander C. Walkowski Institute for Geoinformatics (IfGI), University of Münster, Münster, Germany, e-mail: walkowski@uni-muenster.de

Rolf Weingartner Institute of Geography, University of Berne, Berne, Switzerland, e-mail: wein@giub.unibe.ch

Monika Wnuk German Remote Sensing Data Center (DFD), German Aerospace Center, (DLR), Wessling, Germany, e-mail: Monika.Wnuk@DLR.de

Andreas Wytzisk 52°North GmbH, Münster, Germany, e-mail: wytzisk@52north.org

Lenka Zajíčková Department of Geoinformatics, Palacky University Olomouc, Olomouc, Czech Republic, e-mail: zajickova.l@email.cz

Gunter Zeug European Commission, Joint Research Centre (JRC), Institute for the Protection and Security of the Citizen (IPSC), Global Security and Crisis Management Unit, ISFEREA, Ispra, VA, Italy, e-mail: gunter.zeug@jrc.it

Sisi Zlatanova Delft University of Technology, OTB, GIS Technology, Jaffalaan 9, 2628 BX Delft, The Netherlands, e-mail: s.zlatanova@tudelft.nl

Introduction

Recent large natural and anthropogenic disasters have clearly shown various shortcomings and failures in existing technologies and policies for efficient early warning and emergency response. Many barriers exist to making data available, providing the most appropriate data and allowing systems to work together. Most currently available geodata are designed, stored and managed by organisations that normally have distinct mandates. Under normal circumstances, these organisations operate largely independently of each other. The presentation of data is adapted to the needs and tasks of particular groups of users. Early warning and emergency management requires different types of systems.

Emergency management begins with an alert for a disaster, continues with management of the critical situation by taking care of people, properties, the environment, etc., and finishes when people, assets and animals are out of danger. There are always specifics in the management of different emergencies: a terrorist attack differs from natural disasters, such as floods, fires, earthquakes, or industrial failures such as a fire in a nuclear plant or the release of dangerous substances. Many aspects remain similar, however: time is critical; the primary emergency responders are the same; the systems share characteristics, etc. In contrast, early warning is more of a monitoring activity that is performed continuously. Data coming from various sensors are constantly evaluated, interpreted, analysed and estimated. The responsible specialists (or early warning systems) alert citizens and the responsible emergency specialists only if anomalies are observed. In a way, early warning is still undervalued by society as national and local governments invest in the development of early warning systems only after large disasters (e.g., tsunamis, earthquakes) have taken place.

Geoinformation techniques have already proven that they can significantly facilitate early warning and emergency response. Nowadays, we can acquire and supply enormous amounts of geoinformation, which underwrite making better decisions throughout the disaster management cycle: in the use of natural resources, in environmental protection or fighting disasters or dealing with their aftermath. The maintenance and accessibility of spatial information has improved as well. Spatial data infrastructures (SDIs) are being developed in many countries, which may be responsible for an increased use of spatial data. There is a growing understanding of the importance of SDIs for early warning and emergency response at very high political levels. Various examples can be found at national and international levels, including GMES (Global Monitoring of Environment and Security) and INSPIRE (Infrastructure for Spatial Information for Europe) in Europe, the activities of the Department of Homeland Security in North America, the International Charter and activities of international institutions such as the World Bank and United Nations (for example the United Nations Office for Outer Space Affairs). International geospatial organisations are contributing to this process by organising conferences and working meetings to discuss new challenges and establish new research agendas. The Ad Hoc Committee on 'Risk and Disaster management' of the Joint Board of Geospatial Information Societies has taken the responsibility for preparing a booklet compiling the best practices for using geo-information in disaster management, which is due for publication by July, 2010.

Clearly, the various systems for early warning and emergency response differ, but they serve the same purpose – saving human lives and reducing property losses. Therefore, the systems must be user-centred and ensure that every person receives the information that will help her/him live and minimise damages to critical infrastructure. Furthermore, it is critical that the two phases use the same information. Both phases should be able to benefit from using the same maps and models, the same sensor data products, and the same visuals. While early warning specialists will rely on the analyses/estimates of in situ sensor information to alert the population, emergency responders will use the predictions and simulations to support better decision making. Although this is not the case at the moment, the editors of this book firmly believe that this will happen in the very near future.

It is often discussed that timely information (able to create a good situational awareness), fast access, unified standards, user-friendly interfaces and appropriate visualisations are some of the factors for success for precise decision making in emergency situations. To be able to develop such systems, many disciplines and technologies must be integrated: sensors, communication and localisation, database management, modelling and analysis, semantics, visualisation and simulation. The role of cartography in particular is very important.

Cartography is one of the disciplines that provides tools and models to improve user perception, orientation, knowledge acceptance and understanding. Visualisation of information is not an isolated element of the information transfer process; it depends not only on the type of data but also on the type of user (gender, age, disability, behaviour, preference, habit, task responsibilities and so on). Many present systems for crisis management use static cartographic tools based on predefined models and visualisations. Methods and approaches used in cartography can greatly help in presenting data by adapting the visualisation with respect to the context of the user, and enhance the decision-making process with visual analytics and simulations. The data must fit user concepts, serve the tasks to be performed and represented with respect to the emergency level. Moreover, a larger range of users must be supported. In addition to systems for specialists (emergency responders in the field and decision makers), the general population should also be taken into consideration. Citizens must be educated in how to use this data (maps and models), whether in analogue or digital form.

We note that cartography can support emerging technologies and the possibilities for customising geoinformation to individuals. Specialists who are developing early warning and emergency response systems often overlook centuries-old cartographic principles (which are aimed at usability and perception) and concentrate too much on technical aspects. They think that as long as the data is displayed on a computer screen, their task is complete. This approach often leads to reduced readability and sometimes to confusion and even misunderstanding. For example, the data may be too dense and unclear on the display, or the symbols and colour composites may be inappropriate and convey the wrong message. Frequently, the tools provided to manipulate maps or 3D models can be clumsy and not suitable for work under stress conditions. Users in emergency response need useful support by maps and not just displays of copious amounts of data that are difficult to see and understand.

The editors believe that change is necessary to encourage the standardisation of symbols for emergency response, context-aware visualisations, visual analytics and tools for interaction and immersion with data. Future maps for emergency management must be better adapted to the individual user. Extended research should concentrate on map usability in both map content and visual controls to make map use more intuitive. The effect of advanced technology is lost when the end users do not understand the information on the display and cannot place it in their mental map.

New cartographic areas are evolving, for example ubiquitous or pervasive mapping including context and adaptive mapping, which are expected to be largely applicable for emergency response. These are based on investigating the personal skills, abilities and preparedness of users to understand and use maps in emergency situations. They can be adapted for all layers of human society: geo-professionals, emergency responders in the field and in command and control centres, emergency managers at various levels, and citizens, young and old, disabled and healthy. The development of digital cartography is strongly influenced by ICT and vice-versa. Digital cartography is enhancing efforts to ensure that cartographic products play a more important role in providing current information and knowledge to society. There is clear cooperation between cartographers and users. A tremendous shift from analogue maps to ubiquitous mapping has been observed. Ubiquitous mapping aims to realise technical solutions for map creation and usage - anytime, anywhere and by/for anyone. Moreover, ubiquitous mapping aims to evaluate and predict the effect that the provision of maps in a ubiquitous manner has on society.

This volume is inspired by, and addresses many of, the topics discussed above. The volume consists of 29 peer-reviewed chapters. These were selected from among the 79 papers presented at the Symposium on Cartography and Geoinformatics for Early Warning and Emergency Management: Towards Better Solutions, held in Prague, Czech Republic, in January 2009. The symposium was jointly organised by the International Cartographic Association Working Group on Cartography in Early Warning and Crises Management (CEWaCM), the Ad-hoc Committee on Risk and JBGIS and the International Society for Photogrammetry and Remote Sensing WG IV/8 3D – Spatial Data Integration for Disaster Management and Environmental Monitoring. The authors of the papers were encouraged to revise, extend and adapt their papers to fit the goal of the book.

The chapters in this volume are organised in three parts: *Geoinformation processing and modelling, GeoInformation Services* and *Advanced cartographic visualisation*. A short overview of the chapters follows.

Geoinformation Processing and Modelling: The 12 chapters in Part I discuss general aspects of using geoinformation in disaster management and address approaches for modelling data. In the first chapter, Annoni et al. provide an extensive discussion of existing data collection and data access technologies and their applicability for early warning and emergency response. Several examples from Europe are used to illustrate the use of these technologies. Altan and Kemper elaborate on the use of geoinformation in all phases of the disaster management cycle, providing appropriate examples from the earthquake - vulnerable area of Istanbul. The remaining nine chapters concentrate on methods and approaches for modelling data. The analysis and preparation of information are particularly critical for early warning and emergency management. Four chapters reflect this aspect. Two chapters present methods for gathering information about a population immediately after a disaster strikes (Zeug et al.) or in advance of a disaster, for better risk estimation and preparation (Freire). Kemec et al. present a framework to help risk managers select data models (3D) in order to better estimate the vulnerability of a given area. Breuning et al. introduce a 3D/4D geodatabase that aims at efficient geodata retrieval. The model developed is tested on landslides. The next two chapters (Condorelli and Mussumeci, and Voženílek and Zajíčková) discuss safe (fast) navigation approaches to support fire fighters. The last four papers present methods for damage estimation. Hahmann et al. use SAR images to detect water bodies (in the case of flooding). Maugeri et al. elaborate on an approach for estimating the hazard to pipelines as a result of earthquakes. Vu and Matsioka present a prototype for quick damage detection based on a Geo Grid system. The prototype has been successfully tested using a QuickBird image of Yingxiu, Sichuan, China. Kranz et al. conclude this Part I with a very detailed overview of the rapid mapping techniques applied for monitoring the evolution of several camps in Darfur between 2002 and 2008.

Geo-information Services: Part II consists of nine chapters that discuss system architectures and applications based on open services and the Web. Reichardt and Reed discuss the heterogeneous challenges in geoinformation and elaborate on the activities of the Open Geospatial Consortium (OGC) towards interoperable solutions. Reznik continues the theme on interoperability of data and services and discusses the results of a comparative study on metadata as described in INSPIRE, ISO and OGC specifications. Two chapters present approaches for enhancing and adapting Web services and systems architectures for emergency management. Based on several existing services, Muller et al. present the results from a successful experiment with a multicriteria evaluation service. Maiyo et al. elaborate on the relevance and importance of Geo Web services for detecting damage and present an appropriate Geo Web service architecture that makes use of satellite images. The next three chapters are devoted to systems for early warning for tsunamis (Raape et al.), managing sensor data (Casola et al.) and landslides (Ortlieb et al.). Muler et al. present an approach to integrate multicriteria evaluation analysis (a kind of a common process model) in SDI. Annunziato et al. express doubts about the success of GSDI for world-wide emergency response activities and present an alternative georeporting system (not intended for SDI). Kozel et al. conclude Part II by reporting their experience in the development, configuration and testing of the Contextual Map Service.

Advanced Cartographic Visualisation: Part III concentrates on research and experiments related to map usability, map perception and map context issues. In the first chapter, Fraser analyses the role of the cartographer and suggests that 'the cartographer leading the mapping for early warning and emergency management must be highly trained, educated and skilled'. The chapter investigates the designers and users of the cartographic products that are developed in six cases: a forest fire and a landslide (Australia), mining (China), a hurricane (USA), a tsunami (Thailand) and a volcanic eruption (Montserrat). Kubicek et al. investigate the structure and tasks of the primary emergency responders (fire fighters, medical service and police) in the Czech Republic and specify data and visualisation tools needed. The authors conclude that the only appropriate approach to meet all of their needs is adaptive mapping, which can be achieved by context-aware services. Two chapters concentrate on cognitive issues. Stachon et al. focus on requirements for users working under stress. The authors present the investigation of different map representations among 74 participants. Bandrova et al. study the knowledge of young people about emergency response procedures, maps and symbols and conclude that they need special maps and symbols. The research was performed in four countries - Austria, Bulgaria, the Czech Republic and Slovakia. The last four chapters present various approaches for improved cartographic visualisation and visual analysis. Jern et al. demonstrate the power of visual analytics through a set of tools developed for the interactive exploration and analysis of data. The authors are confident that such tools will greatly improve the perception and understanding of data. Krisp and Špatenková present an enhanced visualisation (based on kernel density estimation) of fire density. The authors conclude that such approaches give a better overview compared to the traditional dotted approach. Lienert et al. present a framework to support flood management. The framework generates Web-based, real-time visualisations of hydro-meteorological data. The last chapter concentrates on the appropriate symbology for emergency response. The author, Friedmannová, elaborates on the need for specially prepared maps for emergency management and presents a set of symbols designed with respect to user context.

This book is intended for professionals, researchers and students who have interest in access to and visualisation of information. Research areas such as data collection approaches, sensor networks, data processing, 3D reconstruction, database management, spatial analysis and decision support systems are covered in a minor way or not addressed. This book reflects recent challenges, research, and developments in European countries. We hope that this focussed publication will make it even more interesting for colleagues and professionals working in this area of endeavour in other parts of the world.

We are grateful to the contributors for their commitment and hard work. We highly appreciate the support obtained by all of the members of the Scientific Committee and are thankful for their time and valuable comments, which contributed to the high quality of this volume. Our sincere thanks go to Costas Armenakis (Canada), Robert Backhaus (Germany), Budhendra Bhaduri (USA), Rolland Billen (Belgium), Piero Boccardo (Italy), Lars Bodum (Denmark), Volker Coors (Germany), Fatih Döner (Turkey), Marc Erlich (France), Astrid Fichtinger (Germany), Georg Gartner (Austria), Michael Goodchild, (USA), Andy Hamilton (UK), Ramon Hanssen (The Netherlands), Lena Halounova (Czech Republic), Bo Huang (Hong Kong), Umit Isikdag (Turkey), Wolfgang Kainz (Austria), Norman Kerle (The Netherlands), Mike Kevany (USA), Gottfried Konecny (Germany), Corinne Lampin-Maillet (France), Robert Laurini (France), Jiyeong Lee (South Korea), Songnian Li (Canada), Yuan Li, (China), Steve Liang (Canada), Karin Mertens (Belgium), Darka Mioc (Canada), Dev Niyogi (USA), Ferjan Ormeling (The Netherlands), Chirs Phillips (UK), Wolfgang Reinhardt (Germany), Carsten Roensdorf (UK), Massimo Rumor (Italy), Vaclav Slaboch (Czech Republic), Josef Strobl (Germany), Marie Wilhelmsson (Sweden), Laszlo Zentai (Hungary) and Sisi Zlatanova (The Netherlands).

The book itself is the result of a collaborative effort involving 93 researchers located in nine countries in Europe, and Australia, the USA and Japan.

Brno, Czech Republik	Milan Konecny
Delft, The Netherlands	Sisi Zlatanova
Sofia, Bulgaria	Temenoujka L. Bandrova

Part I
Geoinformation Processing and Modelling

Earth Observations and Dynamic Mapping: Key Assets for Risk Management

Alessandro Annoni, Massimo Craglia, Ad de Roo and Jesus San-Miguel

Abstract

The timelines and the accuracy of information provided and used are recognized as being critical when dealing with emergencies. The traditional ways to display much of the information in maps had many limitations, which are now being overcome with major progress in the way data are collected, organized, accessed, and communicated. This chapter provides an overview of current Earth Observation capacity with some examples in relation to forest fires and floods that serve to illustrate the main characteristics including governance and existing limitations. The chapter also discusses how new developments in the way individuals can contribute georeferenced information may be of considerable value in addressing some of the limitations identified, particularly when time is of the essence, as is the case in emergencies.

1 Introduction

The methods used to produce traditional cartography seem insufficient when dealing with emergency response that requires a higher level of dynamism in the creation and dissemination of maps. In order to be really effective in an emergency situation those maps should contain, as far as possible, real-time information and up-to-date data.

Timeliness is crucial and the definitions of 'trust' and 'accuracy' significantly differ from those adopted in traditional cartography. In the emergency phase in fact also data of unknown quality and with a low level of accuracy can play an

M. Konecny et al. (eds.), *Geographic Information and Cartography for Risk and Crisis Management*, Lecture Notes in Geoinformation and Cartography,
DOI 10.1007/978-3-642-03442-8_1, © Springer-Verlag Berlin Heidelberg 2010

essential role, often higher than pre-qualified data and maps. A new framework is therefore necessary to be able to quickly integrate real-time observations including those coming from automatic and human sensors.

The situation is slightly different when dealing with risk prevention or early warning. In these two phases there is more time for the preparation of risks maps. Most of early warning maps are based on models and observations. Their quality is mainly influenced not only by the quality of the static data used, but also by the accuracy of the model and the reliability of the space and in situ observations.

In this specific case an additional element that has to be taken into consideration is the sensitiveness of such maps and in particular those that have a direct impact on citizens (e.g., false alert and panic) and on market (e.g., loss of property value). This obliges mapmakers to take into consideration a wide range of communication aspects (with specific attention to how to visualize/represent sensitive issues). This issue becomes particularly critical when the maps are produced under the framework of a specific legislation and therefore are considered, by definition, a legally certified document.

In this chapter we address some of these aspects focusing mainly on real European experiences. The two examples on forest fires and floods facilitate the understanding of current problems and contribute to identifying future challenges in relation to the production and use of maps in different phases of the risk management cycle.

2 Multiple sources of data to support the different phases of the risk management cycle

There are several aspects that should be considered for the production of maps needed in the different phases of the risk management cycle, including (a) access to static data; (b) availability of earth observations from space or in situ; (c) quality of models to be used for risk evaluation, early warning, and damage assessment; (d) trust and confidence (of observations, data, models, and products); (e) timeliness; and (f) communication.

This chapter focuses on data collection and data access providing an overview of current earth observation capacity with some examples to illustrate its main characteristics including governance and current limitations.

The importance of better information to underpin public policy, business, and social interactions has gained considerable recognition during the last decade. Many governments worldwide have been investing in Internet-based infrastructures to support the sharing and effective use of spatial information and data.

Parallel to these initiatives, private sector-led developments in geobrowsers (e.g., Google Maps, Google Earth, and Microsoft Virtual Earth) have brought geography to hundreds of millions of users as a way to explore the Earth, find and visualize information, and become active participants and producers of data.

The opportunities for increased interactions among citizens, governments, and businesses, as well as the development of a spatially enabled information society have never seemed greater. However, despite billions of sensors collecting data today, data availability and usability still remains a major obstacle together with the difficulty of integrating data from multiple sources. The challenge is in harnessing the many initiatives into a framework that is inclusive and open to new developments, i.e., from one-way interactions from government to citizens, to a participatory framework in which all stakeholders take part in developing a shared understanding and in addressing key issues together. As the interdependencies between governmental and social infrastructures become more complex, their governance requires new ways of thinking and new approaches. We should be prepared to manage the heterogeneity of emerging dynamic systems and be able to transform increasing amount of real-time available data into reliable, accurate, timely, and openly accessible information at the relevant geographic and temporal scales. Risk management is a key test case for this new approach given the complexity of interactions among activities and decisions at multiple levels, local to global, that affect us all.

With these considerations in mind, the following sections review some of the key developments in the way we collect, organize, and share information relevant to risk management. We start from a review of the current state of the art in Europe in relation to fires and floods and then explore some of the new developments that may move us forward.

3 Increasing preparedness and response to forest fires in Europe through the European Forest Fires Information System

The first steps for the development of a European Forest Fire Information System[1] (EFFIS) go back to 1998 when the Member States of the European Union requested the support of the European Commission for the development of a harmonized system for the assessment of forest fire risk in Europe and for the estimation of forest fire damages (San-Miguel-Ayanz et al. 2000).

EFFIS became operational in the year 2000. Since then, EFFIS has been the basis for decision making regarding forest fire prevention and fighting at the European Union level. Comparable fire danger information supports decision making when international collaboration on fire fighting is needed. On the political arena, EFFIS has become the reference system for forest fire information in Europe. It was the basis of European Parliament resolutions following major fire events in the Mediterranean region (e.g., EP Resolutions 2006a, and b), and supported the European Commission Communications on Disaster Reduction and Disaster Prevention of 2008 (European Commission, 2008c) and 2009, respectively.

On average, 60,000 fires occur every year in the European territory burning over half a million hectares of forest area. The development of EFFIS built on the efforts of the European Commission and the Member States in the collection of

[1] http://effis.ec.europa.eu

harmonized forest fire information (EC 1992, 1994). EFFIS, which was the basis of the Forest Focus regulation (EC 2003) in relation to forest fire monitoring, works around a network of 26 European countries which provide base data for the calibration of existing modules and the development of new ones (EC 2007a). The following sections describe two of these modules dealing with the early warning for fire danger and the rapid assessment of fire damages.

3.1 Early warning for fire danger

These modules support the operations of preparedness for forest fire fighting and the intervention to mitigate forest fire damages. They feed the Monitoring and Information Centre[2] of the European Commission which coordinates mutual assistance in forest fire fighting operations in Europe.

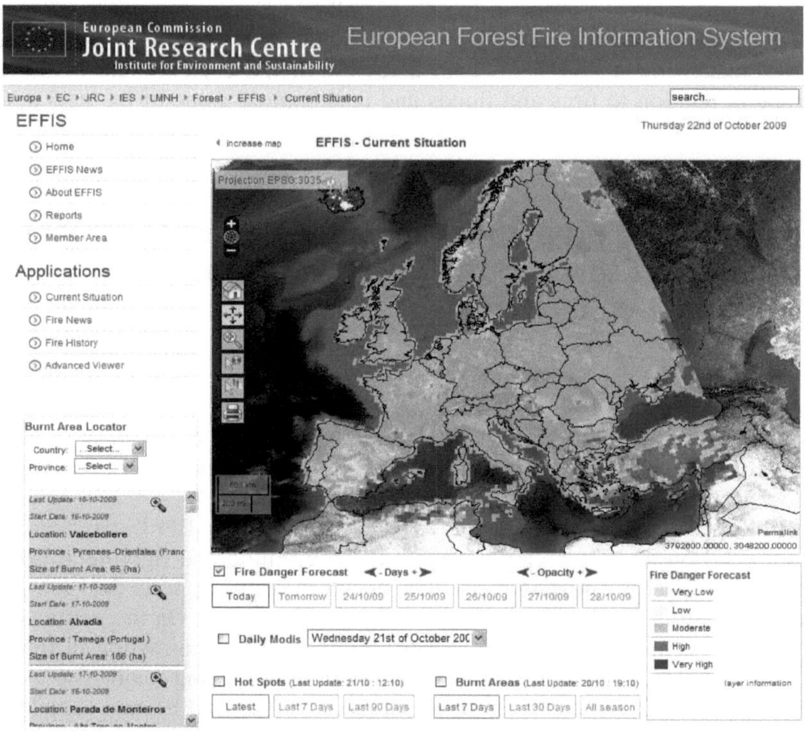

Fig.1. Typical fire danger forecast produced by EFFIS *(source: JRC)*

Fires are a regional phenomenon that affects most of the European countries. Although some countries set up information on forest fires in the early 1990s, this

[2] http://mic.website.env

information was not fully harmonized. Furthermore, national administrations recognized the need to assess fire risk at the pan-European level and to assess the effect of fires throughout Europe in a comparable manner. In those years, the improvements on fire fighting technologies led to an increased cooperation on forest fire fighting, especially in the Mediterranean region. Thus, there was a need to have a comparable assessment of fire risk among the countries which would allow national administrations to decide on the provision of aerial means for fire fighting in other countries. In 2000, EFFIS provided the first harmonized fire danger maps at the pan-European level. In the phase of early warning, EFFIS produces the fire danger forecast up to 6 days in advance(Fig. 1).

This is computed on the basis of calibrated Fire Weather Index with a spatial resolution of approximately 35 km (0.36 degrees). The information is provided as images to the Member States through email and is made available as GeoTIFF data, which allows the incorporation of this information into the national systems. Since EFFIS works on the European ETRS89 Lambert Azimuthal Equal Area coordinate reference system (LAEA 89) (Annoni and Luzet 2001; Annoni 2005), the incorporation into the national systems requires the re-projection of the original images into the national projection systems.

At the pan-European level, fire danger forecast supports operations of mutual assistance and increases preparedness for fire fighting at the European Union level, as the 6-day forecast helps in increasing preparedness when tackling large fire event crisis such as the megafires in Portugal in 2003 or the catastrophic fires in Greece during the summer of 2007.

3.2 Rapid damage assessment of fire damages

Rapid damage assessment of fire damages is critical for quick response to mitigate impacts of the fire events. Information on state of fires and their impact is requested by policy makers and media during and after the fire events. This information is routinely provided to very diverse users. This ranges from the European Parliament and European Commission Institution to national ministries and administrations, including, e.g., the Greek Ombudsman in the case of the casualties in Greece in 2007 where over 60 people were killed. Information is provided to the media whenever agreed by the European Commission institutions and the national administrations.

In the phase of crisis management or rapid response, EFFIS produces up to two daily updates of the perimeters of the fires and the subsequent forest fire damage assessment on the basis of MODIS satellite sensor data with a ground spatial resolution of 250 m. Data are provided through the EFFIS web mapping and web feature services as either images or GeoTIFF data. The perimeters of fires of approximately 40 ha or larger are mapped through a semi-automatic method that includes seed-based identification of hotspots complemented with region-growing algorithms. These fire perimeters are subsequently made available at the EFFIS web interface. This interface permits the overlay of a number of information layers which vary according to the spatial resolution at which the information is viewed in the system.

The system initially provides the perimeters overlaid on the MODIS data. When the user zooms in the area of interest the perimeters are overlaid on Landsat TM data of 25 m spatial resolution. Further zooming enables the display of the burnt areas on Google Earth imagery which allows a high level of detail in most areas in Europe (Fig. 2). Info tools in EFFIS provide online information on the date, the size of the fire, the cover types that were affected by it, and the precise location of the fire in the country/province/commune. This information is updated up to two times a day for active fires, until the burnt area reaches its final perimeter.

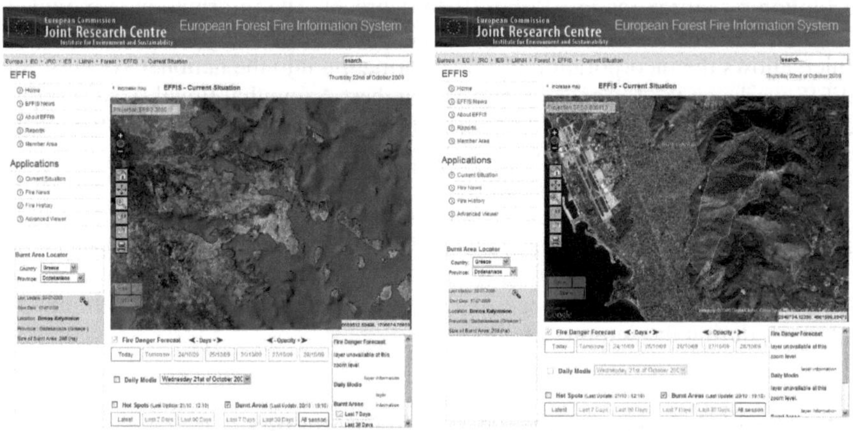

Fig.2. Display of burnt areas on Google Earth imagery *(source: JRC)*

After years of testing (EFFIS is operational since 2000) in which several solutions for data production and storage were tested (description of the storage of data in lat-long for ulterior conversion to a given projection) all the data are currently projected into the LAEA projection. This facilitates the integration with European data sets and the production of harmonized estimates at the European Union level. It should be noted that the EFFIS system is primarily aimed at providing European products. In the case of major events and ad hoc support to the national services, the cartographic products are converted into the national projection systems (Annoni and Luzet 2000; Annoni et al. 2001).

3.3 Communication issues

Information in EFFIS is unique both in the early warning stage and in the rapid damage assessment, as there is no other system providing this at the European level. Information of forest fires is of very sensitive nature. The forecast of extreme danger conditions that were followed by critical fire situations can be interpreted at the national or regional level as a lack of preparedness and can have political implications. The provision of online open information on fire danger was achieved after years of cooperation and negotiations with the EFFIS network. Even now, the provision of fire danger is seen by some as an incentive for arsonists who may use it to produce disastrous fires. However, the knowledge of the level of fire danger by the general public is considered by most as a means to increase awareness and preparedness in the countries.

In the case of the rapid damage assessment provided by EFFIS, the information is as sensitive as in the case of early warning. Although figures are provided online, aggregated estimates of fire damages that are published in the EFFIS newsletters during the fire campaign are only made available to the general public after agreement with the national administrations. In some cases, contradictory figures of damages were utilized by political parties to blame the existing regional or national government and destabilize the political situation. In those cases, EFFIS was used as the neutral source of information that provided reliable estimates across Europe. In major fire events in which countries request the support of the European Commission to mitigate the effects of fires through the European Union Solidarity Fund, EFFIS is used on the European Commission side to provide independent estimates of the areas affected by the fires and the eligible damages.

Next to the existing modules in EFFIS, two important improvements will be available in the near future. One is aimed at the estimation of forest fire emissions and the modeling of smoke dispersion to the nearby populations that can be affected in large fire events. A second improvement refers to the estimation of socio-economic impacts of forest fires. Fires alter drastically the living conditions of the populations and have a direct and strong impact on the local and regional economies. They can alter markets such as those of timber, tourism, and environment-related industries.

4 Increasing preparedness and response to floods in Europe

The following sections illustrate some examples of activities carried out at European level in relation to flood risk prevention, early warning and damage assessment.

4.1 Flood early warning: the European Flood Alert System (EFAS)

Following the devastating floods in the rivers Elbe and Danube in August 2002, the European Commission initiated the development of a European Flood Alert System (EFAS) to increase the flood warning time for transnational river basins (Thielen et al. 2009). EFAS aims at increasing preparedness for floods in transnational European river basins by providing local water authorities with medium-range and probabilistic flood forecasting information 3–10 days in advance, complementing national and regional flood forecasting systems that typically focus on a 1–2 day forecast.

The European Commission Joint Research Centre (JRC) develops EFAS, in close collaboration with the national hydrological and meteorological services. The prototype covers the whole of Europe on a 5 km grid. In parallel, different high-resolution data sets have been collected for the Elbe and Danube river basins, allowing the potential of the system to be assessed under optimum conditions and at a higher resolution.

Flood warning lead times of 3–10 days are achieved through the incorporation of medium-range weather forecasts from the German Weather Service (DWD) and the European Centre for Medium-Range Weather Forecasts (ECMWF).

On a protected website[3] the water authorities can check online the pan-European flood forecasts up to 10 days in advance, which are updated twice a day. Successful alerts are mostly in the range of 4–6 days in advance of the flood.

The flood early warning maps produced (Fig. 3) consist of a gridded map showing which flood alert level is exceeded, e.g., if a moderate/high or major/severe flood is expected. EFAS does not aim to provide accurate flood extent forecasts, but only the location within a river system where problems are to be expected.

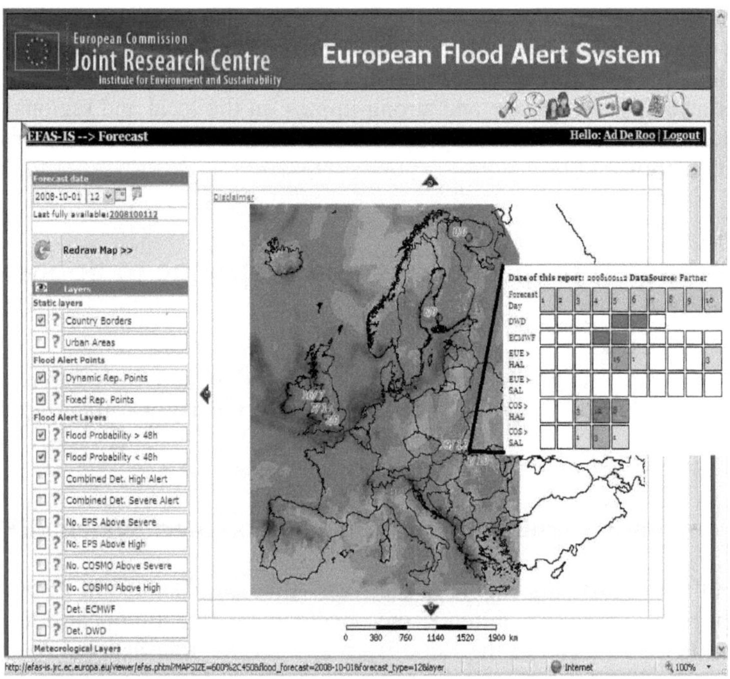

Fig.3. Typical flood early warning maps produced by EFAS *(source: JRC)*

[3] http://efas-is.jrc.ec.europa.eu/index.php

Cartographic problems encountered during the setup and testing of EFAS are numerous, but can be summarized as below:

- Unavailability of an accurate and consistent river network in Europe; this issue is partially solved by the Catchment Characterization and Modeling system (CCM) (Vogt et al. 2007) and the Catchment Information System (CIS) (Hiederer and De Roo 2002) initiatives of the JRC, where now a pan-European network is available, but improvements are still needed.
- Inconsistencies in overlaying Land Cover data (CORINE) with available data on soils from the European Soil Bureau, e.g., open water locations, urban areas; both data sets are necessary data input layers to EFAS.
- Difficulties with obtaining national meteorological and hydrological measurements for calibration, initial conditions, and verification of EFAS. Ongoing JRC initiatives (the EU-FLOOD-GIS project and the ETN-R project) are a step forward (De Roo et al. 2007).

4.2 Flood risk mapping related to European Flood Directive

Another activity carried out by the JRC is pan-European flood hazard and flood risk mapping, partially to be used to compare flood hazard and risk maps to be produced by the European Member States under the obligations of the European Directive on Flood Risk Management (European Commission 2007). One of the main cartographic issues in flood hazard mapping is the accuracy of the Digital Elevation Data used for the flood hazard mapping. Early attempts carried out with GTOPO30 (1 km resolution) and SRTM elevation data (approximately 100 m resolution) prove to be inaccurate for detailed flood hazard assessment. They can serve as a very general overview (Fig. 4).

Recent work using higher-resolution DEMs, e.g., the NEXTmap Europe[4] data (Fig. 5) that aim at pan-European availability, seem to yield much more accurate results, allowing a comparison with maps produced at the national and regional level. Obviously, data quality as obtained with Lidar techniques is the optimum, but this cannot be achieved at Pan-European scale within the next few years.

[4] www.intermap.com

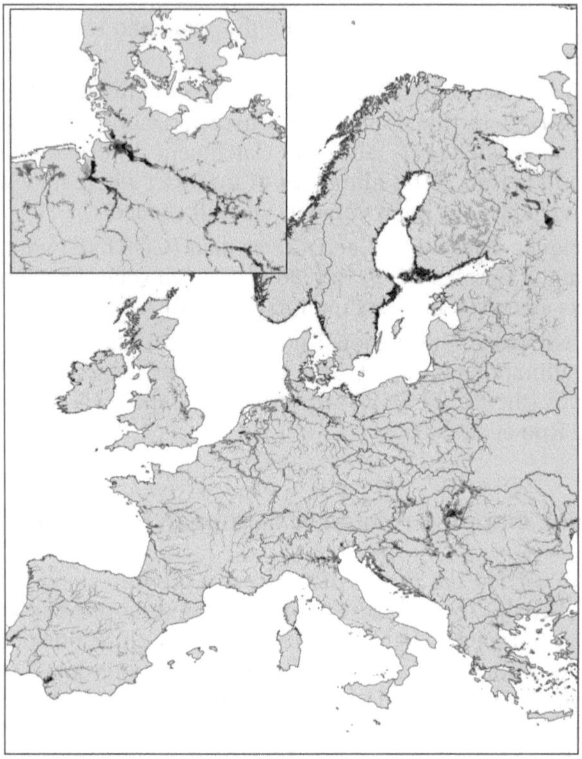

Fig. 4. Pan-European potential flood hazard map based on GTOPO30 elevation data *(source: JRC)*

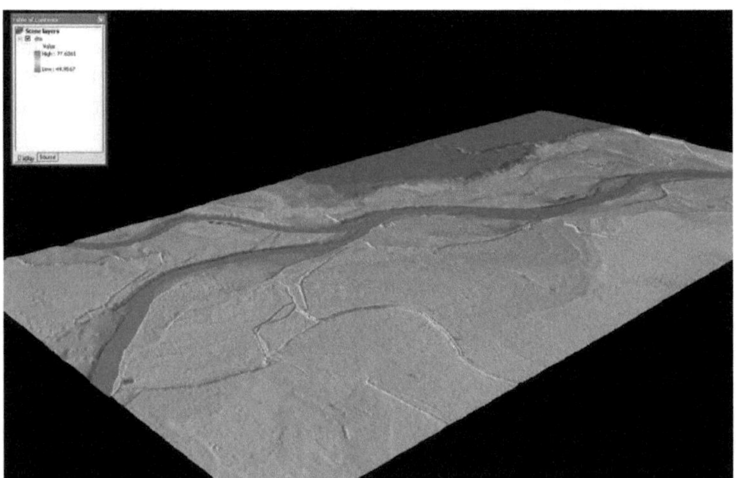

Fig. 5. Flood hazard map for a section of the Po River obtained with NEXTmap DEM data *(source: JRC; DEM data: Intermap)*

4.3 Flood damage mapping related to European Solidarity Fund

Again following the disastrous flooding in the Elbe and Danube in 2002, the European Union established the European Solidarity Fund (EUSF), where countries can request financial assistance from the European Commission following a major disaster in their country, with damages exceeding a certain percentage of the GDP of the country. JRC plays a role in evaluating the requests, by performing a flood damage assessment on the particular flood and comparing it with the claimed damage (Fig. 6). The benefit of this evaluation is in securing correct spending of the EUSF budget and avoiding abuse.

Typical problems encountered here are again the data quality of the DEM used, since that plays a crucial role in estimating water depth. We have used the SRTM DEM, but obviously higher-resolution DEMs would be better. Other problems are the availability of a satellite image showing the flood extent at the right time, e.g., the time of the peak of the flood. In addition, information of flood damage in relation to the water level ('depth-damage-curves') is often of poor quality. To address these issues the following section reviews some recent developments at European and global levels.

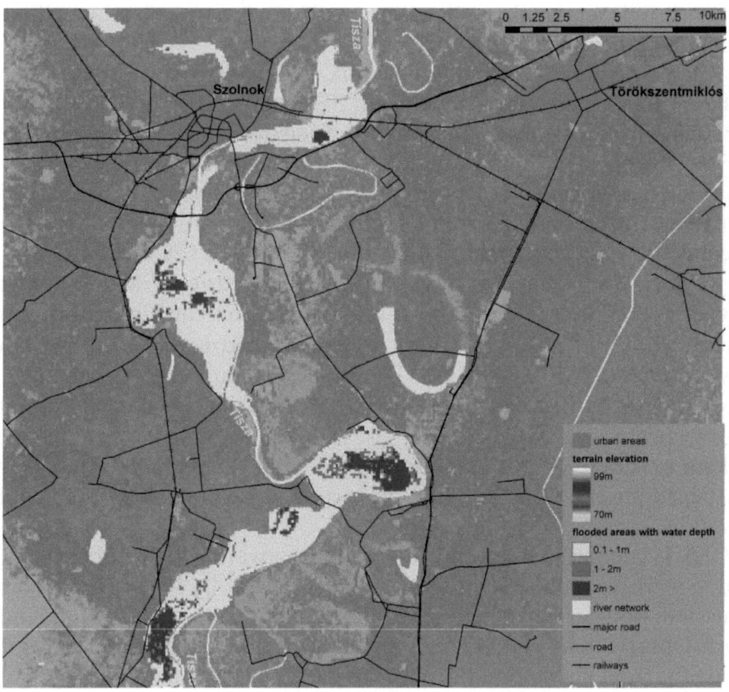

Fig. 6. Flood extent map from the Tisza flood in 2006 *(source: JRC)*

5 Recent developments in geographic information

5.1 Spatial data infrastructures

Many initiatives since the early 1990s have aimed at increasing the availability and accessibility of data through the development of 'Spatial Data Infrastructures' (SDIs). These involve typically the sharing of core data sets of general use, the documentation of existing spatial data sets and services via metadata and related catalogues, and access via distributed Internet-based services under agreed rules and protocols.

In Europe a major recent development has been the adoption of the INSPIRE Directive (EC 2007b), establishing an infrastructure for spatial information in Europe to support community environmental policies and policies or activities which may have an impact on the environment. INSPIRE is based on the infrastructures for spatial information established and operated by the 27 Member States of the European Union. The Directive addresses 34 spatial data themes needed for environmental applications, with key components specified through technical implementing rules.

INSPIRE provides the legal framework and infrastructural layer upon which two other initiatives are being developed: SEIS and GMES.

The Shared Environmental Information System (SEIS)[5] is a collaborative initiative of the European Commission and the European Environment Agency to establish with the Member States an integrated and shared European Union-wide environmental information system linking existing information flows related to European Union environmental policies and legislation (EC 2008a).

The Global Monitoring for Environment and Security (GMES)[6] initiative aims to provide operational information services based on earth monitoring data from satellites and in situ observations on water, air, and land (EC 2008b). While all three initiatives (INSPIRE, SEIS, GMES) pertain to the environmental domain, their complementarities can be portrayed as focusing on data, information, and services respectively.

Linking existing systems and networks to achieve comprehensive, coordinated, and sustained observations of the Earth system is the objective of the Global Earth Observation System of Systems (GEOSS)[7].

GEOSS is a coordinating and integrating network of Earth observing and information systems, contributed on a voluntary basis by 77 nations, the European Commission, and 56 participating organizations of the intergovernmental Group on Earth Observations (GEO), to support informed decision making for society, including the implementation of international environmental treaty obligations.

---

[5] http://ec.europa.eu/environment/seis/

[6] http://www.gmes.info/

[7] http://earthobservations.org

GEOSS aims to achieve comprehensive, coordinated, and sustained observations of the earth system, in order to improve monitoring of the state of the earth, increase understanding of earth processes, and enhance prediction of the behavior of the earth system. GEOSS has several targets, one of them explicitly refers to natural disasters:

> Before 2015, GEO will aim to enable the global coordination of observing and information systems to enable the global coordination of observing and information systems to support all phases of the risk management cycle associated with hazards (mitigation and preparedness, early warning, response, and recovery).

GEOSS will contribute to reducing losses associated with disasters by providing data and information to policy and decision makers for actions associated with disaster preparedness, response, and recovery. One of the best examples of GEOSS in practise is described below.

5.1.1 Argo

There is an increasing concern about global change and its regional impact. Sea level is rising at an accelerating rate of 3 mm/year, Arctic sea ice cover is shrinking, and high-latitude areas are warming rapidly. Extreme weather events cause loss of life and enormous burdens on the insurance industry. Globally, 8 of the 10 warmest years since 1860, when instrumental records began, were in the past decade.

These effects are caused by a mixture of long-term climate change and natural variability. Their impacts are in some cases beneficial (lengthened growing seasons, opening of Arctic shipping routes) and in others adverse (increased coastal flooding, severe droughts, more extreme and frequent heat waves, and weather events such as severe tropical cyclones).

Understanding (and eventually predicting) changes in both the atmosphere and ocean is needed to guide international actions, to optimize governments' policies, and to shape industrial strategies. To make those predictions we need improved models of climate and of the entire earth system (including socioeconomic factors). However, lack of sustained observations of the atmosphere, oceans, and land have hindered the development and validation of climate models.

In 1999, to combat this lack of data, an innovative step was taken by scientists to improve greatly the collection of observations of the oceans. Initiated in 2000, Argo is an international effort to implement a global array of 3,000 free-drifting profiling floats that measures the temperature and salinity of the upper 2000 m of the ocean (Fig. 7). This allows, for the first time, continuous monitoring of the temperature, salinity, and velocity of the upper ocean, with all data being relayed and made publicly available within hours of collection.

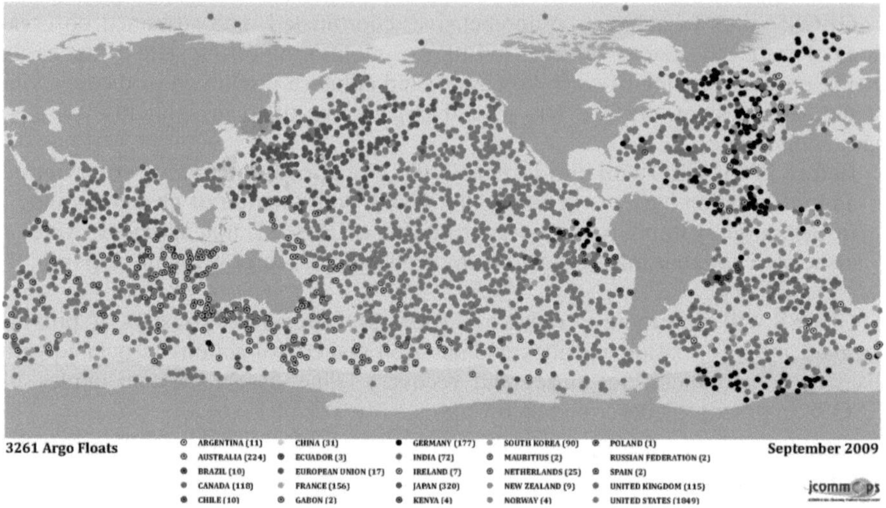

Fig. 7. Argo floats *(source: Argo web site http://www.argo.ucsd.edu/)*

Besides float deployment, Argo has developed two separate data streams: real time and delayed mode. A real-time data delivery and quality control system has been established that delivers 90% of profiles to users via two global data centers within 24 hours. A delayed mode quality control system (DMQC) has been established and 50% of all eligible profiles have had DMQC applied.

Argo is a key example of multinational collaboration to deploy, monitor, and analyze floats and their data. The multinational Argo array is made up of 23 different countries' contributions that range from a single float, to the US contribution, which is roughly 50% of the global array. Funding mechanisms differ widely between countries and involve over 50 research and operational agencies. Each national program has its own priorities but all nations subscribe to the goal of building the global array and to Argo's open data policy.

5.2 Geo-browsers

While the initiatives outlined in the previous section are largely by government organizations, recent developments in the private sector have had a major influence in making geography popular and allowing new forms of data provision. These developments go generally under the name of geo-browsers (e.g., Google Earth, Microsoft Virtual Earth, NASA Worldwind, ESRI ArcGIS Explorer) which use the globe as mechanism to pan, zoom, and fly over the earth's surface to areas of interest. Associated to these 3D representations of the earth are also 2D applications (Google Maps, Microsoft Live Search Maps) that allow users to add and share information via simple application programming interfaces (APIs).

These have enabled a turnaround of the traditional approach in which government agencies (or the private sector) produce data, and citizens (or scientists) are the users, to one in which users themselves become producers. Moreover, the widespread success of these platforms and applications, with hundreds of millions of users, has had the major benefit of popularizing geography among wide strata of the population, as well as enabling companies to add a spatial perspective to their business processes.

5.3 Geosensors

Geosensors can be defined as devices receiving and measuring environmental stimuli that can be geographically referenced. As such they include satellite-based sensors providing multispectral information not only about the Earth's surface (imagery, land cover, vegetation indices, and so on), air-borne sensors for detailed imagery but also for laser scans of physical or man-made structures (LIDAR), and sensors near, on, or under the Earth's surface measuring anything from physical characteristics (pressure, temperature, humidity) and phenomena (wind, rain, earthquakes), to the tracking of animals, vehicles, and people.

Large-scale networks of sensors have been in existence for several decades. What is novel is the web-enablement of these sensors and their networks so that individual sensors can be discovered, tasked, and accessed through web standards (sensor web), and that the networks can exchange information through interoperability arrangements.

A new breed of sensor networks, called wireless sensor networks (WSN), has demonstrated the potential to revolutionize the way we acquire geospatial data. Different from the traditional, large-size, complex, and costly sensor stations, a WSN typically consists of miniature, battery-powered nodes with power-efficient CPUs, short-range radios, and low-cost sensors. The software that runs on the WSN nodes allows them to self-assemble into ad hoc networks, such that the network can be easily deployed (e.g., sensors can be seeded from a low-flying airplane throughout hazardous areas) and data can be relayed across multiple hops and from long distances. This new class of sensing platform will provide an unprecedented volume of real-time geosensor data, along with high spatial and temporal resolution.

5.4 People as sensors

A special type of 'sensors' is that of citizens volunteering geographic information. Goodchild (2007) argues that there is a long tradition of non-specialists contributing to the collection of scientific information such as the case of the Christmas Bird Count[8] or the collection of weather information in the GLOBE programme[9], but that only recently the convergence of greater access to broadband connections, the availability of Global Positioning Systems at affordable prices, and more

[8] http://www.audubon.org/bird/cbc
[9] http://www.globe.gov

participative forms of interaction on the Web (Web 2.0) have enabled vast numbers of individuals to create and share geographic information.

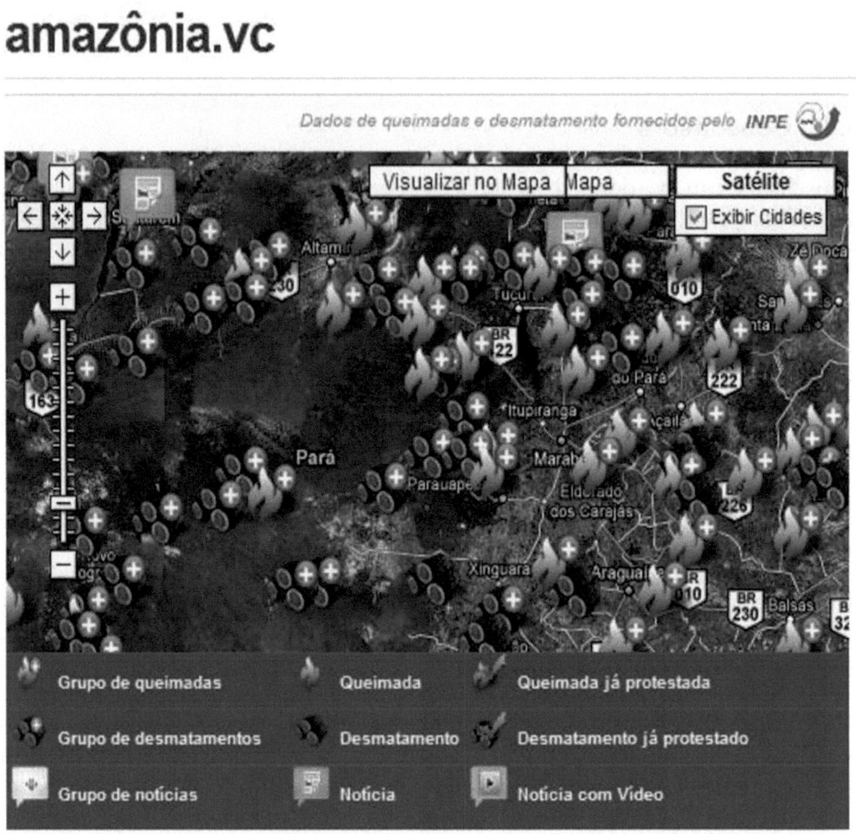

Fig. 8. Reports of instances of illegal logging and forest clearances

Platforms such as Google Maps and Microsoft Live Search Maps have made it possible to publish and make geographically searchable user-generated content to an unprecedented rate. Initiatives such as Wikimapia[10] and OpenStreetMap[11] show how organized volunteered information can challenge traditional data suppliers with good-quality products that are openly accessible to all. As observed by Goodchild, the potential of up to 6 billion human sensors to monitor the state of the environment, validate global models with local knowledge, and provide information that only humans can capture (e.g., emotions and perceptions like fear of crime) is vast and has yet to be fully exploited. An excellent recent example of the opportunities is provided by the Brasilian TV network Globo that opened in

[10] http://www.wikimapia.org
[11] www.openstreetmap.org

September 2008 an interactive website based on Google Maps to monitor the state of the Amazon forest, and provide the opportunity to citizens to report instances of illegal logging and forest clearances (Fig. 8). While having a 'modest' target of 1 million users in the first year, the site[12] already received 41 million reports by individuals in the 3 months to December 2008, making it one of the most successful web sites in Brazil, and obtaining real impact through follow-up legal initiatives and parliamentary enquiries.

Against developments such as the ones described above, current SDIs, and GEOSS, do seem well equipped neither to handle rapidly changing data, nor to consider the major contribution of the public as producers of relevant geographic and environmental data. Aside from technological considerations, there is also a cultural barrier in the perception that only 'official' information, i.e., from certified government sources is reliable, while citizens' information is not. Of course issues of validity, reliability, and trust are crucial, but it is also worth noting here for example that during the forest fires in California in November 2008, in which 230 houses were burned, images and voluntary reports uploaded within minutes on Flickr (Fig. 9) were a key vehicle to alert people of incoming danger and support evacuation, while official information was way too slow, and even Google could not index the relevant information fast enough (Goodchild 2009).

Fig. 9. Tea House Forest Fire: images and voluntary reports uploaded on Flickr

6 Conclusions

This chapter has shown where we are now in respect to risk assessment and forecasting, and highlighted some of the critical cartographic issues including the availability of accurate and timely data.

Governmental or legal initiatives such as the INSPIRE Directive, SEIS, GMES, and GEOSS will be useful mechanisms to unlock existing information under responsibility of public authorities and improve on the gaps and barriers identified. Argo in particular is an excellent example of collaborative efforts at the

[12] www.globoamazonia.com

international scale among different governmental and research organizations to sustain a global operational system.

Despite the progress made, and these examples of best practise, there is a real opportunity to make more of the other recent developments associated with new sensor networks, geo-browsers, and social networks (Web 2.0) allowing individuals to become 'sensors' and providers of timely information as shown in the case of the Tea Fire in California.

A priority therefore is to increase the dialogue and collaboration between government-led and citizens-led initiatives, between 'official' information infrastructures for policy and science, and social networking at the grass roots. It is inconceivable to build an effective and modern risk management system without the active contribution of the citizens who can be affected. Of course quality of the information provided is an issue, but the success of Wikipedia or OpenStreetMap show that if volunteered information is managed through a robust methodology and editorial process, it can achieve the required level of quality.

We must therefore improve our capacity to manage the heterogeneity of emerging dynamic systems and to be more effective in transforming the increasing amount of real-time available data into reliable, accurate, timely, and openly accessible information for risk management and the well being of society.

Acknowledgment

Some theses expressed in this chapter are built on the position paper on Next Generation Digital Earth (Craglia et al. 2008) that came from the specialist meeting on Virtual Globes organized in June 2008 by the Vespucci Initiative for the Advancement of Geographic Information Science and the JRC. We are grateful to Michael F. Goodchild, Gilberto Camara, Michael Gould, Werner Kuhn, David Mark, Ian Masser, David Maguire, Steve Liang, and Ed Parsons who contributed to develop that paper and vision.

References

Annoni A (ed.) (2005) European reference grids: Proposal for a European Grid Coding System, Volume. European Commission, Joint Research Centre, EUR 21494 EN, p. 189.

Annoni A, Luzet C (eds.) (2000) Spatial reference systems in Europe, Volume. European Commission, Joint Research Centre, EUR 19575 EN, 73pp.

Annoni A, Luzet C, Gubler E, Ihde J (eds.) (2001) Map projections for Europe. European Commission, Joint Research Centre, EUR 20120 EN, 131pp.

Craglia M, Goodchild MF, et al. (2008) Next-generation digital earth: A position paper from the Vespucci initiative. IJSDIR 3:146–167. http://ijsdir.jrc.ec.europa.eu/index.php/ijsdir/article/view/119/99.

De Roo A, Thielen J, Maurer T (2007) Harmonised pan-European hydrological and meteorological (meta)-data to support flood studies. In: Schanze J (ed.)

Flood Risk Management Research: From Extreme Events to Citizens Involvement. Leibnitz Institute of Ecological and Regional Development (IOER), Dresden, pp. 122–127.
European Commission (1992) Council Regulation (EEC) No 2158/92 of 23 July 1992 on protection of the Community's forests against fire.
European Commission (1994) Commission Regulation (EC) No 804/94 of 11 April 1994 laying down certain detailed rules for the application of Council Regulation (EEC) No 2158/92 as regards forest-fire information systems.
European Commission (2003) Regulation (EC) No 2152/2003 of the European Parliament and of the Council of 17 November 2003 concerning monitoring of forests and environmental interactions in the Community (Forest Focus).
European Commission (2007) Directive 2007/60/EC of the European Parliament and of the Council of 23 October 2007 on the assessment and management of flood risks.
European Commission (2008a) COM(2008) 46 final 'Towards a Shared Environmental Information System (SEIS)' (http://ec.europa.eu/environment/seis/).
European Commission (2008b) COM(2008) 748 final 'Global Monitoring for Environment and Security (GMES): we care for a safer planet' (http://www.gmes.info/).
European Parliament Resolution (2006a) (P6_TA(2006)0224).
European Parliament Resolution (2006b) (P6_TA(2006)0349).
Goodchild MF (2007) Citizens as voluntary sensors: spatial data infrastructure in the world of Web 2.0. IJSDIR 2:24–32. http://ijsdir.jrc.ec.europa.eu/index.php/ijsdir/article/view/28/22
Goodchild MF (2009, 2–3, March) SDI, the Citizens, and Web 2.0. Presentation at the II Trobada de la IDEC, Barcelona.
Hiederer R, de Roo A (2003) A European Flow Network and Catchment Data Set. Report of the European Commission, JRC, EUR 20703 EN.
San-Miguel-Ayanz J, Vogt J, de Roo A, Schmuck G (2000) Natural hazards monitoring: Forest fires, droughts, and floods – the example of European pilot projects, Surv Geophys 21:291–305.
Thielen J, Bartholmes J, Ramos M-H, de Roo A (2009) The European flood alert system – Part 1: Concept and development. Hydrol Earth Syst Sci 13:125–140.
Vogt JV et al. (2007) Development of a pan-European river and catchment database. In: Peckham R, Jordan G (eds.) Digital Elevation Modelling. Development and Applications in a Policy Support Environment. Springer Berlin Heidelberg, New York, pp. 121–144.

Spatial Information for Disaster Management Using Examples from Istanbul

Orhan Altan and Gerhard Kemper

Abstract

With the passing of each day, the catastrophe risk for urban regions of the world is increasing. Recent events in Northridge USA (1994), Kobe Japan (1995), Marmara Sea, Turkey (1999) and more recently in Sumatra (2004) and Yogyakarta, Indonesia in 2006 are typical examples of what can happen when a major earthquake strikes directly under a densely populated area. Massive human casualties and loss of resources have occurred as a result of the tsunami in South-East Asia. Mega cities created by the rapid urbanization and development in unsafe areas have led to far greater losses than have been experienced in the past. Rapid response in gaining reliable and quick data in these cases is most important for aid management. The Anatolian peninsula is one of the well-known areas endangered by earthquakes. During history many dramatic examples have occurred. In these earthquakes, many people either died or were injured and a lot of damage occurred. In the Sea of Marmara, these earthquakes have also initiated tsunamis which hit the coastline and caused secondary damage. Modern technologies in combination with remotely sensed data in the GIS environment open a wide field for assisting in crisis management. The most important component of any crisis-management system is a crisis preparedness plan where especially our disciplines of photogrammetry, remote sensing, and spatial information science can contribute in many ways. Crisis preparedness plays a key role in protecting the population against disasters. All crisis-management efforts need interdisciplinary cooperation to provide sustainable help for all citizens. We aim to highlight possible contributions by examples of earthquake- and tsunami-risk for Istanbul. Some elements are referred to existing applications already installed or under construction, others are part of our own studies in Istanbul. Crisis-management systems, e.g., can be founded on three columns, the Crisis Preparedness Plan, the Early Warning

M. Konecny et al. (eds.), *Geographic Information and Cartography for Risk and Crisis Management*, Lecture Notes in Geoinformation and Cartography,
DOI 10.1007/978-3-642-03442-8_2,

System, and Rescue and Management Action. International efforts are established to combine and integrate all available data obtained during and after disasters. International cooperation has led to the newly established entity called DMISCO (Disaster Management International Space Coordination), which will be explained in this chapter.

1 Risk-level

Big events like earthquakes and tsunamis do not necessarily cause a high risk potential. Risk for human life depends on the natural conditions in combination with the activities of the population and their society (Heitner, 1969). There are many places where we meet high activity and rapid changes of the environment due to earthquakes, volcanism, tsunamis, weather disasters, and many more. Natural disasters often occur unrecognized in areas apart from the population. Other places bear a high risk to human life. Population growth and the need for agricultural or urban use force populations to make use of these areas. The coastlines are places where fisherman work and live, even though a high risk for tsunamis might exist. Today mainly the urban sprawl has raised the risk level. In the middle ages, Istanbul was situated on the European side of the Bosporus, which is safer than the south-eastern part. Today the city covers the coastline along the Sea of Marmara for a few dozen kilometers and is situated now much closer to the North Anatolian fault. In addition, the densities of urban fabric with houses of several layers enhance the risk level. Strong earthquakes and tsunamis do not appear very frequently. In our fast lifestyle, we forget or ignore such risks. Balancing a risk level is an interdisciplinary task. In the case of earthquakes and tsunamis, we have to cooperate with specialists such as geologists and hydrologists and bring them into contact with city planners and decision makers.

2 Earthquakes near Istanbul

The reason for earthquakes and tsunamis are the movement of geological plates. The North Anatolian fault is one of the biggest and most active tectonic lines. Monitoring of the earthquake history along this fault shows that the epicenters moved from east to west towards the Sea of Marmara. The last strong shock hit Turkey on August 17, 1999 and caused a dramatic disaster. Measured at 7.4 on the Richter scale the tremor was centered between Izmit and Bursa, about 80 km east of Istanbul. This was the most powerful earthquake to ever hit Turkey. More than 15,000 people died, 23,000 were injured, and finally 500,000 were make homeless. Izmit is situated on the North Anatolian fault in the Izmit Bay. This fault leads through the Sea of Marmara just 50 km south of Istanbul's center. The main shockwaves shook and destroyed buildings especially those that were built ignoring rules for safe construction. Most buildings were damaged or destroyed by the S-waves coming a few seconds after the primary ones. The combination of heavy load, big slope, undercut basis of the hills, e.g., by roads and sometimes liquid in the sediments, can lead to landslides as a secondary disaster. A similar effect is liquifaction of the ground that can happen even in flat terrain. A tsunami flew after the earthquake into Izmit bay. With a maximum run-up of 2.5 m along the northern coast of the bay and 1– 2 m on the southern coast, it was a small one and just flooded the area. However, the Sea of Marmara bids a high potential for creating small-and medium-sized tsunamis. Besides the above mentioned, the following usually aggravate the situation. Broken gas-pipelines causing fires that spread quickly in a destroyed city. Industry can initiate environmental disasters like burning oil tanks and petrol leakage. Besides that, criminal activities might begin (see Fig. 1).

Fig. 1. Damage from the earthquake 1999, tectonic subsidence, ground liquification and the tsunami. The ship in the foreground thrown onshore by tsunami wave action. (Kandilli Observatory and Research Institute, www.drgeorgepc.com/Tsunami1999Turkey.html, accessed on 14.04.08)

Sometimes the secondary effects are more destructive than the primary ones. Inside the city of Istanbul many buildings have been destroyed and 3,000 people killed, mainly in the southern part of the mega city and mainly in so-called Gececondo–Districts. Within the last 50 years, Istanbul has grown to a mega city with more than 16 million inhabitants. Rough terrain, forests, and the Black Sea limit the sprawl to the north, so Istanbul expanded to the south on both sides along the coastline of the Marmara Sea. The strongest urban sprawl vector leads to the southeast and already has met the Izmit bay (Altan, and Kemper, 2008). The terrain of Istanbul is hilly and especially along the

Bosporus there are big slopes, a high risk for hang slide as a secondary disaster. The geological survey intensively observes the fault to detect potentially vertical movements. Only strong vertical acceleration might initiate a tsunami wave that grows by approaching the beach. The initial amplitude depends on the shock-intensity, the vertical movement, and the water column over the fault. The path then defines where, at which height, and when a tsunami runs up the coastline. The situation along the Bosporus is critical since the wave can grow by entering this narrow "canyon". The situation as a funnel might increase the height of the waterfront by a factor of 2–3. Reports about tsunamis in Istanbul demonstrate the risk throughout the history (Alpar et al. 2004).

3 Crisis preparedness

A good crisis preparedness plan is the turnkey for any operational crisis-management system and needs highly interdisciplinary work. Usually an inventory of natural and artificial structures and their potential for risk is obligatory and builds its basis. To be effective, they must be part of the city planning in order to take natural risks into account. Like that, it is an important input for the administration. Figure 2. shows a potential run-up analysis of an estimated big tsunami and the population affected. Areas detected as risky have to be treated primarily since we must expect the biggest hit of an earthquake or tsunami there. Organizations that are going to help after the disaster (first aid, fire-fighters, technical Teams…) should be organized without being endangered themselves. GIS data help to detect paths and roads to enter these areas or to evacuate the people. Important for the city and the risk managers is information about the stability of the buildings. To develop a crisis preparedness plan for Istanbul means to cooperate with many disciplines with the assistance of foreign specialists. The integration of all data into a geo-server is essential and must be prepared in a way that decision makers can easily access them for the safety of the society. In a crisis management system, this preparedness plan needs a large amount of data. A good database is the most important criteria for sustainable city planning with respect to risk management as well as the foundation for strategies and management of disasters. Only a complete data collection enables US to set up an early warning system and to organize disaster management. It is important to find acceptance at the population level and to practice behavior in case of earthquake and/or tsunami events.

4 Geo-scientific research

The natural disaster potential must be evaluated by geological and hydrological research. Beside the determination of the potential earthquake centers, the path of the wave-energy must be modeled. In case of an earthquake, the geological structure transports the various waves. In the event of a tsunami, the bathymetric conditions, the vertical water column and the run-up-path are of interest. Geological and hydrological data build the basic layers in a geo-database. Remotely sensed data can assist in detecting significant changes from the air or orbit. Radar data can monitor even very small changes in the terrain that indicate stress in the geological structures. Hyper spectral sensors can assist to in detecting anomalies in the environment, e.g. the emission of thermal heat, gas, or other indicators. This information can also be part of an early warning system. For modeling tsunamis, terrain models of the seafloor, the shore and the coastline must be established. Beside classical hydrological methods e.g. via echo sounder, LIDAR technologies, using water penetrating laser, assist in the off-shore areas for bathymetric measurements. DTM (digital terrain models) and DSM (digital surface models) which include artificial structures are important to compute reliable hydrodynamic run-up simulations. This is very important for tsunami modeling. Aerial surveys use airborne cameras and/or airborne LIDAR sensors and are able to deliver a high-density DTM and DSM. In combination with land use data, risk estimations and generalization of the city into certain risk-levels can be done. Tsunamis cannot be compared with "normal" waves since their energy is extremely high even though the amplitude might be small from the beginning. On the open sea you might not recognize them but their energy shows up when approaching the beach. A typical indicator for a tsunami is the sudden and sustainable falling of the water level where a high front of the tsunami follows. Water can transport material that is then used as "weapons" and increases the destructive force of the wave. Run-up simulation becomes complex when objects or the terrain presses the water into specific directions. The Bosporus builds a funnel within which the water-level can increase several times. The water then runs in a direction not perpendicular to that is the beach; it runs along the shore and hits the objects from the side or even the back. The better the input data the more precise the final model and the results it produces.

5 Risk mapping

Data from the geoscientific survey and research, hydrological models and land use data must be combined with the 3D data in order to achieve spatial risk estimation. The combination of demographic data with urban structural analysis gives a good approximation of a tsunami risk level as shown in Fig. 2. This map was generated using data of the MOLAND Project (Monitoring Land Use/Cover Dynamics) combined with demographic data and terrain models classified for tsunami run-up simulations (Kemper et al. 2005). Even though these estimations are simple, they highlight the risk level. The shown area of Büyükcekmece is covered by residential areas on low terrain 30,000 people can be affect by a tsunami. Such risk maps help city planners and are the basis defining rules for constructing objects in these areas. Those maps support the Crisis Management Team to detect sensitive parts of the city and assist in defining paths for helping the people. Many scientists use GIS combined with remotely sensed data and/or aerial photos to extract the land use map, commonly in combination with spatial or non-spatial ancillary data. Terrain models have various possibilities to contribute to risk mapping. Risk maps help the decision makers to understand the needs for sustainable planning and support an integrated crisis management. Crisis management and the need to redefine city structures more easily can find more acceptance in the population by presenting these risk maps than any other arguments can do. Like that, these maps are of major importance for conveying political decisions needed for a successful crisis management.

Fig. 2. Areas of a certain run-up risk for Tsunamis overlaid with land-use data and population density of residential areas (Kemper et al. 2005)

6 Engineering and architecture

Istanbul has to deal with a difficult "heritage," the so-called Gecekondu areas. These Gecekondu's are illegally built-up areas of residences that were not constructed by engineering rules. Depending on the political situation, they were legalized and then often enlarged by additional stories. These buildings typically are not stable against earthquake shocks and sometimes collapse anyway. Also the foundation anyway is weak especially if the building was enlarged. The technology to construct shockproof buildings is well known in Turkey but only rarely applied. To validate all buildings in Istanbul is an enormous work. The overplanning of the former Gecekondu areas takes place, a good chance for planning new residential areas that consider the risk and make live more safety. There is a need to obtain data of buildings static-stability, their use and their infrastructure. It is good to know how many people remain inside the building at different times of the day. Are emergency exits available and do they really lead to a safer place? Field mapping is very limited for collecting these data. Oblique imaging technologies can assist in the same way as they do for home security analysis. Those images can help engineers and other specialists to validate the building since they enable one to view typical constructive elements of it.

7 City planning and data processing

All information must be integrated in sustainable planning. City planners, administration and decision makers must validate the natural and artificial facts to design a sustainable and long-lasting safer city. City planners deal with various information and must build the interface between geoscientists and the decision makers. They have the knowledge and right feeling about: what can be modified, what is possible in the legal frame and what the right way is to motivate politicians for investing in a "city for tomorrow." As part of the crisis-management system, they are dependent on data other disciplines produce. These data are heterogenic and not ready for easy decision making. Besides traditional development and planning, an entire risk-analysis must be part of enhanced master plans. Other disciplines must be deeply integrated, e.g., geographers, computer scientists and others. Photogrammetry and remote sensing contribute to the GIS application; 3D data and animation in a virtual reality environment contribute to understanding sustainable city development. To simulate different scenarios in 3D helps all involved getting a better understanding of the need for changing existing structures to procedure a safer city. Infrastructural objects such as roads, bridges, pipelines, and dams must be part of a master plan that aims to reduce the risk and show ways to access areas in case of a disaster. Dealing with such amounts of various data needs a powerful server or a server farm with a spatial database. A centralized data server is needed for modeling data and for managing crises. Spatial information sciences use geo-data servers to store these data and to give access to specialists for analyzing, modeling, and to produce new data sets. A geo-data warehouse with a geo-portal manages the access of administrative, scientific, and public parties. GIS and computer scientists continuously optimize the data handling of big data volumes and a large variety of data types via geo-data warehouses. Such data collection enables one to set up an early warning system and to organize a disaster management plan.

8 Decision making and organization

A weak point in all activities is rendering the developed concepts and ideas into real activities and master plans. Good ideas and concepts are lost due to changes of political leadership or through lack of money in the related budget. However, it is easier for decision makers to begin new activities, if the concept is transparent, understandable and meets all aspects in a well-balanced way. Scientists are not good presenters even though they have nice tools to build scenarios and simulations. Too often, we believe that simulations and animations are tools only to attract non-scientists. This technology is able to open doors to the administration and by that; it also hands over a key for accessing the public. If the population is aware of the needs for planning and reorganization of the city, rendering of the plans becomes easier and becomes apart from political competition. This is really sustainable! However, funding is an issue that is on a basis that the scientist can hardly influence.

9 Rescue plans

The Crisis Preparedness Plan must contribute to the rescue planning. Crisis preparedness means to simulate the disaster and to adjust the rescue plans. The geo-scientific data deliver models e.g. estimation of possible destroyed infrastructure by shock waves. Other simulations might deliver run-up simulations and their efficiency on the urban structure. Results are maps that points out where help is needed. The geo-database can then assist in planning the best access to these areas. Where are the roads to access these places, which hospital is the closest, how to get machines and other material there? Where will people go when in panic? What infrastructures can create additional disasters? This extremely interdisciplinary cooperation might result in a "rescue plan" which also can be used to manipulate input variables to improve the city planning and to define rules. A good rescue plan must include the population. They need guidelines on how to behave in a disaster situation and must get training. How should they behave, where should they go and can they assist the rescue teams? Warning must be simple and easily understandable for foreign people. People must be rescued but can become part of the rescue system. Knowing what to do and knowing how to help people reduces panic and makes the rescue easier. Part of this training is the sensitization and information of the citizens. Interactive maps, oblique images and 3D city models with virtual reality simulations assist in learning. Part of the training is to understand natural signatures like pre-earthquake tremors and vibrations or the water run-off at the shore. In addition, there must be a common alarm system.

10 Prediction and early warning

On the basis of the crisis preparedness system described above, prediction and early warning systems need additional information derived from measurements, monitoring or automated sensors. Perfect communication plays the key role. An operational communication, e.g., special channels in the GSM or satellite phones, are essential since wire-based communication technologies frequently are destroyed. Pre-designed data are needed to give in the right alerts at the right time on the right place. Sensors have to communicate with a central organization, but in some cases direct access to the warning is more efficient, e.g., in Japan, fast trains are stopped automatically by an alarm sign and gas pipelines are closed immediately. A prepared alarm-chain has to be activated by the sensors. We have to be aware that theses things have to be tested and trained by the crisis-management teams and by the population to know what must be done if a disaster happens. In 2005, the tsunami disaster at Bandah Aceh had neither a sufficient alarm system nor did the population know how to behave. A big problem of this tsunami disaster was the transport of "weapons" by the water. The transported wooden boards, cars, and many other things made even a 2-m flood extremely dangerous and destroyed more buildings than expected by water only. It must be part of an early warning system to fix such "weapons", e.g., group can it to road blocks, close shops, remove dangerous things inside... An early warning or forecast system must be based on a good preparedness plan. Early warning is a difficult task for earthquake shocks since the reaction time is extremely short and the activity must be designed as an automatic procedure. For Istanbul also the tsunami warning needs an automated workflow since the wave can hit the beach already after 10–30 minutes. Early warning surely has limits by these short timeframes, a good preparedness however can give at least a chance to save lives and prepare for a rapid rescue. Forecast is difficult but can assist to set a first alarm level.

11 Prediction

Earthquake prediction by existing ground-based facilities are unreliable and the earthquakes in recent years point to the need for scientific progress in solving this problem and in employing additional evidence for earthquake prediction. Geodetic science plays an important role in earthquake research (Aksoy 1995). By means of long-term measurements, deformations caused by deformations in the Earth's crust caused by the movement of tectonic plates can be examined. In a recent study Murai and Araki (2005) used data from GPS stations to study the daily change ratio and sudden changes of signs over triangular networks. The big Sumatra earthquake could have been detected and predicted from this data. The evidence of the likelihood of the earthquake was found in the daily change ratio of the triangular area (Singapore–Lhasa–Kunming) in the northing and height coordinates lane 8 days before the earthquake and on the (Indonesia–Singapore–Lhasa) in easting and height coordinates plane 2 and 5 days respectively before the earthquake. Sergey and Kirill (2003) have proposed a concept for a geo-space system to predict and monitor earthquakes and other natural disasters, which is based on monitoring the ionosphere and magnetosphere of the Earth for short-term forecasting of earthquakes. It involves the investigation of the interaction between the ionosphere's F layer variations and variations occurring in the circum-terrestrial environment associated with seismic activity detected by means of ground-based and satellite monitoring. Remote sensing tools can assist too and support geologists to detect stress in the rocks. Radar-sensors in airborne or space-borne platforms can detect even small changes in the surface and indicate stress by using interferometer methods. Rumors in the crustal zone can be interpreted as a warning of shifts in the lithosphere. Changing of vaults, initialized hang slides and other phenomenon show ongoing structural changes. Beside that, "biological sensors" frequently give a short-term prediction for an upcoming event. These "sensors" might assist as a redundant warning system especially in remote areas (see Fig. 3).

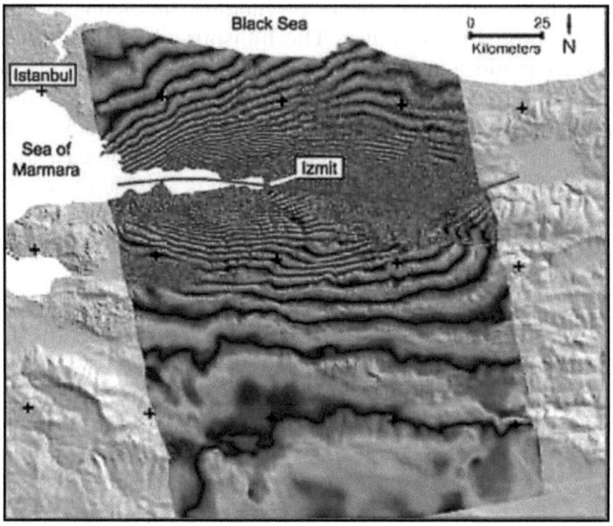

Fig. 3. This interferogram was created from two sets of radar data: The first was obtained before the Izmit earthquake, and the second 35 days after (comet.nerc.ac.uk/images, Accessed on 11.04.2008)

12 Early warning

Real-time observation needs mechanic or electronic devices. Various innovative sensors, already applied in many areas are seismographs, which detect small rumors and vibrations that might indicate an upcoming shock. They are useful to set a pre-alarm but must be handled carefully. Even if the population is trained; every warning creates fear and panic. The main shock must be detected immediately and the alarm chain started automatically. Sensors based on acoustic, accelerative, or other detectors can clearly identify these shocks and validate their strength. In case of an earthquake 100 km south of Istanbul, the P-wave might arrive within 15 seconds, while the S-wave would approach 10 seconds later. A real-time communication is essential in combination with an automated reaction system. Tsunami waves run slower and hit the beach 15–30 Minutes after the shock. Sensors in the Sea of Marmara are just under installation. They provide warning via wireless communication. 15 minutes is a short time, but enables one to start the alarm-chain (see Fig. 4).

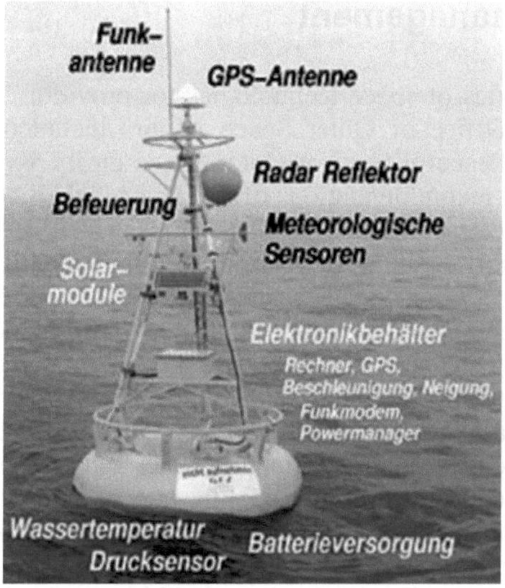

Fig. 4. Tsunami Sensors installed close to tectonic fault zones for detecting Tsunami waves when they start, (http://www.forschung.bmbf.de/de/4879.php accessed on 18.04.08)

13 Automated alarming and communication

Automated actions are needed to deal with the short time frame. In 25 seconds is difficult to estimate as it is a relatively short time frame .e.g. to stop trains and to close pipelines to limit fire disasters. Such actions must be prepared by using the crisis preparedness plan. They must be well tested. The robustness of the alarm communication is extremely important. public alarms can consist of sirens, radio information, warning SMS, and many others. All citizens must be educated to recognize these signs and must be trained in reacting in the right way. Relatively small things can save hundreds of lives. A key issue is communication. Beside communication between sensors and the Crisis Center with its servers, communication is needed to set the alarm and to maintain contact with the rescue teams. Radio transceivers, GSM and satellite phones keep the communication for the crisis management going. Using cellular networks data communication can be possible as long as the transmitters are powered. The power supply at the transmitter and repeater stations are critical factors in case of major earthquakes.

14 Disaster management

Noting the capabilities of space technologies for providing assistance in times of disasters, the UN Office of Outer Space Affairs included in the report of the Committee on the Peaceful Uses of Outer Space on its 5-year review of the implementation of the recommendations of the Third United Nations Conference on the Exploration and Peaceful Uses of Outer Space (UNISPACE III), submitted to the General Assembly at its 59th session, the implemention of an international space coordination body for disaster management (now referred to as DMISCO). At that session the General Assembly agreed "that a study should be conducted on the possibility of creating an international entity to provide for coordination and the means of realistically optimizing the effectiveness of space-based services for use in disaster management and that the study should be prepared by an ad hoc expert group, with experts to be provided by interested Member States and relevant international organizations." It is expected that the resources for the core of the work for the establishment of DMISCO will be undertaken at the United Nations Vienna office (three staff), with contributions in cash from Member States (for facilities, operational costs, and staff), together with in-kind contribution (such as facilities provided by a hosting Member State) and secondments of experts. Additionally, funds will be needed to support the implementation of projects identified in conjunction with National Focal Points (NFPs) and will be defined and secured on a case-by-case basis. DMISCO has been operational since 1st January 2007, and hopefully it will contribute to considerably reducing the impacts of future disasters.

15 Importance of digital archives and space data

With the vast experience gained through operational use of space data, the concept of a space based observation and communication system for disaster management is evolving in different applications. The most important need is to assess the overall requirements of users at various levels and the delivery mechanisms that could effectively provide the services for monitoring, forecasting, warning, assessment, prediction and reduction of natural disasters. The information required by disaster managers in each of the critical phases of disaster management, which includes mitigation and preparedness, response and recovery/relief, consist of

1. database design
2. near real time monitoring/ mapping
3. modeling framework
4. networking solutions
5. multiagency interface.

The success of disaster management largely depends on availability, dissemination and effective use of information. The information needs to include current information on weather, infrastructure (roads, hospital, and administration boundaries), demography etc. to assess the disasters. Currently such data are being generated by multiple users and stored in multiple formats and media, making it difficult to bring the data together to support disaster-management activities. In addition, there is a need to assess the disaster in terms of location, extent and likely impact, so as to plan relief and recovery actions. An integrated system adequately equipped with necessary infrastructure and expertise to constantly monitor the risk profiles on all possible disasters, and maintain a national database, will become relevant. In this context, the GIS technique offers a tool to analyze multiple layers. A pilot study was carried out by the Indian Space Research Organization in 1998–2001, to design a prototype system that will integrate space inputs with conventional data. The study area selected was the Brahmaputra floods in Assam. The system consisted of comprehensive database design, space based near real-time monitoring tools, modeling framework, networking, and a user interface. With appropriate synthesis of these core elements, flood monitoring and damage assessment was carried out. Through the use of networking, the space-based inputs were disseminated to the users. The study has led to a realistic assessment of the gaps in the current system and conceptual framework for a disaster-management system. Earth observation satellites have demonstrated their utility in providing data for a wide range of applications in disaster management. Pre-disaster uses include risk analysis and mapping; disaster warning, such as cyclone tracking, drought monitoring, the extent of damage due to volcanic eruptions, oil spills, forest fires, and the spread of desertification; and disaster assessment, including flood monitoring and assessment, estimation of crop and forestry damage, and monitoring of land use/change in the aftermath of disasters. Remotely sensed data also provide a historical database from which hazard maps can be compiled, indicating which areas are potentially vulnerable. Information from satellites is often combined with other relevant data in geographic information systems (GIS) in order to carry out risk analysis and assessment. GIS can be used to model various hazard and risk scenarios for planning the future development of an area (UN 2004) as demonstrated by the following:

1. Disasters such as floods, earthquakes, forest fires, oil spills, drought, and volcanic eruptions affect large parts of the globe and coordinated international efforts are required to minimize their impacts. Disaster relief requires timely and updated geo-social databases and situational analysis for the various phases of the disaster.
2. Space technology such as remote sensing and meteorological satellites, as well as communications and navigation and positioning systems, can play a vital role in supporting disaster management by providing accurate and timely information and communication support.
3. The utilization of space assets in support of disaster management continues to lag significantly in most parts of the globe and remain as a major challenge; however, there are several international efforts aiming to address the developmental needs and achieve effective utilization of space technology.
4. A considerable gap, however, exists and is likely to remain in all areas of space technology applications for disaster management, including technical, operational, education/training, and organizational areas, unless a global, integrated, coordinated approach is taken. In virtually all countries, there is a

lack of understanding of the benefits of the use of space technologies to support risk reduction and disaster-management activities, especially by the disaster managers and civil protection agencies.

Getting the latest information on the affected area is crucial for managing the rescue teams. Today push broom scanners but also small, medium, and large format digital cameras are combined with precise GPS/IMU orientation systems that allow rapid and fully automated data extraction. These techniques are nowadays small and easy to install and can be adjusted to many aircraft. Any aircraft is able to carry such a system to capture images which are more than documentation only. These data are as that precise as aero triangulated images but enable rapid updating of the database. Many automated tools, mainly designed for remotely sensed data, can extract changes automatically. With only small personal assistance, they can indicate where e.g. destroyed buildings are. Within a few hours, taken imagery and their analysis can be integrated in the data server and become accessible to the rescue and managing teams. Using space-borne data is limited by the repetition rate of the satellite platforms. Airborne hyperspectral sensors can assist perfectly in getting relevant data since they enable a huge combination of spectral bands that can indicate much more than an image alone can do. LIDAR data of the destroyed areas can deliver very dense DSMs, which easily can be validated with the existing DSM data on the server to produce a change map. It is important to produce data as fast but not as accurate as possible. The resolution should meet the requirements only.

16 Conclusions

To be successful in managing disasters, good crisis preparedness, an optimal prediction and early warning and a well-defined management must be established. All efforts must be well distributed over the three columns “preparedness, prediction and disaster management”. Most effective however is the preparedness. If this first and key aspect is not worked out at an early stage all other efforts will not work. Spatial data and their proper handling in geo-data servers provides the chance to combine interdisciplinary work. Today still the various kinds of data sources, the different needs and the various points of view hamper an optimal progress in disaster management. We have many possibilities to achieve data, to store them and to process results. We must take this chance to build up strategies in science as well as in practice in order to address the policy and the decision makers. To solve existing problems is not a technical job, it is a political task. Good early warning systems have strong linkages between different elements. The major players concerned with these elements meet regularly to ensure they understand all of the other components and what other parties need from them. The main element of this chain is information (especially the spatial information) on the disaster area. In former disasters the importance of the need of accurate, timely and information over wide areas has been understood afterwards. A helpful application of geo information technologies requires a solid base of political support, legal frameworks, administrative regulations, institutional responsibility and capacity, and technical training. Early warning systems have to be part of disaster management plans and policies. Preparedness to respond is to be engrained into public awareness.

References

Aksoy A (1995) Kurzer Überblick über die Vermessungsarbeiten in der Türkei. In Altan O, Lucius E (eds.) Neue Technologien in der Geodäsie, Publication House of the Tech Univ. Ist., Istanbul, pp. 1–10.

Alpar B, Gazioglu C, Altinok Y, Yücel ZY, Dengiz S (2004) Tsunami hazard assessment in Istanbul using by high resolution satellite data (Ikonos) and DTM. – XXth Congress of the ISPRS, Istanbul.

Altan O, Kemper G (2008) System architecture for earthquake, Tsunami preparedness and warning. The international archives of the photogrammetry, Remote Sensing and Spatial Information Sciences XXXVII, Part B4: pp. 1061–1071, Beijing.

Heitner KL (1969) A Mathematical Model for Calculation of the Run-Up of Tsunamis. Pasadena, California.

Kemper G, Altan O, Celikoyan M (2005) Tsunami-Gefährdungskarte für Istanbul. – Angewandte Geographische Informationsverarbeitung XI – Agit 2005, Wichmann, Heidelberg.

Murai S, Araki H (2005) Was early warning of Sumatra earthquake possible? Coordinates Magazine July 2005, pp. 8–11.

Sergey P, Kirill B (2003) Ionospheric Precursors of Earthquakes, Springer, Verlag.

UN Office or Outer Space Affairs (2004, 23 December) Final report of the action Team on Disaster Management, Document A/AC.105/C.1/L.273.

Geology and Earthquake Risk, http://www.jpgp.jussieu.fr accessed on 01.04.2008.

Tectonic Structures and Earthquake risk, http://www.jpgp.jussieu.fr accessed on 26.12.2004.

Newspaper Report in Izmit Earthquake, http://arrowsmith510.asu.edu/Exercises/Exercise3/ accessed on 20.03.2008.

Earthquake Risk of Istanbul, http://www.akut.org.tr accessed on 19.04.2008.

Damage of the Earthquake, www.drgeorgepc.com/ Tsunami1999Turkey.html accessed on 14.04.2008.

Tsunami Warning, http://www.enchantedlearning.com/subjects /tsunami/ accessed on 18.04.2008.

Tsunami Sensoring, http://www.forschung.bmbf.de/de/4879.php accessed on 18.04.2008.

Radar-Interferometry, comet.nerc.ac.uk/images accessed on 11.04.2008.

Rapid Population Maps for Crisis Response

Gunter Zeug, Olaf Kranz and Sandra Eckert

Abstract

Population numbers are crucial information in the aftermath of a natural disaster. Questions like how many people are affected, how many survived and how many would need prolonged assistance are key issues for crisis response and reconstruction. However, information and accurate figures on where people live and population characteristics are not always available, especially in developing countries.

A methodology was developed to rapidly estimate population distribution and density in disaster-affected areas. It is based on earth observation satellite imagery with high resolution and the classification of the built-up areas therein conducting a textural analysis. In a second step latest available census data is taken and interpolated on the built-up classes applying binary dasymetric mapping. Result is a population density estimation per built-up area presenting a better picture than known from global data sets like the Gridded Population of the World or Landscan™.

1 Introduction

The response phase after the occurrence of a natural disaster focuses on the basic humanitarian needs of the affected population. For this reason population numbers are crucial information in the aftermath of such an event. Questions like how many people are affected, how many survived and how many would need prolonged assistance are key information for crisis response and the preparation of long-term rehabilitation.

For obtaining a clear picture on the number of inhabitants and their spatial distribution within a country several sources exist: national statistical agencies

M. Konecny et al. (eds.), *Geographic Information and Cartography for Risk and Crisis Management*, Lecture Notes in Geoinformation and Cartography,
DOI 10.1007/978-3-642-03442-8_3,

usually conduct decennial censuses and surveys which are related to country-specific administrative units. Their publication is at lower administrative unit level and often delayed. Considering high population growth rates and migration especially in developing countries underlines the lack of knowledge about a country's effective population situation.

Internationally recognised databases exist like from the United Nations Population Division (UNPD 2006) which combine data from country sources collected through the years and estimate statistically the prospects. Further sources are gazetteers like the World Gazetteer (2008) offering information about country populations, administrative divisions, cities and towns. Sources for the data are official statistics and secondary sources such as yearbooks.

Only few georeferenced data sets exist. The two most widely known global data sets are the Gridded Population of the World (GPW) (CIESIN 2005) and Landscan™ (2008). For the latest version available (GPW 3) administrative boundary data and population estimates were used to produce raster grids showing the estimated number of people residing in each grid cell applying an areal weighting approach which assumes a uniform distribution within each administrative unit. Population density and the area of overlap between administrative unit and grid cell were used to calculate each unit's contribution to the cell population total (Deichmann et al. 2001). Output are several global data sets showing the distribution of population based on national or subnational administrative units with an average spatial resolution of 2.5 arc minutes. Landscan™ provides a global population data set at 30 arc second resolution. Census information is distributed based on likelihood coefficients considering for example proximity to roads, slope, land cover, nighttime lights, and other information as factors (Dobson et al. 2000). The disadvantages of the global data sets are their coarse spatial resolution. In case of GPW, modelling of population distribution is lacking causing a uniform distribution of population within administrative units. Spatial characteristics like distribution in relation to different land cover are not considered. In case of Landscan™ a drawback is its missing publication of the model standing behind the products. Nevertheless, both data sets are invaluable for providing preliminary estimations about affected population in crisis situations. However, there is demand for more precise population density data at a finer scale.

In case of a crisis only little time is available between alert and reaction and thus for preparing baseline and population maps. For this reason a methodology was developed to rapidly estimate population distribution and density in disaster-affected areas applying a binary dasymetric mapping approach. It is based on earth observation satellite imagery with very high resolution and the classification of the built-up areas therein by textural analysis. In a second step latest available census data is taken and distributed to the built-up classes. Result is a population density estimation per built-up area presenting a better picture than known from global data sets.

2 Population modelling

Census data is typically collected spatially aggregated corresponding to administrative units. A simple spatial representation is in form of a vector polygon associated with an attribute reporting the total population count. This mapping approach assumes a homogenous distribution of population within the administrative unit not considering any geographic phenomenon influencing the distribution of population in space. This uniform distribution is unlikely to happen in the real world.

Several models have been developed to spatially disaggregate the uniform distribution of data and allocate them to finer scale. Among these areal interpolation and surface modelling are the best known. Areal interpolation refers to the allocation of data from one set of geographical units to a second set of independent units whereas surface modelling refers to the allocation of a data set available in vector format to a regularly spaced grid (Deichmann 1996). Both strategies know different methodologies for conducting the disaggregation.

- Simple areal weighting is based on a proportional allocation of population from administrative units to grid cells considering the proportion of the area a grid cell covers (Deichmann et al. 2001, Li et al. 2007). This method assumes also a uniform distribution of people within the source zone. Examples applying simple areal weighting are the GPW 2 and GPW 3 data sets (Deichmann et al. 2001).
- A more realistic representation of population distribution is considering ancillary data to differentiate between populated and unpopulated space. This dasymetric mapping approach has been successfully applied in several cases (Langford and Unwin 1994, Mennis 2003, Holt et al. 2004). The binary dasymetric mapping uses a binary land use classification (populated versus unpopulated land) as ancillary information. Population data is only allocated within the populated land use class assuming a fixed population density inside the populated land. A weakness of the model is that more complex land use situations are not considered in the model (Li et al. 2007). Langford et al. (1991), for example, use Landsat TM data for obtaining a land use classification as ancillary data to build predictive models regressing population densities for the different land use classes. Langford and Unwin (1994) use satellite imagery to determine residential and non-residential classes on which population is redistributed among the residential pixels.
- A more advanced method is three-class dasymetric mapping that respects different population densities within the different land use classes. Eicher and Brewer (2001) redistribute county level population to sub-county level by applying predetermined percentages for each land use class. Differences and improvements between binary and three-class dasymetric mapping are investigated by Langford (2006). Mennis (2003) introduces an empirical sampling technique to assess the relationship between ancillary data and population.
- The limiting variable method is applying simple areal weighting to all inhabitable zones in an area followed by thresholding of maximum population densities for particular land use zones. If a polygon density exceeds its threshold it is

assigned with the corresponding land use threshold value. The remaining data are removed from the zone and evenly distributed to the remaining zones. Eicher and Brewer (2001) apply this technique in a comparative study.

- Pycnophylactic modelling is known as a method to smooth a population grid by iteratively applying a moving filter that replaces each grid value with a weighted average of its neighbouring values keeping the initial total population density for the whole area of interest (Deichmann 1996). The method was introduced by Tobler (1979) and applied on several examples.
- Smart interpolation is another grid-based technique creating first a grid of population potentials followed by an allocation of population numbers available at administrative unit level. The population potential modelling is based on weightings per grid cell considering several factors influencing the inhabitability of the single cells like the size of settlements, their location to infrastructure, land use classes, uninhabitable areas and others (Deichmann 1996). Examples were conducted by Schneiderbauer and Ehrlich (2005) for rural Zimbabwe and Mubareka et al. (2008) for Iraq.

Studies have been conducted about the accuracy of the approaches with different results (Li et al. 2007, Eicher and Brewer 2001, Mennis and Hultgren 2006). The method applied in our example and explained in the following section is based on binary dasymetric mapping.

3 Study area and project background

The study was conducted within the EU-FP6 project LIMES (Land and Sea Integrated Monitoring for European Security) developing and applying satellite-based technologies in the field of security for Europe. The exercise was carried out for the South-Western coast of Cyprus as part of the Assessment Mission Course (AMC). The AMC is jointly planned and conducted by the German Agency for Technical Relief (THW) and the medical non-governmental organisation Johanniter Unfallhilfe on behalf of the European Commission. It has the aim to train disaster managers of the EU-Member States and other international organizations in planning and conducting assessment missions. The exercise was based on an earthquake scenario with a following tsunami wave, together with secondary threats and major incidents as consequences of the earthquake. Participants were asked to perform damage and need assessments in urban and rural areas as well as at industrial and logistically important sites. Key sectors assessed have been search and rescue, shelter, water supply and sanitation, food and nutrition, transport and logistics, health, environment, and law and order (European Virtual Academy 2006). The LIMES cluster for humanitarian affairs coordinated by the Center for Satellite Based Crisis Information (ZKI) at the German Remote Sensing Data Center (DFD) at DLR supported the disaster relief workers with up to date satellite images and satellite maps. The Joint Research Center as cluster partner created and provided the population maps as part of the overall contribution.

The study area for population mapping extended over a 45-km strip along the South-Western coast of Cyprus reaching approximately 15 km inland.

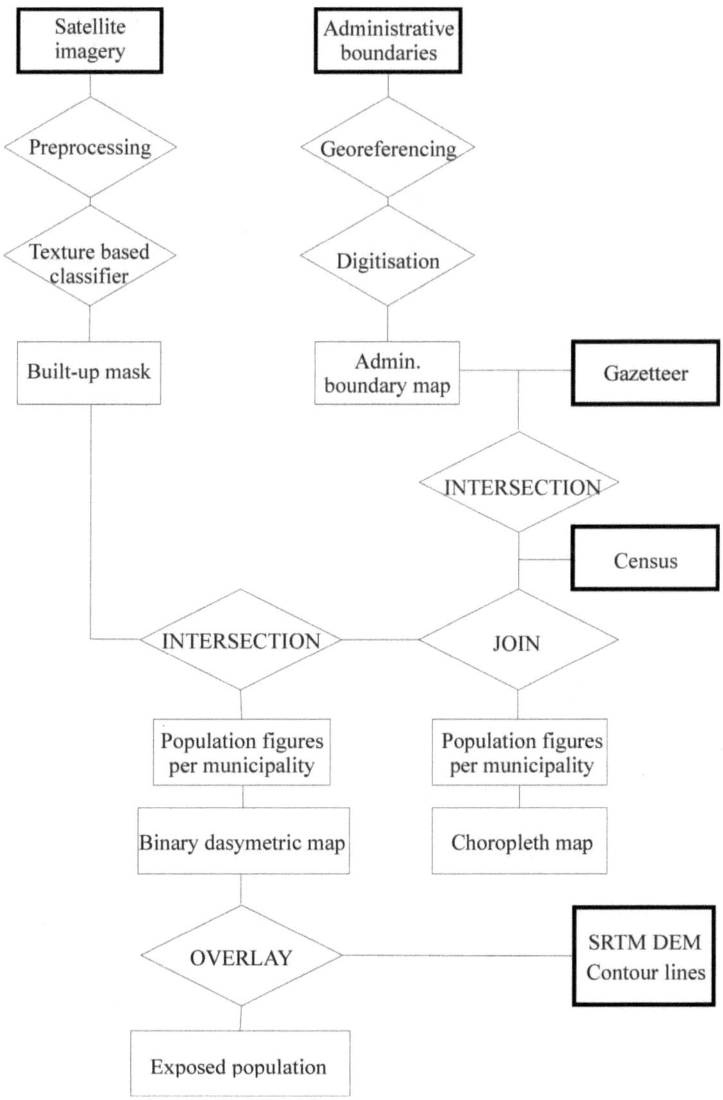

Fig. 1. Processing flow

4 Data and methods

As latest available census, the figures on 'population by municipality' from 2001 were used (Statistical Service of Cyprus 2001). The aim of the study was a proof of concept. For that reason no population projections and adjustments for the year 2008 were carried out. For European Union Member States the NUTS database (Nomenclature of Territorial Units for Statistics) provides a geocode standard for referencing the administrative divisions of countries for statistical purposes. The unit boundaries are available in vector format (Eurostat 2008). However, for Cyprus no third or higher level (municipality) administrative boundary layer is available from NUTS. Within the short time of the exercise the data could not be obtained from another source. For that reason a municipality map in raster format found in the Internet was used. The further processing required also the application of gazetteer information for which the GEOnet Names Server (GNS) provided by the US National Geospatial-Intelligence Agency's (NGA) was used (NGA 2008). The classification of occupied and unoccupied space was based on multispectral Quickbird imagery from 2008. For an analysis of exposed population for the expected tsunami wave elevation data from the Shuttle Radar Topographic mission (SRTM) was applied (CGIAR-CSI 2008). The main data processing and analysis was conducted in three stages and is represented in Fig. 1:

1. Creation of new target zones using remote sensing
2. Creation of administrative boundary map layer and integration of census data in source zones
3. Areal interpolation transforming source zone data to new target zones, defined by built-up classes extracted from satellite imagery

The use of remote sensing data in relation to population estimations is well known (Harvey 2002, Lo 2003, Mesev 2003). Its use has the advantage that data is available for most areas of the world, it is cheap to cover large areas and it has the potential to use data that was recorded close to the date the latest available census was conducted (Langford et al. 2008).

For determining occupied space as input for the binary dasymetric mapping a methodology developed by Pesaresi and Gerhardinger (2009) was applied. The derivation of a built-up surface layer is based on fuzzy rule-based composition of anisotropic textural co-occurrence measures considering also the different directional component of the textural signal. The method was successfully applied in several test cases both on optical (Pesaresi et al. 2008, and Pesaresi and Gerhardinger et al. 2009) and radar imagery (Gamba et al. 2008).

The determination of urban areas can successfully be derived by calculating textural measures derived from the gray-level co-occurrence matrix (GLCM) (Haralick et al. 1973). As built-up structures in an image are anisotropic at the scale of the urban area and the GLCM is sensitive to rotation it is necessary to consider the number of displacement vectors around each single pixel to fully exploit the anisotropic invariance. This is followed by the combination and integration of the resulting texture channels. According to previous studies (Pesaresi 2000) the textural measure of contrast has been selected as one of the most efficient in the discrimination of built-up and non-built-up. For the combination of the resulting data sets it was demonstrated that *min* and *max* operators are superior to standard averaging of the individual channels. For the derivation of a binary built-up mask the delineated textural classes are adjusted and standardised by applying threshold levels. A visual inspection was carried out for an accuracy check of the resulting built-up layers. Figure 2 shows a subset of the original Quickbird imagery used for the analysis.

For the creation of an administrative boundary layer the municipality map was manually georeferenced by draping the map over the earth surface available in Google Earth. In a following step the map boundaries were digitised. The resulting administrative boundary map did not contain the municipality names, IDs or population figures as attributes. For this reason an intersection with the gazetteer point data set was applied to determine the right municipality names. These were used to join the resulting administrative unit boundaries with the census database containing the population figures for 2001. This information was used to create a choropleth map based on population totals as well as a population density map considering the total municipality areas.

For the dasymetric mapping the population figures, now available on the base of administrative units, had to be redistributed to the primarily derived built-up area mask. This was done applying the following formula:

$$P_t = \sum_{s=1}^{s} \frac{A_{tsp} \times P_s}{A_{sp}} \tag{4.1}$$

where P_t is the estimated population in the target (built-up) zones, A_{tsp} is the overlapping area between source and built-up zones, A_{sp} is the area of the source zone and P_s is the population at the source zone.

The resulting map represents the number of population per built-up land of each municipality. The map was overlaid with contour lines at 10-m intervals derived from the SRTM elevation model for a visual comparison of built-up areas and their corresponding population numbers below 10 m and 20 m elevation.

5 Results and discussion

As a result four different population maps were created. Two choropleth maps of the study area were done using population totals and population density as attribute. Figure 4 shows a subset of one of these maps. The maps were created for a comparison between population totals and densities of choropleth and dasymetric map. Two other dasymetric maps were created showing allocation of population totals on built-up areas only. An example is given in Fig. 3. The contour lines based on SRTM were used as additional map feature for visualisation purposes. In a second dasymetric map the population attributes were changed. Due to their availability from the statistics the population figures were given on housing unit level showing the numbers of population with permanent residence and temporary residence.

This was seen as important information in the aftermath of a disaster as the coastal area around Paphos is a famous touristic spot. Dependent on each season this information can give additional knowledge about the number of exposed people.

It could be shown that even with the time constraints during the AMC exercise population maps could rapidly be produced applying the dasymetric mapping approach using ancillary information derived from satellite imagery. Compared to other available data sets like GPW 3 or Landscan™ (see Fig. 5) the resulting maps have higher spatial resolution and thus give a better picture of the population situation which is highly necessary for the planning and implementation of relief actions.

Fig. 2. Subset of the Quickbird satellite imagery used for the delineation of built-up areas. (Original QuickBird Imagery © DigitalGlobe, Inc., 2008. Distributed by Eurimage)

Nevertheless some limitations have to be mentioned as well:

- The low accuracy of the available municipality map which was only available as non-georeferenced image file without any attribution influences the overall map accuracy. The allocation of attributes was achieved by intersecting the digitised map with gazetteer points. This procedure was successful most of the times. Only three municipality units could not be identified. The missing accuracy of the administrative map has also to be considered looking at the population density figures due to their relation to the unit areas.
- The calculation of the built-up surface layer required a visual accuracy check due to several misclassifications especially in agriculturally used areas.
- The overlay with SRTM elevation data should illustrate the increasing knowledge in relation to a tsunami event by combining population figures

and topography. For the use of the SRTM data set with 90-m resolution it must be considered that SRTM represents a surface model of the earth not adjusted to building heights or infrastructure.

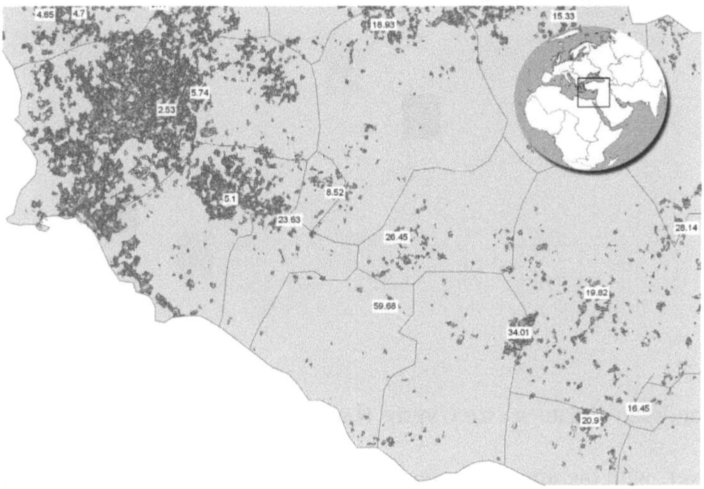

Fig. 3. Dasymetric population map

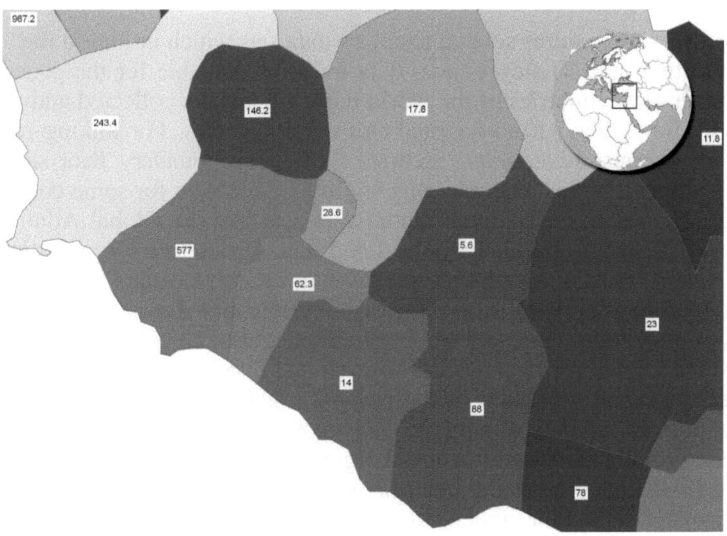

Fig. 4. Choropleth population map

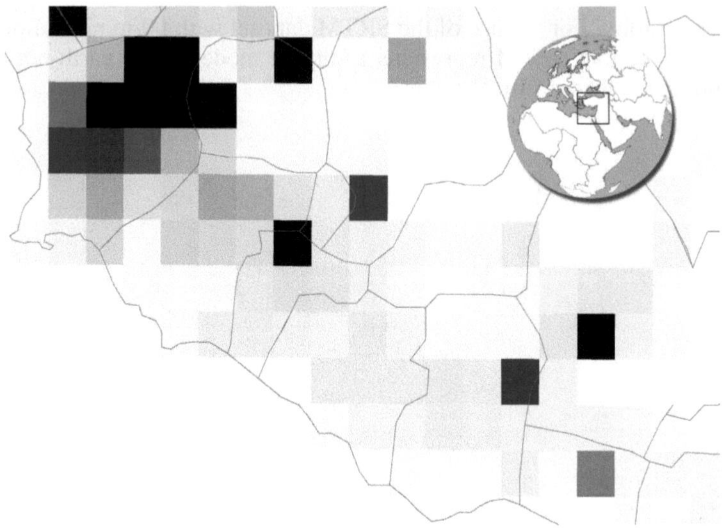

Fig. 5. Landscan™ data set for Cyprus (Landscan 2008)

However, with the approach presented it could be demonstrated that population maps with high spatial resolution can be produced within the short time given in the aftermath of a disaster event. Moreover, the limitations reflect the difficult working conditions and challenges of remote sensing and mapping related to crisis response.

The processing requires several ancillary data sets which influence the accuracy of the resulting maps. Latest census data must be available for the generation of the maps. In most countries of the world those figures are collected and published (even with time delay) from national statistical agencies. For linking population figures to administrative units, accurate GIS-ready boundary data sets are required. As shown in this example, this might be a problem for some countries and a careful assessment of available global data sets like the Global Administrative Unit Layer (GAUL) provided by the Food and Agriculture Organisation of the United Nations (FAO) should be considered in future research. The classification of built-up zones for a redistribution of the population figures requires satellite imagery. Today those data sets are frequently acquired after crisis events as their use is well accepted for damage assessments and disaster management in general. However, a general limitation of the proposed method is given by non-available VHR satellite image data sets due to bad weather conditions and cloud coverage which is especially the case for tropical regions. The latter case could be overcome applying SAR imagery for the determination of the built-up zones. Yet, besides their advantage of weather independency Gamba et al. (2008) report about drawbacks due to slant-range to ground-range image reprojection which reduces the possibility to detect small settlements.

The application of a textural based classification algorithm demonstrated to be a robust method for the determination of built-up areas as it provided fast and accurate results.

The use of a dasymetric mapping approach provides a better picture than commonly used choropleth maps or gridded data sets with coarse resolution. A comparison of differing population densities between the two map types illustrates this impressively.

For future research we consider to include demographic changes between the time a census was conducted and the disaster event by applying population projection models. Another task will be to integrate information on the different use of the built-up areas, e.g. residential areas, business areas, tourist spots and others. A further need is to include time as an important factor for risk assessments starting with different population exposures during day and night, weekdays and weekends or high and low tourist seasons.

References

Center for International Earth Science Information Network (CIESIN), Columbia University; and Centro Internacional de Agricultura Tropical (CIAT) (2005) Gridded Population of the World Version 3 (GPWv3). Palisades, NY: Socioeconomic Data and Applications Center (SEDAC), Columbia University http://sedac.ciesin.columbia.edu/gpw (19.12.2008).

CGIAR-CSI SRTM 90m Digital Elevation Data (2008) http://srtm.csi.cgiar.org/ (28.12.2008).

Deichmann U (1996) A review of spatial population database design and modeling. Technical Report 96-3. National Center for Geographic Information and Analysis, Santa Barbara. http://www.ncgia.ucsb.edu/Publications/Tech_Reports/96/96-3.PDF (25.12.2008).

Deichmann U, Balk D, Yetman G (2001) Transforming population data for interdisciplinary usages: from census to grid. Documentation for GPW Version 2. http://sedac.ciesin.columbia.edu/plue/gpw/GPWdocumentation.pdf (23.12.2008).

Dobson JE, Bright EA, Coleman PR, Durfee RC, Worley BA (2000) LandScan: A Global Population Database for Estimating Populations at Risk. Photogramm Eng Remote Sensng 66(7):849–857.

Eicher CL, Brewer CA (2001) Dasymetric mapping and areal interpolation: Implementation and evaluation. Cartogr Geogr Inf Sci 28(2):125–138.

European Virtual Academy (2006) Basic Information on the Assessment Mission Course (AMC) Republic of Cyprus. http://www.evanetwork.net/download/documents/brochures/Info_AMC.pdf (27.12.2008).

Eurostat GISCO – Geographic Information System of the European Commission (2008) http://epp.eurostat.ec.europa.eu/portal/page?_pageid=2254,62148876,2254_62153824&_dad=portal&_schema=PORTAL (28.12.2008).

Gamba P, Pesaresi M, Molch K, Gerhardinger A, Lisini G (2008) Anisotropic rotation invariant built-up present index: applications to SAR data. Proceedings of the IEEE International Geoscience & Remote Sensing Symposium July 6–11, Boston, MA, USA.

Haralick RM, Shanmugan K, Dinstein I (1973) Texture features for image classification, IEEE Trans Syst Man Cybern 3:610–621.

Harvey JT (2002) Estimating census district populations from satellite imagery: some approaches and limitations. Int J Remote Sens 23:2071–2095.

Holt J, Lo CP, Hodler TW (2004) Dasymetric estimation of population density and areal interpolation of census data. Cartogr Geogr Inf Sci 31:103–121.

LandScan™ (2008) Global Population Database. Oak Ridge, TN. Oak Ridge National Laboratory. http://www.ornl.gov/sci/landscan (19.12.2008).

Langford M (2006) Obtaining population estimates in non-census reporting zones: an evaluation of the 3-class dasymetric method. Comput Environ Urban Syst 30:161–180.

Langford M, Higgs G, Radcliffe J, White S (2008) Urban population models and service accessibility estimation. Comput Environ Urban Syst 32:66–80.

Langford M, Maguire DJ, Unwin DJ (1991) The areal interpolation problem: estimating population using remote sensing in a GIS framework. In: Masser I, Blakemore M (eds.) Handling Geographical Information: Methodology and Potential Applications. Longman, London. pp. 55–77.

Langford M, Unwin DJ (1994) Generating and mapping population density surfaces within a geographical information system. Cartogr J 31(1):21–26.

Li T, Pullar D, Corcoran J, Stimson R (2007) A comparison of spatial disaggregation techniques as applied to population estimation for south East Queensland (SEQ), Australia. Applied GIS 3(9):1–16.

Lo CP (2003) Zone-based estimation of population and housing units from satellite generated land use/land cover maps. In: Mesev V (ed) Remotely Sensed Cities, Taylor & Francis, London, pp. 157–180.

Mennis J (2003) Generating surface models of population using dasymetric mapping. Prof Geogr 55:31–42.

Mennis J, Hultgren T (2006) Intelligent dasymetric mapping and its application to areal interpolation. Cartogr Geogr Inf Sci 33(3):179–194.

Mesev V (2003) Remotely Sensed Cities. Taylor & Francis, London, pp. 372.

Mubareka S, Ehrlich D, Bonn F, Kayitakire, F (2008) Settlement location and population density estimation in rugged terrain using information derived from Landsat ETM and SRTM data. Int J Remote Sens 29(8):2339–2357.

NGA GEOnet Names Server. http://earth-info.nga.mil/gns/html/index.html (10.05.2008).

Pesaresi M (2000) Texture analysis for urban pattern recognition using fine-resolution panchromatic satellite imagery, Geogr Environ Model 4(1):47–67.

Pesaresi M, Gerhardinger A (2009) Enhancing the discrimination accuracy of the built-up index in semi-desert areas with scattered vegetation – the Al Geneina case. Proceedings of the 2009 Joint Urban Remote Sensing Event. May 20–22, Shanghai.

Pesaresi M, Gerhardinger A, Kayitakire F (2008) A Robust Built-Up Area Presence Index by Anisotropic Rotation-Invariant Textural Measure. IEEE JSTARS 1(3):180–192.

Schneiderbauer S, Ehrlich D (2005) Population density estimations for disaster management. Case study rural Zimbabwe. In: van Oosterom P, Zlatanova S,

Fendel EM (eds.) Geo-information for Disaster Management. Springer, Delft, pp. 901–921.

Statistical Service of the Republic of Cyprus (2001) Census 2001 – Population per Municipality. http://www.mof.gov.cy (10.06.2008).

Tobler WR (1979) Smooth pycnophylactic interpolation for geographical regions. J Am Stat Assoc 74(367):519–530.

The World Gazetteer (2008) http://www.world-gazetteer.com (19.12.2008).

United Nations Population Division (2006) World Population Prospects: The 2006 Revision Population Database. http://esa.un.org/unpp (19.12.2008).

Modeling of Spatiotemporal Distribution of Urban Population at High Resolution – Value for Risk Assessment and Emergency Management

Sérgio Freire

Abstract

Knowing the spatiotemporal distribution of population at the local scale is fundamental for many applications, particularly in risk analysis and emergency management. Because of human activities, population counts and their distribution vary widely from nighttime to daytime, especially in metropolitan areas, and may be misrepresented by census data.

This study uses a dasymetric mapping approach to refine population distribution in Portugal. The most recent census enumeration figures and mobility statistics are combined with physiographic data to allocate nighttime population to residential areas, and workplaces and workforce are georeferenced to model daytime distribution.

Main results represent expected maximum daytime population and maximum nighttime residential population for each 25-m grid cell in the study area. Since the same spatial reference base is used to allocate population, day and night distributions are directly comparable, as is demonstrated in sample applications in the context of emergency management. Verification and validation procedures and demonstrations indicate that the approach suits the objectives.

M. Konecny et al. (eds.), *Geographic Information and Cartography for Risk and Crisis Management*, Lecture Notes in Geoinformation and Cartography,
DOI 10.1007/978-3-642-03442-8_4,

1 Introduction

1.1 Population distribution and emergency management

Natural or man-made disasters, either resulting from natural hazards, technological accidents, or terrorism, usually occur without warning and can potentially affect people from a local to a continental scale. Risk is usually defined as a function of hazard probability and vulnerability, the latter resulting from a combination of exposure and ability to cope (UNDP 2004). Misestimating one of these components necessarily affects the accuracy of the overall risk assessment and mapping.

Human life is unquestionably the most important *asset* to protect, and the distribution and density of the overall population is a rather basic geographical indicator. Therefore accurately estimating population exposure is recognized as a key component of catastrophe loss modeling, one element of effective risk analysis and emergency management (FEMA 2004, Chen et al. 2004, NRC 2007). However, until recently, assessment and mapping of human vulnerability has been lagging behind hazard analysis efforts (Pelling 2004), and potential loss derived from exposure has served as a quantitative proxy for risk (Lerner-Lam 2007).

Also, updated and detailed mapping of population distribution is important for decision support in practically every phase of the emergency management cycle, if produced at appropriate spatial and temporal scales (Sutton et al. 2003), e.g., central to the planning stage is the assessment of population exposure and vulnerability in the hazard zone; mitigation includes reducing vulnerability and exposure (namely by displacing population or other measures); preparedness may involve fitting and placing means and resources according to vulnerability, for more efficient response in case of disaster; during and after the event, locating and estimating victims is essential to tailor response and rescue efforts, including allocating emergency personnel, hospital beds, etc.; and finally, estimating all activities and people affected, even if indirectly, facilitates the recovery process.

Despite recent efforts by Dobson (Dobson 2003, 2007) at devising a population estimation technique that could be employed in real time once a disaster occurs, for planning and simulation purposes and to ensure a timely response, adequate population distribution data should be produced and made available beforehand whenever possible.

1.2 Population in space and time

The increased availability of digital spatial data combined with improved capabilities of Geographic Information Systems (GIS) have allowed for the development of several global population distribution databases, such as the GPW, HYDE, and LandScan (Tobler et al. 1995; Goldewijk and Battjes 1997; Dobson et al. 2000). However, their spatial resolution is still too coarse to adequately support analysis at the local level and most do not represent the dynamics of population distributions. Dobson acknowledges that "even finer resolutions are needed for many types of disasters", namely those that can "impact areas as small as a neighborhood, city block, or single building" (Dobson 2002). Data sets may be produced globally, but people are always affected locally.

Due in part to the complex nature of population as a geographical variable, several approaches have been adopted to estimate their spatial distribution, including statistical modeling (correlation), surface modeling, and cartographic methods (Fisher and Langford 1996; Wu et al. 2005). However, many of these methods require assumptions that over-simplify the reality or disaggregate population totals based on heuristic or empirical parameters. Additionally, obtaining positive values and preserving the total volume of people—Tobler's "pycnophylactic condition" (Tobler 1979)—are basic requirements to produce realistic representations. Dasymetric mapping is a cartographic technique, originally used for population mapping, which aims at limiting the distribution of a variable to the areas where it is present, by using related ancillary information in the process of areal interpolation (Eicher and Brewer 2001).

Population distributions are not static in time, varying over daily, seasonal, and long term time scales (Sutton et al. 2003) due to a number of human activities, such as work and leisure. For emergency planning in urban areas, it is the population variation during the daily cycle that is particularly important to be able to estimate the number, age, and socioeconomic classes affected by an impact (Alçada-Almeida 2009). In Portugal, as in most countries, existing population distribution maps and vulnerability or exposure analyses are normally based on census data, often aggregated at the commune level (e.g., Oliveira et al. 2005). Census figures register where people reside and usually sleep, although their spatial distribution varies widely between night and day, especially within metropolitan areas. Due to daily commuting alone, the daytime population of municipalities in the metro areas of Lisbon and Porto can differ by more than 50% of the official census figures (INE 2003). Also, the availability of a total count for census or administrative areas creates problems for analysis (e.g., MAUP, ecological fallacy), being generally assumed that the distribution is constant and exhaustive in those areas, or represented by their centroid.

Making population distribution data available as a high-resolution raster database facilitates rapid GIS analysis at the local level and for any zoning. Therefore, when disaster strikes or is imminent, knowing how many people are likely to be in the affected area can be invaluable information for adequate emergency response and evacuation planning. The LandScan project (Dobson et al. 2000) successfully incorporated the temporal dimension in population distribution by mapping "ambient population", a temporally averaged measure of population density that accounts for human activities (such as sleep, work, study, transportation, etc.). This representation may be more adequate for certain applications (such as emergency management) than residence-based population density. However, LandScan's spatial resolution (30 arc-seconds), while adequate for national to global analyses, is insufficient for most practical uses in Portugal. Furthermore, ambient population corresponds to a compromise between daytime and nighttime distributions that strictly represents neither period. Therefore, population distribution databases having higher temporal and spatial detail are being developed for the territory of the USA (McPherson and Brown 2003, Bhaduri et al., 2002).

This study is aimed at developing and testing a data-based model to map the nighttime and daytime population distributions in Portugal at high resolution to enable local-level analysis. This effort correlates with recent recommendations to improve vulnerability analyses (Cutter 2003, Balk et al., 2006, Birkmann 2007; NRC 2007).

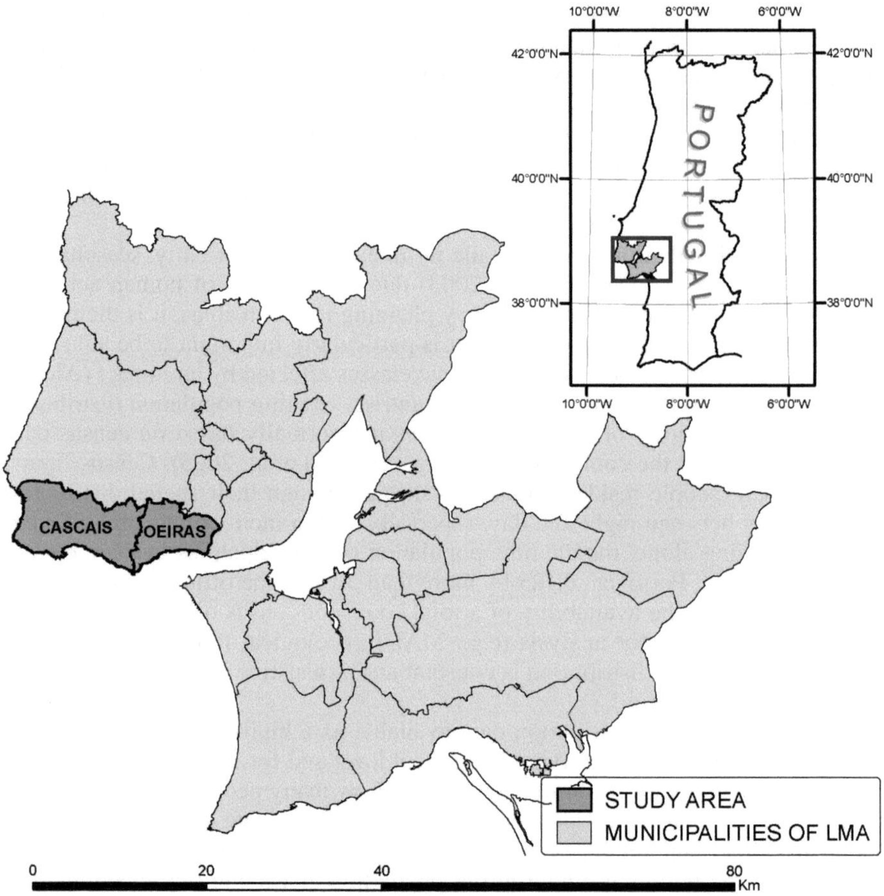

Fig. 1. Location of the study area in Portugal and in the Lisbon Metro Area (LMA)

2 Methods

Pre-processing and modeling of geographical data were conducted in ESRI® ArcGIS 9.1, a GIS application. GIS offers the necessary tools and flexibility to implement raster or vector-based dasymetric methods, and was used to verify, correct, and integrate geographic data sets, and for modeling, analysis, and mapping the results for presentation.

Because some data sets were made available on a municipal basis, each municipality was modeled separately. For each municipality, 25-m raster grids were created representing different types of population distribution. The raster structure provides uniform and flexible units that facilitate the reaggregation for any zoning, thus being suitable for modeling and analysis. A spatial resolution of 25 m was adopted to approximate the size of a single-family residence (half block). Also, when the model was tested for sensitivity to cell size, an increase in resolution to 12,5 m yielded marginal gains in the accuracy of results.

2.1 Study area

The official administrative limits of the municipalities (*concelhos*) of Cascais and Oeiras in 2001 constitute the study area for this research. These are 2 of the 18 municipalities that comprise the Lisbon Metropolitan Area (LMA), the main metropolitan area in Portugal (Fig. 1).

Cascais and Oeiras occupy 97 and 46 km^2, respectively, and have a combined resident population of 332 811; this results in an average population density of 2332 inhabitants/km^2, well above the national average density of 112 inhabitants/km^2. However, population density varies widely throughout the study area, from high density in multistory residential apartments to low density in rural areas. Even at the census block group level, some polygons are quite large and do not reflect their uneven population density.

This area was selected for several reasons: (a) its characteristics, namely with regard to urban and suburban character, and strong economic activity, (b) the availability and access to input data, and (c) personal familiarity with the area, which facilitates data verification and field work.

2.2 Data sets

In this study, the spatial detail of census zones whose counts are to be disaggregated should be met in scale, resolution, and accuracy by ancillary data sets used for disaggregation. Input variables used for modeling include both physiographic and statistical data. In the first group are census tracts, street centerlines, and land use and land cover (LULC), while the second includes census counts (INE 2001), data on workforce by workplaces, and commuting statistics (INE 2003) for the study area. These data were obtained from various sources and in different formats which are listed in Table 1.

Table 1. Main input datasets used for modeling nighttime and daytime population

Data set	Source	Date	Data type
Street centerlines	Private	2004	Vector polyline
LULC (COS90; CLC2000)	Public	1990; 2000	Vector polygon
Census block groups	Public	2001	Vector polygon
Census statistics	Public	2001	Database (MS Access)
Workplaces and employment	Public	2001	Table
Commuting statistics	Public	2001	Table (O/D matrix)

In general, temporal consistency among data sets was very high, with the exception of street centerlines whose reference date was 3 years subsequent to the model target date (2001). For this reason and owing to the importance of this data set in the model, it was decided to modify it in order to better represent the reality of the target date. The LULC data originated from two maps and was also corrected and improved upon within the study area.

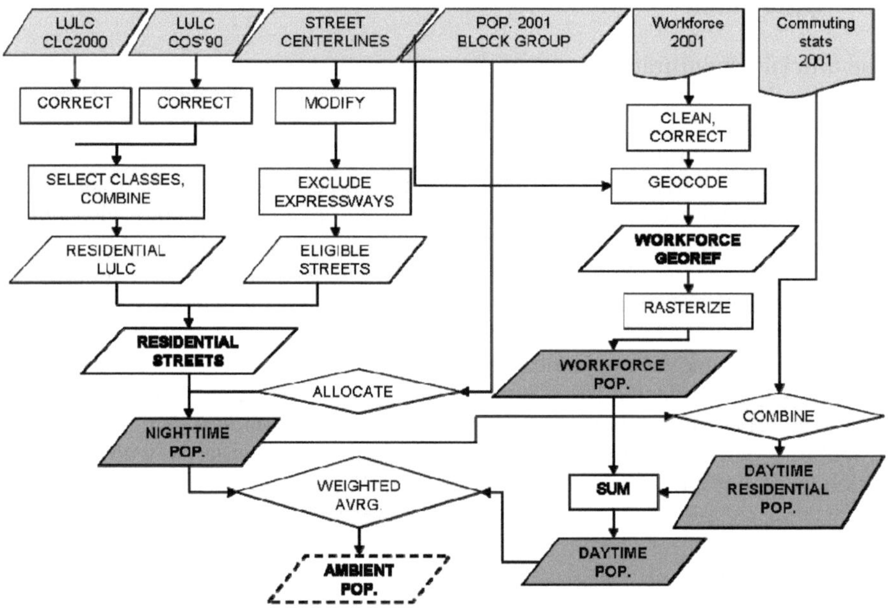

Fig. 2. Flowchart of main tasks involved in the model: input data are noted in *light gray*, secondary products in *bold*, and main results noted in *bold* and *darker gray*

2.3 Model

The modeling of population distribution is based on dasymetric mapping using street centerlines as spatial reference unit to allocate population counts. The most recent statistical and census data (2001) provide the population counts for each daily period, while physiographic data sets define the spatial units (i.e., grid cells) used to disaggregate those counts. This general approach was successfully implemented by the Los Alamos National Laboratory (New Mexico, USA) to map daytime and nighttime population distributions in the United States at 250-m resolution (McPherson and Brown 2003), and is adapted and applied to Portugal.

An overview of tasks involved in this modeling process is presented in Fig. 2. Input data are noted in light gray, secondary products previously unavailable in Portugal are noted in bold, and main results are noted in bold and darker gray.

The map of nighttime population distribution was obtained by using a grid binary dasymetric mapping method to disaggregate residential population from census zones to residential streets. First, available digital LULC maps were improved, relevant classes selected, and combined, in order to identify residential land use. Street centerlines were also modified in order to better represent the road network existing in 2001. Then, freeways are removed from consideration and the resulting eligible streets are combined with residential land use classes from LULC data to obtain residential streets. These are subsequently rasterized at 25 m and the population from census block groups (source zones) are interpolated to the respective residential street cells (target zones) using areal weighting.

The daytime population distribution results from the combination of two components (as illustrated in Fig. 2): (a) the daytime population in their places of work or study—the workforce population surface, and (b) the population that remains home during the day—the daytime residential population grid. The latter is obtained by multiplying the nighttime distribution by the percentage of resident population who, according to official statistics (INE 2003), do not commute to work or school. In the absence of other information, it is assumed that non-commuters remain in their residences in the daytime period. This implies that this study is not including the potential effects of leaving home for shopping and several other activities on daytime population distributions.

The workforce population surface was created by georeferencing 4,316 workplaces and schools and respective workforce and students in the study area; 1,395 of these were georeferenced manually using ancillary data and field work. The remainder workplaces were geocoded to the street centerlines in ArcGIS using their addresses.

Total daytime population surface results from the sum of workforce population with daytime residential population on a cell-by-cell basis. The ambient population distribution is estimated by computing a weighted average of nighttime and daytime distributions, considering the proportion of nighttime and daytime periods occurring in a typical 7-day weekly cycle.

3 Results

Main results consist of raster surfaces of nighttime (residential) population, daytime residential population, daytime worker and student population, total daytime population, and ambient population. These have the number of people in each 25-m cell in 2001, thus representing population density by 625 m^2. Nighttime and daytime distributions represent maximum expected density on a typical workday, assuming that everyone is at home at night and all workers and students are in their workplaces and schools, and the remainder in their residences. Although still a simplification of reality, it is preferable than computing a zonal average by census polygon alone.

3.1 Verification and validation

Validation of daytime population distribution was limited by unavailability of compatible reference data sets for Portugal. However, this limitation can be considered less relevant since daytime population distribution originates from a combination of the nighttime distribution surface (subject to formal validation) with mostly official statistical data, as opposed to being derived from heuristic or empirical weights. Still, the daytime population distribution was subject to verification in several ways: (a) input data (especially workplaces' addresses) were verified through cross-checking with other sources and field work, regarding location of workplaces (firms), (b) results were verified with high-resolution imagery to confirm positional accuracy of distributions, and (c) it was checked that the total number of workers provided and other statistics (census and mobility) were not contradictory.

The nighttime population distribution was subject to a formal accuracy assessment process, using the higher-resolution census blocks as reference (i.e., ground truth) in a correlation analysis. Cell values in modeled distributions were aggregated by census block in ArcGIS and compared against the respective census count. Correlation coefficients (Pearson's r) of 0,84 were obtained for the municipality of Cascais and 0,79 for the municipality of Oeiras. Even when the 36 blocks that coincide with block groups are not considered, the coefficients decrease only slightly to 0,79 and 0,75, respectively. This indicates that model performance is rather good, in light of its high resolution and considering the large number of samples used for validation (2,433 census blocks in Cascais, 1,456 in Oeiras).

4 Sample applications

Three fictitious case study scenarios are presented as sample applications of model results in the context of risk analysis and emergency management. Situations are examples of technological hazards, natural disasters, and terrorist attack. These scenarios illustrate the usefulness of improved population distribution mapping for planning or response to actual events.

4.1 Case study A: technological hazard (airborne toxic plume release)

A truck transporting highly toxic chemical products is involved in a serious accident on the off-ramp of a busy freeway in Oeiras, at 1100 hours (AM). An airborne toxic plume is released and the dominant winds slowly push it southwards. Figure 3 represents this scenario, showing the plume and the daytime population distribution overlaid on orthophotos from 2004.

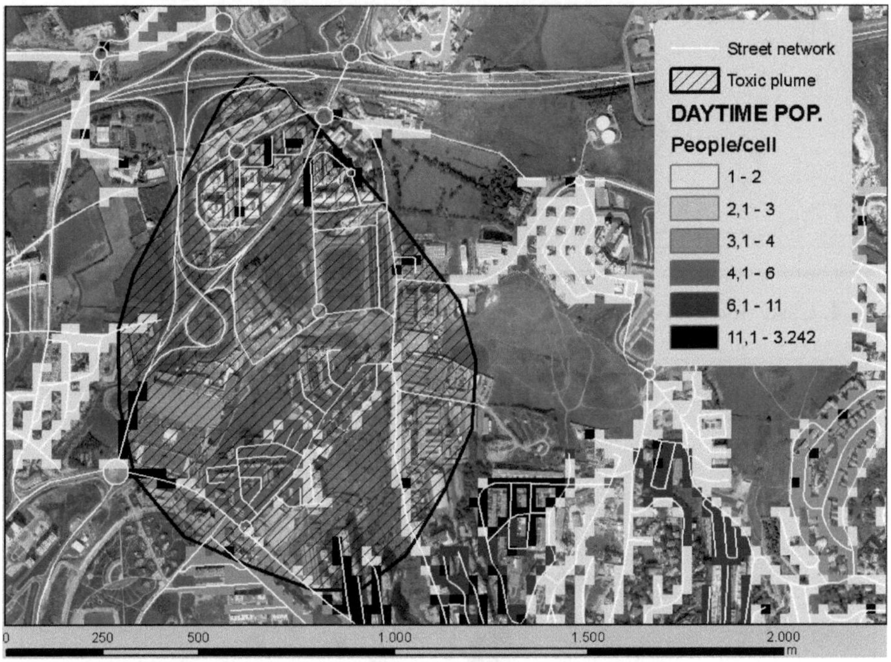

Fig. 3. Case study A: airborne toxic plume release in Oeiras

4.2 Case study B: natural disaster (earthquake)

The Lisbon Metro Area is located in a high seismic risk area, having suffered destructive earthquakes in the past (e.g., 1755). Around 1500 hours a strong earthquake strikes the town of Cascais, causing major damage around downtown and along the coast. Figure 4 represents this scenario, showing the area most seriously affected by destruction and collapse of buildings, and the daytime population distribution overlaid on 2004 orthophotos.

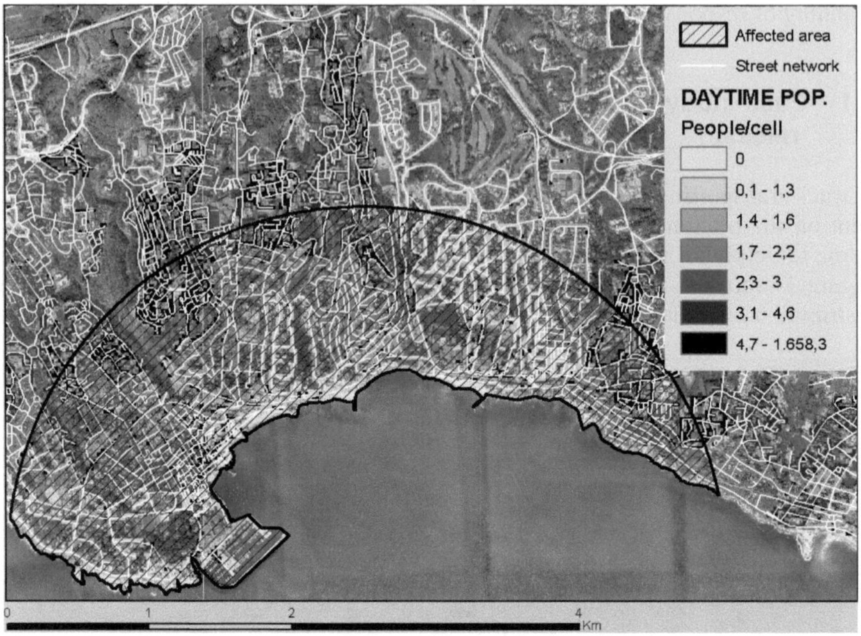

Fig. 4. Case study B: earthquake affecting downtown Cascais

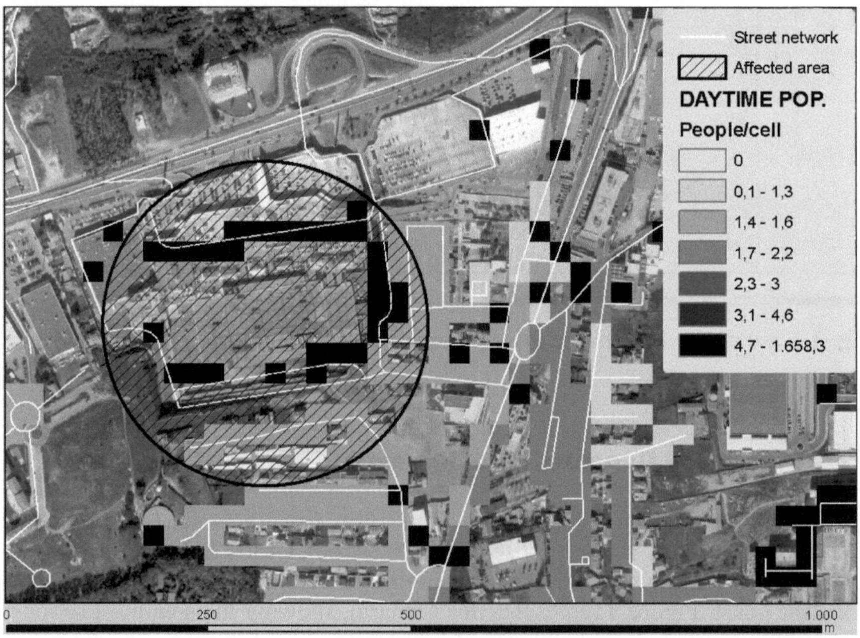

Fig. 5. Case study C: terrorist attack in a Cascais shopping center

4.3 Case study C: terrorist attack (bombing of shopping center)

Around 1630 hours a powerful explosive is detonated in a large shopping mall in Cascais, causing great destruction and a fire. Figure 5 represents this scenario, showing the area most affected by the blast, and the daytime population distribution overlaid on 2004 orthophotos

Based on the scenarios presented above, it is relatively straightforward to use GIS analysis to calculate the affected (exposed) population in each event. Table 2 quantifies and compares exposure based on direct use of census data versus use of appropriate model results.

Table 2. Total exposed population in each case study

Scenarios	Census data (Block group)	Model results (Daytime pop.)	Difference [persons (%)]
A	2144	3950	1806 (84)
B	22841	28944	6113 (26)
C	86	2151	2065 (2400)

In all the scenarios there are important differences in the figures of total exposed population using census data compared to more appropriate model data. In these daytime scenarios, use of census data indicates significant under-estimation of affected population, more dramatic in Scenario C. This could possibly lead the authorities to under-prepare the emergency response, resulting in an increase of more serious victims or unnecessary suffering. Also, use of daytime distributions to simulate and plan for daytime events also improves assessment of exposure and may contribute to better planning and mitigation measures.

4.4 Case study D: planning of best route for hazardous materials transportation

Scenario D illustrates the use of spatiotemporal population distribution in a planning stage as a criterion for truck routing of hazardous materials between the A5 freeway exit and a processing facility, in an industrial area of the municipality of Oeiras. Consideration of population distribution in this context allows two fundamental issues to be addressed simultaneously: what is the best period (day or night) and route in order to minimize population exposure along the way?

In the example, population distribution in a 50-m wide buffer along the streets is considered as the cost impedance to select the route between two locations which minimizes overall population exposure. Best routes were computed for the nighttime period and for the daytime period, considering each period's respective population distribution resulting from the model as the sole criterion (Fig. 6).

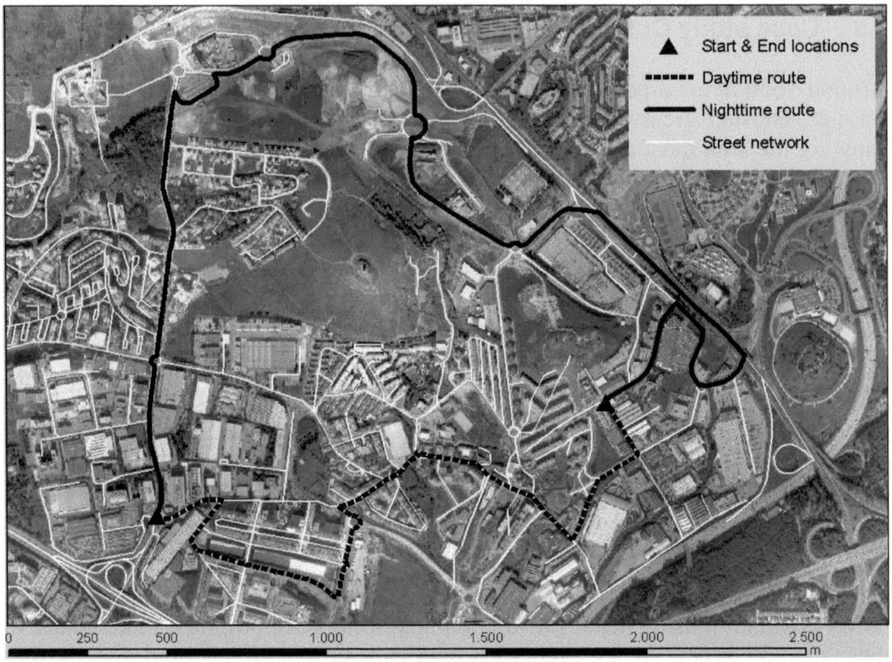

Fig. 6. Case study D: best route considering the population distribution

Table 3 quantifies and compares overall population exposure along each route and in each period.

Table 3. Total exposed population along each route and period

		Exposed population		
Best routes	Length [km]	Nighttime	Daytime	Difference [persons (%)]
Nighttime	4.9	345	3352	3007 (872)
Daytime	2.7	1001	2734	1733 (173)

Results show that consideration of spatiotemporal population distribution for routing leads to definition of quite different routes, having quite different lengths. Although the best route for the nighttime period is 4.9 km long, only 345 people are within the affected zone, whereas in the daytime period 2,734 people would be exposed along a shorter 2.7-km route. Additionally, total exposure along the best route for the nighttime period varies widely between night and day, increasing by 3,007 people to surpass the daytime's best route exposure.

Therefore, in order to minimize overall population exposure it is the nighttime period and longest route that should be selected for truck routing of hazardous materials in the study area.

5 Conclusions

As the population is not static, it was demonstrated that emergency management activities would greatly benefit from the use of reliable population distribution data with increased spatiotemporal resolution for estimation of human exposure and vulnerability. GIS enables both the development and spatial analysis' applications of these data sets.

The adopted approach, based on official data, allows the mapping of nighttime and daytime population distribution at high spatial resolution, to support local-level analysis. This method also efficiently accommodates people that work at home, by not considering that all active population leaves their residences during the workday. The main value of these results includes the increased spatial resolution of nighttime distribution, the fact that both nighttime and daytime distributions share the same spatial reference basis, and that daytime distribution is better approximated. Furthermore, the combination of both distributions yields an approximate representation of ambient population. The dasymetric mapping technique and zonal interpolation meet the requirements for disaggregation of population counts—obtaining positive values and preservation of total mass—so the final distribution matches official statistics and counts.

Given the availability of input data sets, this approach could be applied to all municipalities (19) comprising the Metropolitan Areas of Lisbon and Porto, that together contain 40% of Portugal's population.

Subsequent versions would benefit from a number of improvements: better modeling of "distributed activities" (e.g., cleaning, security), and accounting for people present in transportation networks, in hospitals and prisons, or involved in leisure and shopping activities; increased temporal segmentations of population distribution, so as to represent differences on a weekly basis (workdays vs. weekend) or on a seasonal basis (winter vs. summer); and the use of statistical sources beyond census demographics to consider tourism influx in areas and periods where that activity is significant. In the context of exposure and risk analysis, it would also be useful to model people who are indoors versus those involved in outdoor activities.

Acknowledgment

The kind support of GeoPoint Lda. throughout this project was greatly appreciated. The author also thanks the Lisbon Metropolitan Area for providing data on school facilities, from the *Sistema Metropolitano de Informação Geográfica – Grande Área Metropolitana de Lisboa.*

References

Alçada-Almeida L, Tralhão L, Santos L, Coutinho-Rodrigues J (2009) A multiobjective approach to locate emergency shelters and identify evacuation routes in urban areas. Geogr Anal 41:9–29.

Balk DL, Deichmann U, Yetman G, Pozzi F, Hay SI, Nelson A (2006) Global mapping of infectious diseases: methods, examples and emerging applications. In: Hay SI, Graham AJ, Rogers DJ (eds.) Advances in Parasitology, vol. 62. Academic Press, London, pp. 119–156.

Bhaduri B, Bright E, Coleman P, Dobson J (2002) LandScan: Locating people is what matters. Geoinformatics 5(2):34–37.

Birkmann J (2007) Risk and vulnerability indicators at different scales: Applicability, usefulness and policy implications. Environ Hazards 7:20–31.

Chen K, McAneney J, Blong R, Leigh R, Hunter L, Magill C (2004) Defining area at risk and its effect in catastrophe loss estimation: A dasymetric mapping approach. Appl Geogr 24:97–111.

Cutter S (2003) GI science, disasters, and emergency management. Transactions in GIS 7(4):439–445.

Dobson JE (2002) War is god's way of teaching GIS. Proceedings of the Merrill Conference on Science at a Time of National Emergency (http://www.merrill.ku.edu/publications/2002whitepaper/dobson.html).

Dobson JE (2003) Estimating populations at risk. In: Cutter SL, Douglas BR, Wilbanks TJ (eds.) Chapter 5.5 in Geographical Dimensions of Terrorism. Routledge, New York and London, pp. 161–167.

Dobson JE (2007) In harm's way: Estimating populations at risk. technical paper in NRC [National Research Council]. Tools and Methods for Estimating Populations at Risk from Natural Disasters and Complex Humanitarian Crises. National Academy Press, Washington, DC, pp. 161–166.

Dobson JE, Bright EA, Coleman PR, Durfee RC, Worley BA (2000) A global population database for estimating population at risk. Photogramm Eng Remote Sensing 66(7):849–857.

Eicher CL, Brewer CA (2001) Dasymetric mapping and areal interpolation: Implementation and evaluation. Cartogr Geogr Inf Sci 28:125–138.

FEMA (2004) Using HAZUS-MH for Risk Assessment. Technical Manual, FEMA 433. Washington, DC: Federal Emergency Management Agency. Available at www.fema.gov/HAZUS.

Fisher PF, Langford M (1996) Modeling sensitivity to accuracy in classified imagery: A study of areal interpolation by dasymetric mapping. Prof Geogr 48:299–309.

Goldewijk K, Battjes JJ (1997) A hundred year (1890–1990) database for integrated environmental assessments (HYDE, version 1.1). Report no. 422514002, National Institute of Public Health and the Environment (RIVM), Bilthoven, The Netherlands.

INE (Instituto Nacional de Estatística) (2001) Recenseamento Geral da População e da Habitação. Lisboa.

INE (Instituto Nacional de Estatística) (2003) Movimentos Pendulares e Organização do Território Metropolitano: Área Metropolitana de Lisboa e Área Metropolitana do Porto 1991–2001. Lisboa.
Lerner-Lam (2007) Assessing global exposure to natural hazards: progress and future trends. Environ Hazards 7:10–19.
McPherson TN, Brown MJ (2003) Estimating daytime and nighttime population distributions in U.S. cities for emergency response activities. Preprints: 84th AMS Annual Meeting, AMS, Seattle, WA, 10pp.
NRC (National Research Council) (2007) Tools and methods for estimating populations at risk from natural disasters and complex humanitarian crises. Report by the National Academy of Sciences, National Academy Press, Washington, DC, 264 pp.
Oliveira CS, Mota de Sá F, Ferreira MA (2005) Application of two different vulnerability methodologies to assess seismic scenarios in Lisbon. Proceedings of the 250th Anniversary of the 1755 Lisbon Earthquake. November 1–4, Lisbon, Portugal.
Pelling M (2004) Visions of risk: a review of international indicators of disaster risk and its management. UNDP Bureau for Crisis Prevention and Recovery, Geneva.
Sutton P, Elvidge C, Obremski T (2003) Building and evaluating models to estimate ambient population density. Photogramm Eng Remote Sensing 69(5):545–553.
Tobler WR (1979) Smooth pycnophylactic interpolation for geographical regions. J Am Stat Assoc 74(367):530–535.
Tobler WR, Deichmann U, Gottsegen J, Maloy K (1995) The Global Demography Project. (Technical Report No. 95-6). National Center for Geographic Information and Analysis. UCSB. Santa Barbara, CA, 75pp.
UNDP (United Nations Development Program) (2004) Reducing Disaster Risk: A Challenge for Development. UNDP Bureau for Crisis Prevention and Recovery, New York, 146pp.
Wu SS, Qiu X, Wang L (2005) Population estimation methods in GIS and remote sensing: a review. GISci Remote Sensing 42:80–96.

A Framework for Defining a 3D Model in Support of Risk Management

Serkan Kemec, Sisi Zlatanova, and H. Sebnem Duzgun

Abstract

A conceptual framework is proposed that defines the most appropriate 3D visualisations for different types of natural disasters in urban environments. Based on the disaster type, the needed level of detail for a 3D model is derived, which is then linked to the time needed to process the data and obtain this level of detail. The levels of detail are compliant with the 3D international standard CityGML. The framework is designed to serve risk managers and to help them make a better selection of 3D model representations to perform their tasks. After a brief introduction on the relations between types of disasters, data needed to manage the disasters and different users involved in the risk management process, the chapter elaborates on the parameters according to which types of hazards are classified. The framework is demonstrated for an earthquake case in Eskisehir, Turkey. The paper concludes with a discussion of the advantages and disadvantages of the given framework, as well as an outline of future research.

1 Introduction

Natural disasters have caused many casualties and injuries, as well as economic, cultural and structural damage, over the history of civilisation. Given that more than half of the world population currently lives in cities and that many economic assets are concentrated in urban areas, naturally, the vulnerabilities increase due to the growing complexity of urban processes.

Natural disaster risk is a function of hazards, elements of risk and vulnerability. Risk is defined as the expected losses, including lives, personal injuries, property

M. Konecny et al. (eds.), *Geographic Information and Cartography for Risk and Crisis Management*, Lecture Notes in Geoinformation and Cartography,
DOI 10.1007/978-3-642-03442-8_5, © Springer-Verlag Berlin Heidelberg 2010

damages and economic disruptions, due to a particular hazard for a given area and time period (WMO 2002). Risk assessment is one of the key elements of a natural disaster management strategy as it allows for better mitigation and preparation. It provides input for decision making, and it increases risk awareness among decision makers and other stakeholders (U.S. HHS 2002). Previous studies have shown that the presentation of hazards, vulnerability, coping capacity and risk in the form of digital maps has a higher impact than traditional analogue information representations (Martin and Higgs 1997). Digital maps are increasingly being used by disaster managers. Many authors believe that 3D visualisations have the potential to be an even more effective communication tool (Kolbe et al. 2005, Marincioni 2007, Raper 1989, Zlatanova et al. 2002); 3D graphical representations significantly reduce the amount of cognitive effort and improve the efficiency of the decision-making process (Kolbe et al. 2005, Zlatanova 2008). However, to achieve an appropriate 3D visualisation, two aspects must be ensured: appropriate presentation and appropriate tools for interaction.

The use of 3D spatial data for the whole disaster management process is a new, but quite attractive, topic in geoscience. There have been several studies on the use of 3D geographic information for modelling hazard phenomena and corresponding urban environments. Uitto (1998) proposed a framework using GIS for urban disaster management, considering the disaster vulnerability concept. Herold et al. (2005) outlined a framework for establishing an online spatial disaster support system for disaster management. The framework integrates GIS, spatial databases and the Internet for disaster management concepts. Zlatanova et al. (2007) discussed an emergency response framework. Technical necessities of multi-risk emergency situation response systems were evaluated from a 3D spatial information perspective, and they proposed a system architecture that covers data management and communication for problem areas. These studies, however, have not provided an adequate methodology for defining the level of 3D urban modelling for various types of natural hazards.

This paper presents a framework that establishes a link between the disaster type (e.g., flood or earthquake) and several components of 3D urban visualisation. The framework takes into consideration issues such as the resolution of the 3D model, the time/effort needed to create a model, and the availability of software and source data. This chapter is organised into five sections. The next section provides the needed background information. Section 3 introduces the framework. Section 4 demonstrates how the framework is applied. The last section elaborates upon the next steps for extending and testing the framework.

2 Background

Generally, 3D urban modelling is a holistic process of conceptualisation, data capture, sampling and data structuring and depends on the aim of visualisation (Raper 1989). In risk management, 3D modelling is very much dependent on the type of disaster being represented and the type of users involved in the risk management process. This section introduces related topics such as hazard classification, users and 3D urban objects of interest for risk analysis. Since 3D urban objects can be modelled with different resolutions, types of geometry and attributes, aspects of 3D semantic modelling are also briefly introduced.

2.1 Hazards

The development of a framework for 3D visualisation of various natural disasters first requires a classification of natural disasters depending on their characteristics. Several authors have provided classifications of the types of disasters depending on various parameters of hazard and risk characteristics (Shaluf 2007, Kaplan 1996, Mitroff 1988). Among them, the classification given by Burton et al. (1993) is used as a starting point in the proposed framework. Burton et al. (1993) provided six parameters for assessing the potential impact of a natural disaster. *Frequency* reflects the time interval in which a natural disaster occurs. For example, earthquakes may occur with a log-normal frequency, while landslides may occur seasonally. *Duration* is the period of time over which a disaster continues. Hence, it has a wide range, from seconds to years. For example, earthquakes are very short, while some landslides (e.g., creeping slopes) are long-term processes. *Spatial Dispersion* refers to the pattern of distribution of a hazard over the geographic area in which the hazard can occur. This parameter ranges from small to large. *Speed of Onset* is an important variable since it establishes the warning time. Most extreme disasters such as earthquakes, mud flows and flash floods give virtually no warning. Other disasters, such as creeping slopes, drought and desertification, act slowly over a period of days, months or years. *Areal Extent* is the spatial density of the disaster over the whole earth (e.g., earthquake zones are limited as they are governed by the tectonic plates). *Temporal Spacing* refers to the sequencing and seasonality of the disaster events. Some disasters are random, like volcanoes, while others have seasons, such as hurricanes, tropical cyclones and floods. These parameters give sufficient background to find the proper 3D spatial object representation needed for our framework. In the proposed framework, five parameters *(frequency, duration, spatial dispersion, speed of onset* and *areal extent)* are taken into account as they were found to be sufficient for generating a disaster prevalence index. This forms the first component of the proposed framework. Burton's last parameter, *temporal spacing*, is not applicable for the framework since a relation of the temporal spacing to the spatial and temporal definitions in the 3D urban models could not be clearly defined. The last parameter to be used in the framework is described in Sect. 3.1.

2.2 Users

As different user groups or decision makers may require different types of 3D urban models and functionalities, the next parameter to be identified for the framework is the characterisation of decision makers in risk management. Correct user group determination is also important for determining the functional content of the 3D urban model. Different users are involved in different phases of disaster management. For example, fire brigades, ambulances and police might be the main responders in the emergency phase, while urban planners and risk management specialists might be the users in the preparation phase (Zlatanova et al. 2007 and Zlatanova 2008). The introduction of a fundamental classification of users is beyond the scope of this paper. The users considered in this study are general users such as financial institutions (e.g., World Bank, insurance industry), academia (e.g., universities), the private sector (e.g., industrial organisations), governmental organisations (e.g., governors, municipals), civil society organisations (e.g., Red Crescent), international financial institutions and other public bodies.

2.3 3D urban modelling

The 3D urban modelling is another important factor to be considered in the development of the framework. Although various aspects of 3D modelling can be accounted for in the proposed framework, the type of objects to be modelled, their representation and the resolution, or Levels of Detail (LoD), play the most critical roles in 3D urban modelling for natural disaster situations. Thus, the proposed framework utilises previous knowledge of semantic models to define the objects to be used and the required resolution for urban modelling applications. In practice, the LoD concept can be directly related to the resolution; therefore, it is adopted in the proposed framework.

Various initiatives exist to define objects of interest in urban areas. Among them, CityGML is one of the few 3D urban modelling concepts that considers 3D semantics, geometry and topology in a generic sense (i.e., it is not application oriented). The most important objects are *Building* objects, which provide representations of thematic and spatial aspects of buildings; *Relief*, which is simply the terrain; *Transportation*, which represents objects of all modes of transportation, for example, roads, tracks, srailway or squares; and *Land Use*, which describes areas of the earth's surface dedicated to a specific land use. Several other objects, such as *City Furniture*, *Vegetation* and *Water*, can be also useful for risk management. *City Furniture* objects are immovable objects like lanterns, traffic lights, traffic signs, advertising columns, benches, delimitation stakes or bus stops. *Vegetation* is used to represent solitary tree objects, plant covers and surface or plant canopy. *Water* objects represent the thematic aspects and 3D geometry of seas, rivers, canals, lakes and basins. The only limitation of the current version of CityGML is the lack of underground objects. However, ongoing research and developments are considering extensions in this direction (Emgard and Zlatanova 2008). These can be included later in the framework.

These object classes compose the *object pool* in the framework. A set of objects important for any particular disaster can be obtained from this pool.

The most interesting concept in CityGML is the notion of an LoD, which defines the resolution with which 3D objects have to be modelled. The LoD is best developed for buildings. LoDs range from LoD0 to LoD4. LoD0 is the 2.5D level, over which an aerial image or a map may be draped (Kolbe et al. 2005). For buildings, LoD1 defines a box model, while LoD4 defines the inside of buildings. Naturally, the resolution increases from LoD0 to LoD4 (Gröger et al. 2006). The LoD concept is quite generic and suitable for small-to-large area applications. The LoD as developed in CityGML is adopted as a starting point in the proposed framework.

2.4 Sensor products and 3D reconstruction

Once the conceptualisation of 3D models is established (i.e., with respect to the types of objects to be modelled and their resolution), the next requirement to be satisfied for the development of the framework is the generation of the corresponding 3D models. Data used to generate 3D urban models can be obtained from passive sensors, active sensors or a combination of both (Hu et al. 2003, Kerle et al. 2008). Passive sensor methods are usually cost effective if large-scale 3D urban modelling is needed and require a combination of aerial and terrestrial images. Active sensors can provide, in a short period of time, large amounts of 3D data, but the 3D reconstruction models need to mature. Most urban modelling applications require integration of different data sources and sensor products. This integration can be in the form of simple overlay or multi-data information extraction (increasing the dimensionality using DSM and 2D images, increasing spatial resolution using high spatial resolution panchromatic images with multispectral images in relatively low spatial resolution and multi-criteria analysis to assess the natural disaster risk by considering various elements at risk) (Kerle et al. 2008, Tao 2006).

Sensor products and methods for reconstruction are important factors in the proposed framework. Risk managers have to be aware of how much effort and money is needed to obtain an effective 3D model for visualisation. For example, current technology is relatively limited in terms of radar systems (less than 45), as compared to more than 130 laser scanners and thousands of optical systems. In practice, this means that a method based on products of laser scanning or optical sensors is more likely to be better matched with the resources of a specific municipality. The selection of products is also dependent on the desired LoD. For instance, 3D models textured with images always require the use of optical sensors. The efforts required for creating a detailed 3D model (e.g., buildings in LoD3) differ significantly from the efforts needed for obtaining an LoD1 model. The proposed framework aims to help risk managers in municipalities make the most appropriate decisions regarding the resolution of the 3D model.

3 The framework

The framework consists of the following four groups of parameters: (1) hazard assessment, (2) user/elements at risk, (3) data and process requirements and (4) needs for visualisation (Fig. 1). *Hazard Assessment* refers to the determination of the three characteristics of the 3D urban model, which are *Hazard Characteristic Medium, Indoor/Outdoor Resolution* and *Data Representation.* Hazard Characteristic Medium is the vulnerability value of any model object (e.g., of a building in an earthquake case or a sea-water object in a tsunami case). Indoor/Outdoor Resolution defines the abstraction levels of each modelling object. Low spatial resolution would mean a low LoD, while high spatial resolution would mean a high LoD. Data Representation involves the data and procedures needed for a specific model. Here, the alternatives to 3D data representations such as boundary (surface) or volume approaches (e.g., voxel) should be evaluated.

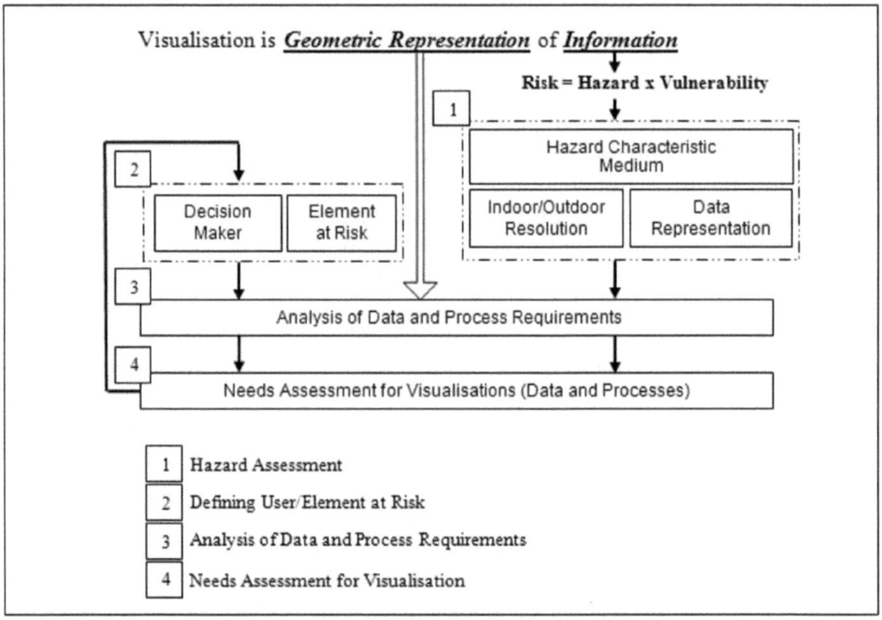

Fig. 1. Steps and workflow

The next group of parameters is defined as the *User/Elements at Risk.* A user might be interested in different sets of risk elements, which depend on the components of the urban environment. For example, an insurance company may have interests only in buildings, while utility companies might be mostly concerned with the effect on their networks. This is to say that the objects to be considered (and included) in a particular 3D model have to be selected with respect to the user.

When urban objects and the elements at risk are defined, they are fed into the analysis of data and process requirement stages to specify efforts needed for establishing model objects with data and process efforts.

3.1 Hazard assessment

This section introduces the final parameter called *Indoor Penetration (to the Built Environment).* Some natural hazards exert their fatal effects on built environments by the intrusion of hazard-related material into the built structure. This material can be soil, mud or rock in landslides or water in floods and tsunamis. The indoor penetration parameter is used to determine the indoor LoD, together with the spatial dispersion. For instance, in many disaster cases such as tsunamis, floods and landslides, it would beneficial to have 3D indoor models with floor distributions (or even apartments) to better estimate those parts and floors of the building that will be affected.

The six hazard assessment parameters, adapted from Burton's hazard classification (see Sect. 2.1.) and the newly introduced parameters are used to achieve two basic outputs, the first being the so-called *prevalence index* of different hazard types. The *prevalence index* of different hazard types is obtained from the parameters of *frequency, duration, spatial dispersion, speed of onset* and *areal extent.* Figure 2 presents a graphical representation of the prevalence index. As can be seen, a hazardous event that occurs frequently, with a long duration and slow speed of onset, over a widespread and large area causes the most pervasive effect to the urban environment. For example, specific urban earthquakes can be a more prevalent hazard than landslides.

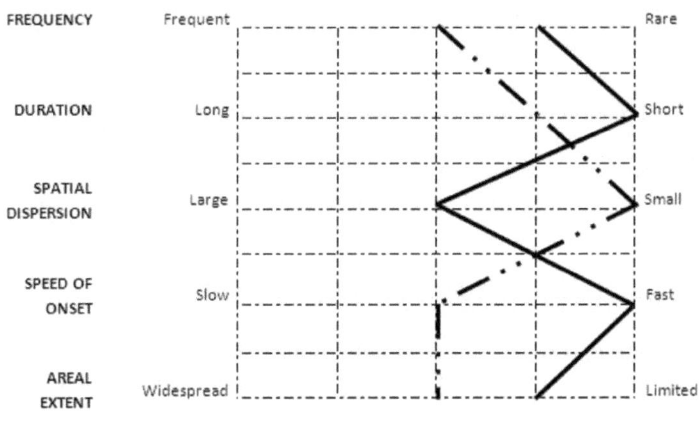

Fig. 2. Hazard prevalence level of earthquake and landslide cases (the *flat line* represents an earthquake, and the *dotted line* represents a rapidly moving landslide)

Each parameter is normalised to obtain a value on a scale from one to five. Each of the five parameters that form the shape of the prevalence function is related to a hazard type. The cumulative evaluation of these parameters constitutes the hazard-related part of the prevalence approach, which aims to relate the hazard type and spatial characteristics of the desired 3D urban model.

The hazard prevalence value is used with the urban area extent and the population density parameters of the target city to evaluate the LoD level of the desired 3D urban model. If a hazard occurs in an urban area that has a high areal extent and/or low population density, the needed model detail level decreases.

To determine the resolution (or LoD), hazard characteristic medium and data representation, one or more hazard assessment parameters can be used. Indoor and outdoor resolutions (or LoD) are obtained from the spatial dispersion and intrusion parameters. The hazard characteristic medium is basically identified from the parameter type of natural hazard. The data representation is established by using a combination of the physical mode and duration parameters.

Further formalisation of the link between the hazard type (defined by the six parameters) and the derived LoD can be found in Kemec et al. 2009.

3.2 Users and elements at risk

Hazards can affect physical assets and/or humans, and therefore, the risk can be considered to be human-related or nature/infrastructure-related. In an urban area, citizens, cultural heritage (e.g., buildings and natural phenomena), assets, infrastructure (e.g., roads, utility networks and rivers), and private and public housing may be vulnerable to a specific disaster and are considered to be elements at risk. Furthermore, different risk management specialists (stakeholders) might be interested in specific elements at risk related to their professional area. For example, a water engineer may be interested only in the risk of dam breakage and will not be involved in an estimation of the vulnerability of citizens. The most likely social issues will be considered by a social expert. Therefore, it is important to know who the users (user groups) are and what elements at risk they are interested in. The identified elements at risk have a direct influence on the required 3D model and visualisation. These issues are, however, beyond the scope of this paper.

3.3 Analysis of data and process requirements

This section focuses on finding a proper representation for each model object. Each object is pointed out on a three-dimensional object representation scale composed of three axes: *Level of Detail (LoD)*, *Level of Data Processing (LoP)* and *Hazard Type (HT)* (Fig. 3).

With respect to hazard type, a specific level of detail is required, which determines the level of processing. This relation may be represented as a point, line, area or volume in the three-dimensional scale. The *prevalence level* found from the hazard assessment is used to rank different hazard types. Hazards that have a low prevalence level are placed close to the origin, and those with a high prevalence are placed far from the origin. The resolution obtained from the hazard assessment is used for relating the HT and LoD. The LoD axes comply with the LoD as defined in CityGML. Further improvements and refinements of these levels will be needed, however. Currently, LoD4 is the only level for indoor, and this might be insufficient for the purpose of the framework.

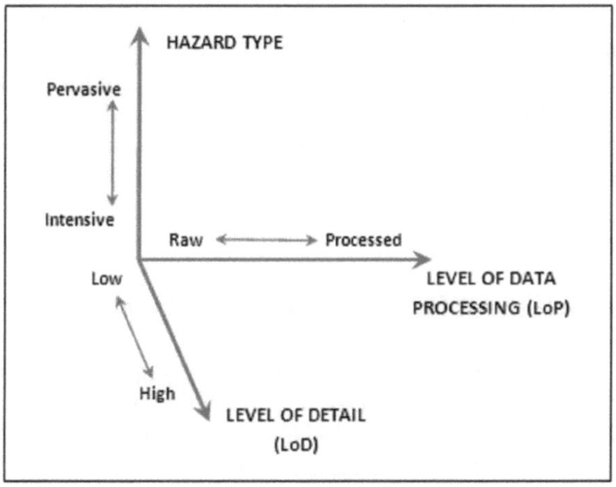

Fig. 3. The three axes used to define the 3D model: Hazard Type, Level of Detail and Level of Data Processing

The LoP axis represents the effort required to generate the needed geometrical representation or to obtain the available data. In our data visualisation environment, there are three main components: *urban objects (above* ground), *terrain objects (on the ground)* and *hazard medium*. The integration of these three components constitutes the resulting model. Each of these components has process steps, model components and related spatial data analysis techniques. For urban objects, the needed process steps could start with digitising (if the model generator has paper copy maps) or building footprint extraction (the generator has only aerial or high-resolution images in this case). Land cover classification and building height detection from stereo pairs (in this case, we have building footprints but there is no height data) may also be beneficial for urban object determination. Terrain objects generation could be done by using contour data, stereo pairs or photogrammetric field surveys. The accuracy and compulsory process cost increases for each method, respectively. The last component, and the context of our model, is the hazard medium. The process needed for the hazard medium originates from the characteristics of specific hazards or the aim of the visualisation. The hazard medium could be dynamic or static, and thus, the process needs increase, respectively. These examples may clarify the LoP definition; some municipalities may need to have a 3D extrusion model of a given area. To create such a model and provide it to the corresponding specialists, raw data may be needed to process flood inundation areas (if data and software are available). However, if indoor model of several buildings is required, the process needs more effort to create such an output. Processing efforts (represented by the axis LoP) start with creating the needed 3D representation with data collection. Then, 2.5D terrain representations with draped aerial or satellite images can be considered but require more effort, and therefore, these come after the raw data. 3D extrusion such as façade texturing with ground images, 3D object generation and integration with the surface, automatic or semi-automatic roof construction, detailed façade modelling by using ground point clouds and detailed indoor modelling can be at the end of this axis. Processes that need personal experience and immature functionality require more effort and take place at the right side of the LoP axis. Automatic methods for creating an output are located on the left side.

3.4 Needs assessment for visualisation (data and processes)

In this section, the available data and outputs are compared with the results obtained from the data and process requirement analyses. The required new data and model can be discussed with respect to the resources of the municipality. For example, the visualisation of a given hazard may require a 3D model with LoD4 resolution. However, the municipality may only have high resolution stereo aerial images. To come up with the desired product, they will need additional data such as ground images, building data that contain detailed indoor information, software for image processing of stereo images, 3D geographic modelling software and visualisation software to combine detailed 3D geographic objects and other spatial outputs like terrain models. Moreover, this modelling process will reveal that there will be a need for a considerable amount of human resources for the operation of this advanced software. It is up to the decision makers to find the best balance between the required model and the available resources.

4 Illustration of the framework within an earthquake case study

For an illustration of the framework, an earthquake case in Eskisehir, Turkey, is used. In this case, a visualisation application must be built for the users in the Eskisehir municipality. The users are municipality staff, such as urban planners, cartographers and sociologists. They need a clear view of the distribution of vulnerable regions throughout the city. The Eskisehir municipality has an urban information system infrastructure, so they have quite some data and software (planning, cartographic and GIS software). In this case, a 3D urban model environment is used to visualise the previously calculated social, physical and accessibility vulnerability indexes of each building object (Kemec and Duzgun 2006). Eskisehir is one of the important centres of industry in Turkey. A number of dams and two universities are located within and near the city. Due to its rapid development, Eskisehir has become a popular location for new investments. It is an industrial city in central Turkey with a population of over 500,000. The greatest part of the settlement in the city is located on alluvium. The largest earthquake (Ms: 6.4) occurred in 1956. The pilot area is a part of the city centre and has various kinds of city development textures including low-rise, historic buildings and high-rise apartments. There are nearly 400 buildings in the case study area.

Step 1. Hazard Assessment: In this case, as only one hazard is considered, the result will be a horizontal cross-section. The resolution, hazard characteristic medium and data representation are defined as follows: earthquakes affect a large area, and there is no intrusion. This means that an extruded building model at LoD1 from CityGML is suitable for this case. Moreover, earthquakes occur beneath the earth's surface, and the aim of the model is a visualisation of the vulnerability index values of each building. Because it is a data-driven process model, the used data representation is B-rep.

Step 2. Defining User/Elements at Risk: As mentioned above, the users are municipality staff, and they can be both technicians and non-GIS experts. In general, they need relatively realistic 3D urban models. The basic elements at risk are considered to be buildings and socioeconomic structures of the people living in the buildings. To construct a realistic urban environment and express the aim of the visualisation model, the objects are defined as Buildings, Relief, City Furniture (e.g., utility poles and street lamps), Transportation and Vegetation.

Step 3. Analysis of Data and Process Requirements: The model objects defined in the previous step are represented with LoD1 of CityGML with some modifications. LoD1 for buildings, according to CityGML, does not contain façade image mapping, but the users request realism; therefore, it will be included. The model objects and their representations in the case study are clearly defined with respect to CityGML as well. Figure 4 represents the results of the data and process requirement analysis for this case. The normal letters denote object representations, and the italic letters denote initial data types.

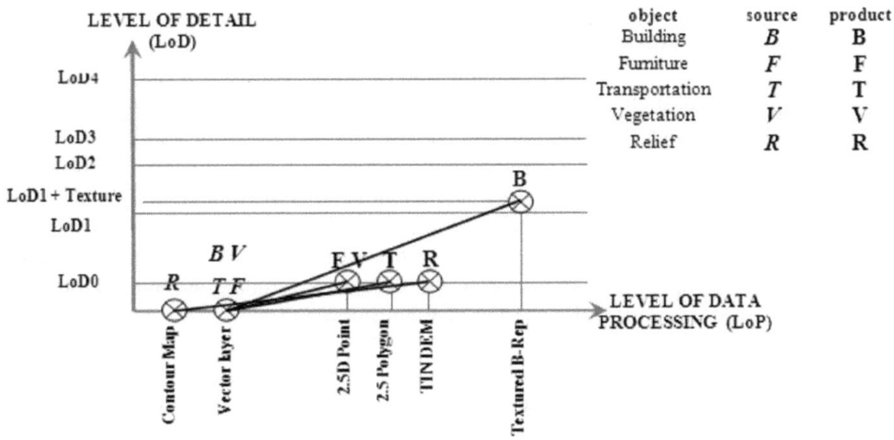

Fig. 4. Cross-section of framework for an earthquake case

Step 4. Needs Assessment for Visualisation: The available data for urban modelling are as follows: vector layers from the Eskisehir Municipality Digital City Information System, street and building footprint layers and 1/25000 digital contour maps. To construct building objects, façade images and building height data are needed, which are collected by a field survey. Process steps to develop an urban model are 2.5D DEM generation from digital contour maps, generation of LoD1 façade textured B-rep buildings by using building height information and ground images, draping of city furniture, tree points, road data and building models with a terrain model, relating tabular index data (hazard characteristic medium in this case) to the building objects (Fig. 5).

Fig. 5. General view from the 3D city model generated for an earthquake case

5 Conclusions

The proposed framework draws a link between the disaster type and the needed 3D model for an appropriate 3D visualisation of vulnerabilities in risk assessment. To be able to establish this link, parameters are defined for hazard type (frequency, duration, spatial dispersion, speed of onset, areal extent and indoor penetration) and 3D visualisation (representation, level of detail and level of processing). The framework can be used as a tool by risk managers (and other stakeholders) to determine the necessary data types for 3D models to be used in a particular risk assessment (for simulation and/or visualisation). The main expected benefit of this proposed framework is the ability to create understandable, yet well-balanced 3D models to be used as risk communication tools.

This study shows that technological developments in CityGML, 3D data sources, data processing and 3D visualisation can be easily adapted to the framework. The initial applications of the framework are very promising and well-accepted by risk managers. However, the tests have revealed that further developments and/or refinements along all of the axes will be needed. LoDs, as currently defined by CityGML, may not be sufficient for all types of disasters. For example, in some cases, only information about levels of a building is required, without the need to create the LoD4. A combination of CityGML LoDs may be more appropriate. For example, low outdoor (LoD1) resolution with low indoor resolution (floor level) can be used for a hazard application that has a large spatial dispersion and an intrusion to the built environment (e.g., a tsunami). Future work will investigate these possibilities in detail.

The level of data processing demands further elaboration as well. Currently, the level of data processing is established by a rough estimation of the resources needed to obtain a specific model. Further investigations are needed to define the appropriate parameters that influence the data-processing efforts. These parameters will most likely be specific for a municipality or data-processing unit.

More research and development are needed to formally define the relations and dependencies between users and elements at risk. A set of criteria should be made, according to which 3D objects can be selected from the object pool.

References

Burton I, Robert WK, Gilbert FW (1993) The Environment as Hazard, Guilford Press, New York.

Emgard L, Zlatanova S (2008) Implementation alternatives for an integrated 3D information model. In: Van Oosterom PJM, Zlatanova S, Penninga F, Fendel E (eds.) Advances in 3D Geoinformation Systems, Lecture Notes in Geoinformation and Cartography, Springer-Verlag, Heidelberg, pp. 313–329.

Gröger G, Kolbe TH, Czerwinski A (2006) CityGML Implementation Specification, Developed by the Special Interest Group 3D (SIG 3D), OGC Document Number 06-057r1.

Herold S, Sawada M, Wellar B (2005) Integrating geographic information systems, spatial databases and the internet: A framework for disaster management, Proceedings of the 98th Annual Canadian Institute of Geomatics Conference June 13, Ottawa, Canada.

Hu J, You S, Neumann U (2003) Approaches to large-scale urban modelling. IEEE Comp Graph Appl, Published by the IEEE computer Society November/December :62–69.

Kaplan LG (1996) Emergency and Disaster Planning Manual, McGraw-Hill, New York, NY.

Kemec S, Duzgun S (2006) 3D visualization for urban earthquake risk in geohazards. In: Nadim F, Pöttler R, Einstein H, Klapperich H, Kramer S (eds.) ECI Symposium Series, Vol. P7 http://services.bepress.com/eci/geohazards/37.

Kemes S, Zlatanova S, Duzgun S (2009) Selecting 3D urban visualisation models for disaster management: A rule-based approach. In Proceedings of TIEMS 2009 Annual Conference, June 9–11, Istanbul, Turkey, pp. 99–110.

Kerle N, Heuel S, Pfeifer N (2008) Real-time data collection and information generation using airborne sensors. In: Zlatanova S, Li J (eds.) Geospatial Information Technology for Emergency Response, Taylor & Francis, London, pp. 43–74.

Kolbe TH, Gröger G, Plümer L (2005) CityGML-interoperable access to 3D city models. In: Oosterom P, Zlatanova S, Fendel EM (eds.) Proceedings of the International Symposium on Geo-information for Disaster Management March 21–23, Delft, Springer Verlag.
Marincioni F (2007) Information technologies and the sharing of disaster knowledge: The critical role of professional culture. Disasters 31(4):459–476. Blackwell Publishing, USA.
Martin D, Higgs G (1997) The visualization of socio-economic GIS data using virtual reality tools. Trans. GIS 1(4):255.
Mitroff II (1988) Crisis management: cutting through the confusion. Sloan Manage. Rev. 29(2):15–20.
Raper JF (1989) The 3D geoscientific mapping and modeling system: A conceptual design. In: Raper JF (ed.) Three dimensional Applications in Geographical Information Systems. Taylor and Francis, London, pp. 11–20.
Shaluf IM (2007) Disaster types. Disaster Prevention Manage. 16(5):704–717.
Tao CV (2006) 3D data acquisition and object reconstruction for AEC/CAD. In: Zlatanova S, Prosperi D (eds.) Large-Scale 3D Data Integration Challenges and Opportunities. Taylor and Francis, London, pp. 39–56.
Uitto JI (1998) The geography of disaster vulnerability in megacities. Appl Geogr 18(1):7–16.
U.S. Department of Health and Human Services (2002) Communicating in a Crisis: Risk Communication Guidelines for Public Officials, Department of Health and Human Services, Washington, D.C.
WMO (World Meteorological Organization) (2002) Guide on Public Understanding and Response to Warnings, WMO, Geneva.
Zlatanova S (2008) SII for Emergency Response: The 3D Challenges. In: Chen J, Jiang J, Nayak S (eds.) Proceedings of the XXI ISPRS Congress, Part B4-TYC IV, July, Beijing, pp. 1631–1637.
Zlatanova S, Holweg D, Stratakis M (2007) Framework for multi-risk emergency response. In: Tao CV, Li J (eds.) Advances in Mobile Mapping Technology, Taylor&Francis, London, ISPRS Book Series, pp. 159–171.
Zlatanova S, Rahman AA, Pilouk M (2002) 3D GIS: Current status and perspectives, In: Proceedings of the Joint Conference on Geo-spatial theory, Processing and Applications, July 8–12, Ottawa, Canada, p. 6. CDROM.

DB4GeO, a 3D/4D Geodatabase and Its Application for the Analysis of Landslides

Martin Breunig, Björn Schilberg, Andreas Thomsen, Paul Vincent Kuper, Markus Jahn, and Edgar Butwilowski

Abstract

The analysis and preparation of information are still particularly critical points in the early warning chain of natural hazards. Also there is a strong demand for analyzing geological hazards such as earthquakes, landslides, and tsunamis. Therefore the management of heterogeneous and large sets of geodata may become one of the key issues in early warning and emergency management. Geodatabases may support the analysis of landslides significantly, as the responsible decision makers are usually confronted with huge amounts of original and interpreted data. The 3D/4D geodatabase introduced in this chapter provides efficient geodata retrieval by its service-based architecture which enables an easy-to-use internet-based geodata access. Concepts and implementation issues for the management of spatial and spatiotemporal data are presented. An application scenario for early warning demonstrates the applicability of the proposed approach. Finally a conclusion and an outlook on future research are presented within a distributed software environment for early warning geotools.

1 Introduction

During the last years a strong demand for 3D/4D geodatabases has become evident (Bode et al. 1994, Balovnev et al. 2004, Breunig and Zlatanova 2006, Zlatanova and Prosperi 2006) e.g., for the analysis of surface and sub-surface geoprocesses. However, hitherto there is no ready-to-use 3D/4D geodatabase system on the market. Therefore research is necessary that enables web-based 3D/4D geodata

M. Konecny et al. (eds.), *Geographic Information and Cartography for Risk and Crisis Management*, Lecture Notes in Geoinformation and Cartography,
DOI 10.1007/978-3-642-03442-8_6, © Springer-Verlag Berlin Heidelberg 2010

access and that brings 3D/4D geodatabase concepts nearer to geographers. For geographers, the special requirements to 3D/4D geodatabases are as follows:

- combined spatial, temporal, and thematic data access to geo-objects;
- database support to analyse geoprocesses such as landslides;
- archives for very large data sets.

Following the tradition of GeoStore, an information system for geologically defined geometries (Bode et al. 1994), and GeoToolKit, a library of 3D geometric data types (Balovnev et al. 2004), both developed at the Collaborative Research Centre 350 at Bonn University, we are currently developing DB4GeO, a service-oriented geodatabase architecture (Breunig et al. 2004, Bär 2007). Our early prototype systems GeoStore and GeoToolKit, both implemented in C++ programming language, have been based on a proprietary object-oriented database management system (Object-Store®). Furthermore, they did not support complex services and did not allow web-based access. GeoToolKit also did not have strong semantic support for geometries, i.e., thematic objects had to be implemented in a separated data model. DB4GeO is written in Java programming language and has been designed for service- and web-based data access right from the beginning. It is implemented on top of an open source object-oriented database management system (db4o®). DB4GeO is developed as an extensible toolkit for 3D/4D geo-database services to be used for different application fields such as analysis and early warning of geological hazards (Breunig et al. 2007, Glade et al. 2007) and emergency management (Zlatanova et al. 2007). DB4GeO provides suitable data types and geometric geodatabase services to support data analysis by complex geodatabase queries. Furthermore, combined thematic and spatiotemporal database queries are supported by its data model. The easy-to-use internet-based interface enables the direct service access and usage from other geotools. DB4GeO is embedded in a simple service infrastructure that may also be used as an integration platform for 3D and 4D geoscientific data.

2 A scenario for the early warning of landslides

One of the emerging applications for DB4GeO is the support for early warning scenarios such as examined in (Breunig et al. 2007). Within the framework of an early warning project (Breunig et al. 2007, Geotech 2009) a part of the hillsides in the Isar valley in the South of Munich (see Fig. 1), next to Pullach and Neugrünwald, has been selected for exemplary landslide simulations (Trauner and Boley 2009) in cooperation with the Bavarian State Office for Environment (Bayerisches Landesamt für Umwelt). In this area, the height difference of the slope reaches up to almost 40 m and the potentially endangered human infrastructure is located close to the edge of the slope. In the early and the mid-1970s there have been landslides in this area. As a reaction to these events and because of the risk potential several measuring devices were installed by the Bavarian State Office for Environment. Furthermore, the soil layers of the slope were investigated through numerous outcrops and boreholes.

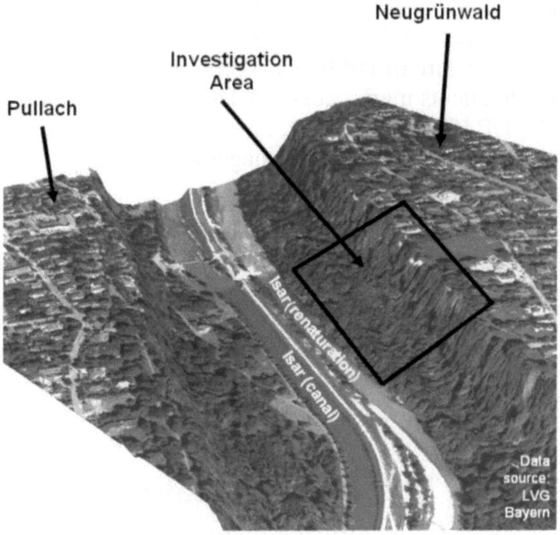

Fig. 1. Application area "Isarhänge Grünwald"

Today, after more than 30 years of investigations, extensive knowledge of the subsoil structure and the failure mechanism are available and can be used in the present project.

The requirements of this early warning scenario to a geodatabase are the following:

- Storage of original and interpreted geological data;
- retrieval of spatially and temporally selected geological data;
- geometric selection on the digital elevation model (DEM);
- integration of heterogeneous data;
- archiving of simulation results.

These requirements are well met by a geodatabase that provides spatial and spatiotemporal services for the analysis of landslides.

3 DB4GeO – the concepts

3.1 Service-based architecture

DB4GeO has been designed with a service-based architecture right from the beginning and is exclusively implemented in the Java programming language, i.e., its interface provides Java-based services. Presently, REST (Fielding 2000) is used as architectural style, i.e., REST style web services are used in enabling remote interaction with the geodatabase. In (Bär 2007) also a mobile database client has been realized, that interacts with versioning of 3D geometric objects on the server side. The system architecture of DB4GeO is presented in Fig. 2. On the client side, GIS clients or mobile clients may access 3D data managed by the DB4GeO server. On the server side, DB4GeO may be accessed exclusively via its services, which are divided into operations and version management.

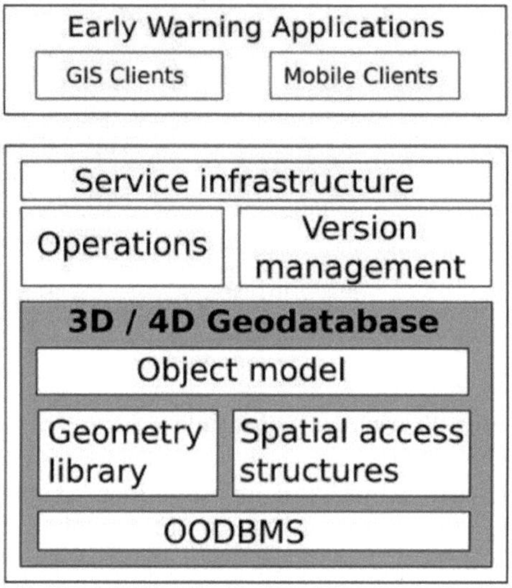

Fig. 2. System architecture of DB4GeO

The central part of DB4GeO is its 3D/4D geodatabase which is based on a geometry library and the R-tree based spatial access structures. The latest version of DB4GeO is implemented upon the open source object-oriented database management system db4o (Paterson et al. 2006, DB4O 2009).

3.2 Examples of complex DB4GeO services

3D geodatabase services should contain 3D spatial operators. Therefore DB4GeO offers primitive web services, such as

- distance between 3D geoobjects;
- intersection between vertical plane and a set of geoobjects;
- calculating boundary elements in each dimension;
- testing topological relations between 3D geoobjects and simplexes.

However, some applications such as light mobile clients, also need the capability to visualize 2D results of 3D spatial operators. Often small mobile devices do not have enough main memory and processing power to compute complex 3D models. That is why projection of a 3D model onto a plane is an important operator for geodatabases.

Figure 3 shows the principle of the so-called 3D-to-2D service, realized in DB4GeO: First, the location of a vertical profile section is defined on a 2D map. Secondly, the stored geometries are retrieved from a corresponding window in 3D space. Then the vertical intersection of the section plane with the 3D geometries is computed. Using a 3D buffer surrounding the plane, sample data are retrieved and projected onto the profile plane. Finally, for purposes of visualization and further analysis, the result may be projected onto a 2D vertical "paper plane" carrying the profile section (see Fig. 3).

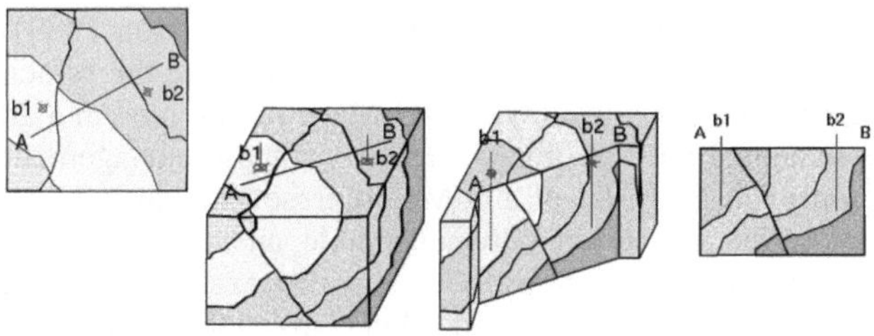

Fig. 3. Principle of the 3D-to-2D geodatabase service

3D geometric operations are provided by the DB4GeO libraries, i.e., spatial data types and spatial access methods are made available. An example for the application of the 3D-to-2D service is demonstrated in Fig. 10. Another example for DB4GeO's services is the "4D-to-3D service" which selects single snapshots of moving 3D geoobjects. With this service a set of geoobjects can be analyzed during a time interval.

3.3 The 3D/4D data model

The object-oriented data model of DB4GeO has been designed to support thematic, spatial, spatiotemporal, and combined thematic and spatiotemporal database queries. Thus it is possible "to attach" thematic attributes to geometric objects (see also Fig. 7). The implementation of the data model is currently based on simplicial complexes, i.e., on points/nodes, lines/edges, triangle nets/meshes, and tetrahedron nets/meshes, respectively. The data model has also been extended to simplicial complexes varying with the parameter time. As is well known, simplicial complexes are a special form of cell decomposition, which are used for the modeling of geoobjects in geoscientific applications. A special feature of the cell decomposition is the consideration of topology. The decomposition is done with cells of the same type. For instance, a surface decomposition can be achieved with triangles (2-simplex) and volumetric decomposition with tetrahedrons (3-simplex). Simplexes in DB4GeO are topologically classified in the usual way, i.e., towards their dimension: 0-simplex = node, 1-simplex = edge, 2-simplex = triangle, 3-simplex = tetrahedron. The implementation of this data model is shown in Fig. 6. Advantages of this special case of cell-complexes are that it is technically easy to implement and geometric algorithms can be executed on primitive data types. The drawback is the large memory consumption.

3.4 Plug-in for topology

Besides thematic and geometric geodata retrieval, in future also topology may be retrieved by DB4GeO. To achieve this, a topological plug-in is being designed for DB4GeO. The plug-in is based on Generalized Maps (G-Maps) implemented with cellular complexes (Lienhardt 1989, Lienhardt 1994, Brisson 1989) and has been examined in detail by (Thomsen et al. 2007). To illustrate the principle of this topological data structure, Fig. 4 shows 3D-G-Maps represented in a graph structure. It may be implemented with an object-oriented database management system.

Each cell-tuple (N_i, E_i, F_i, S_i) of a G-Map represents a particular incidence relationship, the so-called involution, between a node, an edge, a face, and a solid, and references neighboring relationships by exchanging exactly one component: either a node, or an edge, or a face, or a solid. This Cell-Tuple Structure (Brisson 1989) or Generalized Map (Lienhardt 1989) provides a cellular complex with the combinatorial structure of an abstract simplicial complex, and completely describes the topological relationships within the cellular complex. The graph structure in Fig. 4 can easily be transferred to a relational database structure. To traverse this structure, orbits are generated using combinations of different types of "involutions" α_0, α_1, α_2, α_3, respectively. By definition, an orbit $orbit_{i..j}(ct_0)$ consists of the subset of cell-tuples that can be reached from ct_0 using any combination of $\alpha_i,\ldots,\alpha_j,\ldots$ The components of dimension k where k is not contained in the set of indices i,…,j remain fixed, e.g. if $ct_0=(n,e,f,s)$, then $orbit_{012}(ct_0)$ leaves solid s fixed, and returns all cell-tuples of the form $(*,*,*,s)$.

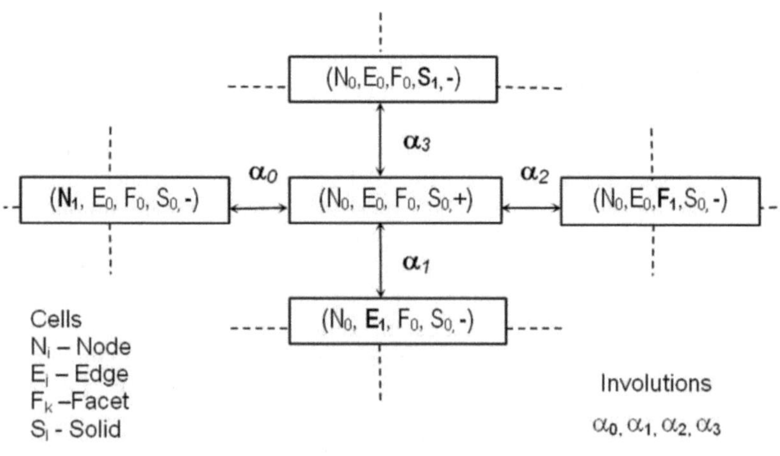

Fig. 4. Graph structure of a G-Map

The main advantages of G-Maps in contrast to other spatial representations such as boundary representation are

- they support the dimension-independent topological access to spatial and spatiotemporal data;
- they are based on a clear mathematical theory;
- they provide a unique topological data model for 2D and 3D space.

Another reason to have selected G-Maps is its strong acceptance in geological solid modeling (Mallet 2002). They are often used for sub-surface reservoir modeling.

G-Maps are currently used in the GoCAD®[1] 3D modeling and visualization tool (Mallet 1992) that is particularly used by oil and gas companies. Research on G-Maps and their application in 3D-modeling (Mallet 2002) and visualization is also focused at the Laboratoire SIC of Poitiers University, where the G-Map viewer and editor Moka is available as free software (GNU LGPL Version 3) (Moka 2006).

3.5 Version management

As shown in Fig. 2, the version management of DB4GeO is realized "on-top" of the data model at a separate level. There are additional services for the management and request of object versions (Bär 2007). A conversion option between version model and data model allows the use of all other database functionality for versions of database objects. The decisive advantage of the "integration on-top" approach is that the existing database management system does not need to be changed at all. Thus a separate development and maintenance of DB4GeO and the version management is enabled. Using DB4GeO in "normal mode," i.e., without using version management, is enabled as well as using it with version control. Existing applications of DB4GeO that do not use the version control, remain independent in their functionality. Applications that intend to use version management, however, have to implement these innovations and work together with the available services for version management.

A typical application of DB4GeO's version model is shown in Fig. 5. Here geometries of a landslide movement are changing in time being managed by DB4GeOs version model supporting the analysis of geological moving crevasses. Internally this is realized by generating an alternative workspace for each alternative version of an object. Hitherto the management of the versions is only implemented as linear history.

[1] GoCAD is distributed by Paradigm.

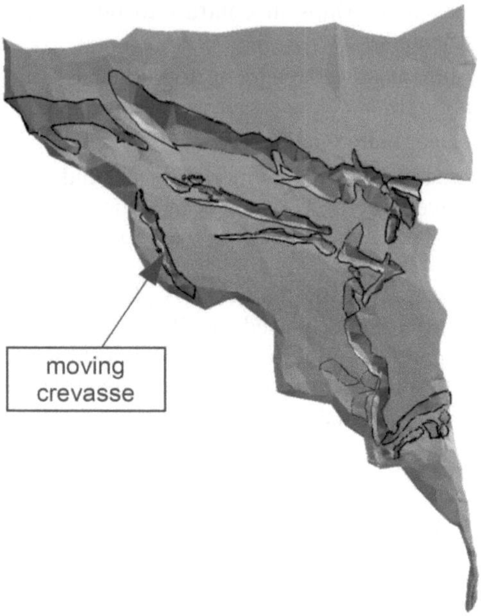

Fig. 5. Geological moving crevasses managed by DB4GeO's version model

4 Implementation issues

4.1 Services and internet-based service infrastructure

DB4GeO has been designed for the embedding into web-based infrastructures. Therefore a REST-architecture (Fielding 2000) is used. REST is deceptively simple—it is based on three fundamental design decisions:

- Agreement on a global addressing scheme, Unified Resource Identifiers (URIs), to identify resources;
- Agreement on a limited, generic application interface to manipulate resources. HTTP provides this generic interface via its "verbs," which include GET, POST, PUT, DELETE;
- Agreement on common formats, particularly in our case GOCAD®, to represent resources. Over time, additional formats have been added to export from DB4GeO, including GML and VRML.

Here are some examples of URIs:

1. http://server/projects/
2. http://server/projects/p1/area1

3. http://server/projects/p1/area1/all.gml
4. http://server/projects/p1/area1/all?3dto2d(CuttingPlane)
5. http://server/projects/p1/area1/sandsoil?getTimeStep(STEP)

The result of the first call is a list of available projects managed by the database. The second call gives an overview of the geoobjects that are part of the 3D geomodel stored in the geodatabase for the application area "area1" of the project with the name "p1." The third call creates a GML representation of this 3D geomodel. The result of the fourth call is the result of the 3D-to-2D-service described in Sect. 3.2. Input parameter is a vertical plane ("CuttingPlane"). Finally the last call returns a geoobject of type "sand soil" at a given time step. Spatiotemporal operators are described below.

4.2 3D/4D data model

To establish the independent adoption to other OODBMS, an abstraction layer was implemented which provides abstract interfaces dealing with connection and management of single objects and object sets. The interfaces are accessed through factories. Each OODBMS that may be used has to implement those interfaces and factories.

Spatial data sets usually belong to a specified project and a space within the project. A project may consist of multiple spaces. A space is responsible for efficient and consistent data storage for objects of equal dimension, coordinate system, and integrity constraints such as "polygons must not contain holes." Furthermore, data sets can be stored in temporal workspaces for storing intermediate results, e.g., to keep data consistent within service chains.

4.2.1 Spatial data types

The geometry library of DB4GeO consists of geometric and analytic objects. The relevant classes are *Vector3D*, *Line3D*, *Plane3D*, *Point3D*, *Segment3D*, *Triangle3D*, *Tetrahedron3D,* and *Wireframe3D*. The simplex types are collected and implemented in the *geom package* (Fig. 6). A *Point3D* is the geometric representation of a point in 3rd dimension. A *Segment3D* models a bounded line in 3D space with references to the start and end *Point3D* objects. A *Triangle3D* object is modeled by three *Point3D* objects and a *Tetrahedron3D* with four Point3D objects.

The class *ScalarOperator* makes methods available to compare scalars with a given accuracy constant (space feature). All geometric comparisons are based on this principle to avoid errors due to imprecise representations of floating-point numbers. The class MBB3D defines a minimal bounding box of 3D objects.

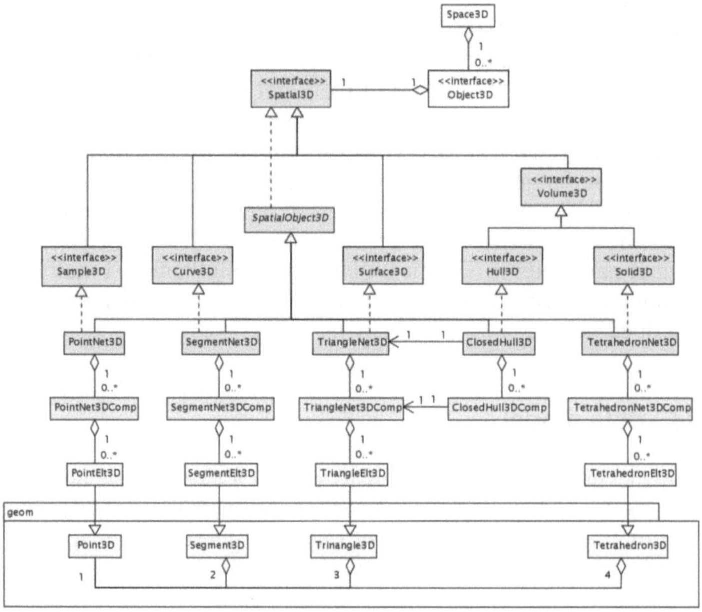

Fig. 6. Geometric model of the database objects in UML notation

Based on the geometry library a geometric model was implemented (Fig. 6). Its structure is symmetrical to the dimension of the geometric objects. Putting <Simplex> in place of the k-dimensional *Point3D*, *Segment3D*, *Triangle3D*, or *Tetrahedron3D*, the model can be described as follows: The class *<Simplex>Net3D* has a number of *<Simplex>Net3DComp* components. The components have no topological coherence to each other. Each *<Simplex>Net3DComp* models the specific k-dimensional simplicial complex through an adjacent disjoint set of 3D elements, i.e., *<Simplex>Elt3D* objects. There are no isolated simplexes within those components, i.e., they are topologically connected. *<Simplex>Elt3Ds* are finally direct subclasses of the supported simplex types collected by the geometric library. A special case of the implementation is the support of solids. The hull representation is established by 2-dimensional simplicial complexes implemented in *ClosedHull3D* and *ClosedHull3DComp*. Those classes wrap around *TriangleNet3D* and *TriangleNet3DComp*. Area calculations are passed to these simpler classes and volume calculations are implemented in the wrapping classes.

4.2.2 Basic spatial operations

The database core implements only basic spatial operations. All further operations such as input/output, constructing, or sophisticated geometric operations between geometric objects are not implemented in the core.

The basic operations between simplexes may be divided into the following types:

Basic operations between simplexes, such as

- testing topological relations between two simplexes with their bounding boxes,
- testing topological relations between two simplexes with their exact geometries,
- intersection calculations between two simplexes,
- projection calculations on planes,
- testing strict equivalence and geometric equivalence,
- testing validity, regularity, correct shape,
- calculation of face-simplexes,
- calculation of specific indicators such as area, volume, center, angle, etc.

Basic operations between 3D geoobjects, such as

- testing topological relations between 3D geoobjects and simplexes,
- calculating boundary elements in each dimension,
- calculation of specific indicators such as total area, total volume, number of boundary elements, number of simplexes per dimension etc.

4.2.3 Semantic model for spatial data types

The "semantic" (thematic) model (Fig. 7) implements the attributes, which may be attached to the geometric objects. The theme is defined by the class *ThematicObject3D* which implements the abstract class *Thematic3D*.

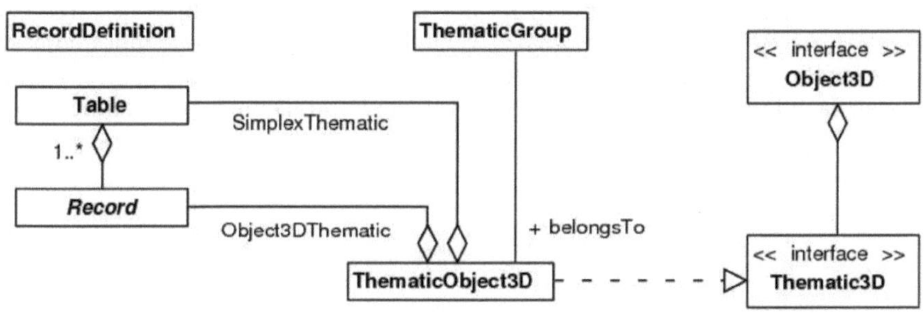

Fig. 7. Thematic model of the database objects, showed in UML notation

Geometric objects are members of a thematic group in which attributes are defined. Floating point numbers, integers, Boolean values, strings and vectors are supported. Likewise an indexed table can be defined in runtime to include a collection of data sets for thematic purpose.

4.2.4 Spatiotemporal data types

4D objects in DB4GeO are structured into components, which have a unique sequence for every single geometry object, such as a triangle. The spatiotemporal model of the database objects is shown in Fig. 8.

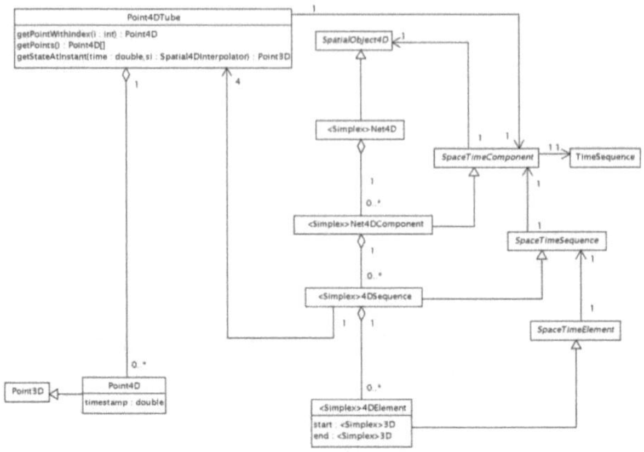

Fig. 8. Spatiotemporal model of the database objects, shown in UML notation

A spatiotemporal component C consists of a set $C = \{seq_1, seq_2, \ldots, seq_n\}$ of sequences which temporal discretizations are equal (Fig. 9). The elements of C build a contiguous and topological invariant net.

A sequence $S = \{ste_1, ste_2, \ldots, ste_n\}$ consists of a series of n spatiotemporal states (*ste*). Every element ste_{m+1} is a continuation of the element ste_m. Each of these 4D element states (*ste*) consists of two 3D objects and a time interval. The first 3D object describes the start state and the second 3D object describes the end state of an object during this time interval. Every point of a 4D object is stored in a *Point4DTube* $PT = \{point4d_1, point4d_2, \ldots, point4d_n\}$. Each of these tubes consists of n Point4D objects and describes one point of the simplex in the specified time interval, each Point4D object is realized with one Point3D object and a time step.

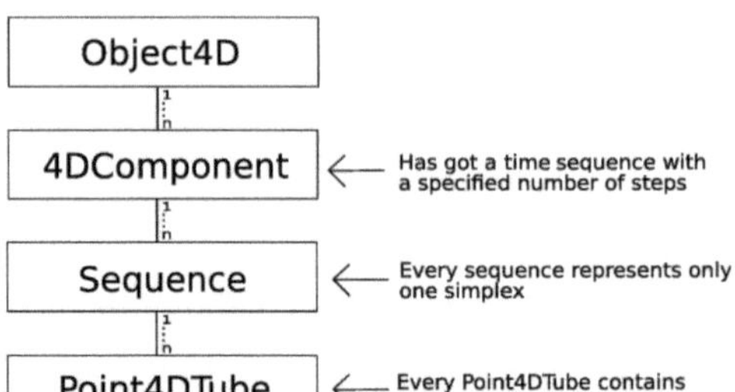

Fig. 9. Structure of 4D objects

Time dependency is one of DB4GeOs key features. Thus besides the request of spatial data we offer the ability to access temporal data in a simple and effective way. We describe the operators of the data types separately.

4.2.5 Basic spatiotemporal operations

DB4GeO offers operations on spatiotemporal geoobjects, such as:

- snapshot(instant): computes and returns an object which represents the state of this spatiotemporal geoobject at a specified instant with the help of linear interpolation.
- averageSpeed(): returns the average speed of a 4D geoobject. It is the quotient of way/time.
- intersects(BoundingBox box): determines, whether this object‘s trajectory crosses a given bounding box.
- isContained(BoundingBox box): determines, whether this object is contained in a given box for at least one arbitrary instant of its existence interval.
- isCompletelyContainedInBox(BoundingBox box): determines, whether this geoobject’s trajectory is completely contained in a given box, in other words, if it is enclosed in it during its existence time.
- getTrajectoryMBB(): returns the minimum bounding box of this object's trajectory.

The REST requests for these six sample queries run as follows:

1. http://server/projects/p1/area1/object?getSnapshot(INSTANT)
2. http://server/projects/p1/area1/object?getAverageSpeed
3. http://server/projects/p1/area1/object?intersects(MBB3D)
4. http://server/projects/p1/area1/object?isContainedInBox(MBB3D)
5. http://server/projects/p1/area1/object?isCompletelyContainedInBox(MB B3D)
6. http://server/projects/p1/area1/object?getTrajectoryMBB

4.2.6 4D-to-3D service

In DB4GeO, spatiotemporal geoobjects are accessed in a simple and effective way. But how do we get the geoobjects at a specified time to step out of the database? To handle this request we have developed the “4D-to-3D service.” With this service we offer access to the following queries:

- “Get one specified geoobject at a specified time step.”
- “Get all geoobjects contained in a space4D at a specified time step.”
- “Get one specified geoobject and all its time representations in one response.”
- “Get all geoobjects and all their time representations in one response.”

The REST requests of these four sample queries run as follows:

1. http://server/projects/p1/area1/object?getTimeStep(STEP)
2. http://server/projects/p1/area1/all?getTimeStep(STEP)
3. http://server/projects/p1/area1/object.gml
4. http://server/projects/p1/area1/all.gml

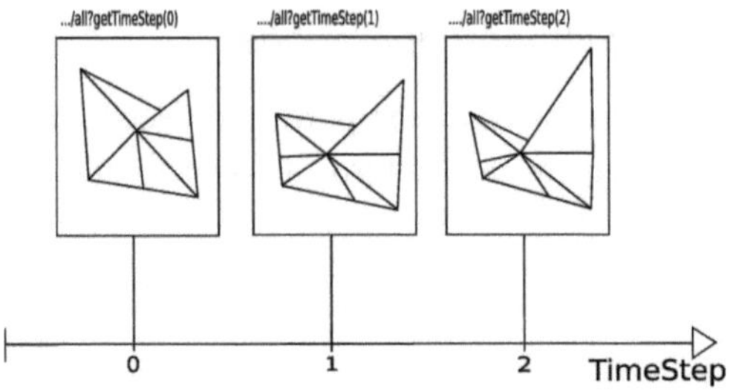

Fig. 10. Results of the 4D-to-3D service at different time steps

As we can see in Fig. 10, the access to the temporal data is very simple and easy structured. With every request we get the 3D geometry of the object at the specified time step. The 4D-to-3D service has explicitly been designed to retrieve snapshots of moving 3D objects.

In our further research we are dealing with the change of the topology of 4D objects over time. One topic is the refining of the meshing of existing data objects over a period of time. Another research topic is the possibility to extend an existing time interval, at the moment the data model is restricted to a predefined static time interval.

4.2.7 Indexes for supporting spatial and spatiotemporal operators

To support the spatial operators efficiently, a fast access is necessary on single elements of simplicial complexes, such as spatially specified triangles of a triangle net. The *<Simplex>Elt3D* objects are indexed in the class *<Simplex>Net3DComp* and inserted into a spatial access structure via the SAM interface (spatial access method). There are two implementations available in the prototype of DB4GeO. The class *RStar* is based on the R*-Tree structure (Beckmann et al. 1990). Furthermore, as an additional spatial access method there is the *Octree* class available. The octree is used to partition a three-dimensional space by recursively subdividing it into eight octants. Its advantage lies in fast insertion times, because no re-sort has to be done in this data structure. It is also planned to integrate a spatio-temporal access method (STAM) into DB4GeO (Rolfs 2005).

4.3 Storage of geoobjects

The users of DB4GeO will mostly work with complex objects. For this reason we wanted a database that is easy to use and high in performance. Our decision was to use an object-oriented database, because of its easy handling with complex objects and the proximity to the object-oriented programming language Java in which DB4GeO is written. Another reason mentioned by Mueller (2000) is that we need an OODBMS "...because of the high level of aggregation and the number of topological relations between the spatial object classes in geoscientific applications...." Furthermore, we intended to provide a database which is on the State of the Art, well documented and modifiable. Our choice fell on the free software database db4o (Paterson et al. 2006, DB4O 2009).

Because db4o meets our requirements, it seems to fit perfectly in our concept. The handling of storing objects in the database is quite simple. You just need to open a file (which will be created if it does not already exist) and initiate an *ObjectContainer*. To store one or more objects in a space, you call the set method of the *ObjectContainer*, specifying the object you want to store. DB4O stores "plain" objects – i.e. such objects that do not need to know about, or be altered to work with persistence mechanisms (Paterson et al. 2006, p. 6).

Example of opening a database and start working with it:

```
/** open the db4o database **/
ObjectContainer db = Db4o.openFile(databasename);

/** save a project in the database that may contain 1..n spaces **/
db.set(project);

/** get the project out of the database **/
ObjectSet result = (ObjectSet) db.get(new NormalPro-
ject(projectName, false));
```

Obviously, it is quite easy to store data in the db4o database. There is no need to deal with any table schemata.

To increase the performance and to avoid memory problems, we had to take care of the particular storage handling of db4o. We had to adapt the *updateDepth* and *activationDepth*, because it is not always desirable to activate all possible objects, as in our situation this could result in a very large graph being loaded into main memory. That is why we decided to activate every part of the graph for itself when it is necessary to work with it.

Example of the ActivationDepth handling:

```
/** activate the objects of the space to make them available (up to level 16 of the object graph) **/
db.activate(space.getObjects() , 16);

/** save only the object you want to save, not the whole project or space **/
db.set(space.getObjectByID(objects[i].getOID().getObject()));
```

With these extensions, the storage of arbitrary single objects and set of objects is provided by passing the maximal depth – e.g., the level in object hierarchies – in the graph during activation of the corresponding space.

5 Using DB4GeO in the early warning scenario

DB4GeO is being tested in the introduced early warning scenario "Isarhänge Grünwald." The following heterogeneous data are managed by DB4GeO:

- digital terrain model (resolution of 2 m),
- drilling profiles (borehole data),
- approximated surface model in 3D space.

The approximate surface model has been designed manually with the 3D modeling tool GOCAD® and then stored in DB4GeO. Thus the basic 3D database operators (*intersects*, *isContained*, *boundingBox*) and services (3D-to-2D service) are applied to the surface model.

Figure 11 shows the approximate surface model of the application area and the result of the 3D-to-2D service to our application scenario "Isarhänge Grünwald."

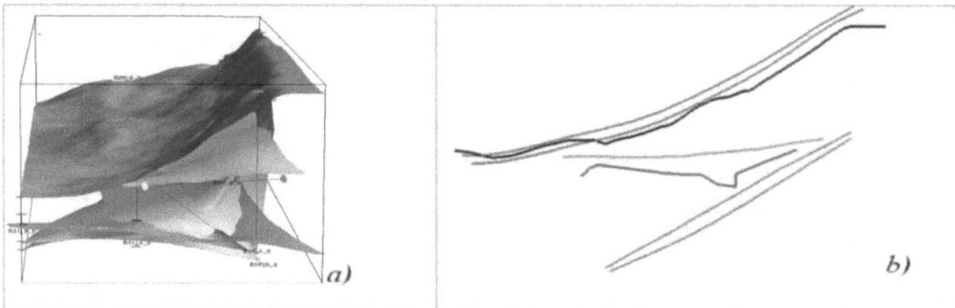

Fig. 11. **(a)** Approximate surface model of application area "Isarhänge Grünwald" **(b)** Result of 3D-to-2D service

A typical *workflow*, i.e., geodatabase session in this scenario is described as follows:

1. Export the 3D model from the 3D modeling tool and import it into DB4GeO.
2. Use the DB4GeO services (e.g. 3D-to-2D service) for data retrieval.
3. Work with the spatiotemporal client application analyzing the temporal change of landslides.

For example, the following types of analysis are supported by our geodatabase architecture:

- Operation *snapshot(instant)* of the geodatabase architecture returns the status of a slope at a given time point or time interval. Part of our future research is the implementation of advanced operations concerning the interpolation of objects between two snapshots.
- Operation *averageSpeed()* of the geodatabase architecture supports movement aspects of landslide analysis returning the average value for time series of measured location data, e.g., for a given crevasse.
- Operation *getTrajectoryMBB()* of the geodatabase architecture returns the area of interest (3D box), e.g., the region in which a boulder has moved during a landslide.

Together with simulation tools, the spatial and spatiotemporal services of DB4GeO are supporting the analysis of landslides. For advanced early warning functionality such as "triggering" a client application when the content of the geodatabase has changed, DB4GeO is being coupled with suitable services such as GeoRSS (http://georss.org). This functionality will be important for future works.

6 Conclusion and future research

In this chapter we have described the concepts of DB4GeO, an object-oriented 3D/4D geodatabase. Particularly its service-based architecture and 3D/4D data model including spatiotemporal operations enable DB4GeO to support geographers by providing web-based access and by archiving 3D/4D data. Its application to an early warning scenario of landslides near Munich, Germany, has been demonstrated by operations of the geodatabase architecture.

In our future works database applications for early warning of landslides will profit from accessing (key, value)-pairs of finite element models used in numerical simulations. Therefore the geodatabase has to be coupled with finite element model libraries. Obviously, further "plug ins" or Java-wrappers have to be made available to couple DB4GeO with other external software libraries. Coupling DB4GeO with advanced time-dependent services will allow early warning applications to react to changes in the geodatabase and to trigger special geoservices. Undoubtedly DB4GeO would also profit from interoperating with a distributed software environment for early warning, consisting of services for statistical analysis (Gallus et al. 2007) and linguistic information extraction or mechanic geotechnical simulation (Trauner and Boley 2009).

Acknowledgments

This is publication no. GEOTECH – 1229 of the R&D-Programme GEOTECHNOLOGIEN funded by the German Ministry of Education and Research (BMBF) and German Research Foundation (DFG), Grant 03G0645A. We would like to thank the anonymous reviewers for their valuable comments that helped definitely to improve the manuscript.

References

Balovnev O, Bode Th, Breunig M, Cremers AB, Müller W, Pogodaev G, Shumilov S, Siebeck J, Siehl A, Thomsen A (2004) The story of geotoolkit – An object-oriented Geo-database kernel system. Geoinformatica 8:1, Kluwer Academic Publishers, 5–47.

Bär W (2007) Verwaltung geowissenschaftlicher 3D Daten in mobilen Datenbanksystemen. Ph.D. Thesis, University of Osnabrück, 166p.

Beckmann N, Kriegel H-P, Schneider R, Seeger B (1990) The R*-tree: an efficient and robust access method for points and rectangles. Proceedings ACM SIGMOD, Atlantic City, NY, pp. 227–239.

Breunig M, Bär W, Thomsen A (2004) Advancement of geoservices – Services for geoscientific applications based on a 3D-Geodatabase kernel. Geotech-nologien Science Report No. 4, Information Systems in Earth Management, Potsdam, pp. 35–39.

Breunig M, Broscheit B, Reinhardt W, Ortlieb E, Mäs S, Boley C, Trauner F-X, Wiesel J, Richter D, Abecker A, Gallus D, Kazakos W, Bartels A (2008) Towards an information system for early warning of landslides. Informatics and Industrial Ecology Proceedings 22nd Conference on Informatics for Environmental Protection "EnvironInfo 2008", Lüneburg, Shaker Verlag, pp. 476–481.

Breunig M, Reinhardt W, Ortlieb E, Mäs S, Boley C, Trauner FX, Wiesel J, Richter D (2007) Development of suitable information systems for early warning systems. Geotechnologien Science Report, Potsdam, pp. 113–123.

Breunig M, Zlatanova S (2006) 3D Geo-DBMS. In: Zlatanova S, Prosperi D (eds.) Large-scale 3D Data Integration – Challenges and Opportunities. CRCPress, Taylor & Francis group, London, pp. 88–116.

Brisson E (1989) Representing geometric structures in d dimensions: Topology and order. In Symposium Computational Geometry. ACM, 218–227.

DB4O (2009) http://www.db4o.com/, accessed on 26.2.2009.

Fielding RTh (2000) Architectural styles and the design of network-based software architectures, Ph.D. thesis, University of California, Irvine.

Gallus D, Abecker A, Richter D (2007) Classification of Landslide Hazard/ Susceptibility in the Development of Early Warning Systems. SDH/ SAGEO, Montpellier, France, Springer.

Geotech (2009) http://www.geotechnologien.de, accessed on 5.6.2009.

Glade T, Becker R, Bell R, Burghaus S, Danscheid M, Dix A, Greiving S, Greve K, Jäger S, Kuhlmann H, Krummel H, Paulsen H, Pohl J, Röhrs M (2007) Integrative landslides early warning systems (ILEWS). – In: Stroink L (ed.) Geotechnologien. Science Report No. 10. – 136 p., Kick-Off-Meeting 10 October 2007, Technical University Karlsruhe, 89–100.

Lienhardt P (1989) Subdivision of n-dimensional spaces and n-dimensional generalized maps. 5th annual ACM Symposium on Computational Geomentry. Saarbrücken, Germany, 228–236.

Lienhardt P (1994) N-Dimensional generalized combinatorial maps and cellular quasi-manifolds. J Comp Geom App 4(3):275–324.

Mallet JL (1992) GOCAD: A computer aided design programme for geological applications. Turner AK (Ed.) Three-Dimensional Modelling with Geoscientific Information Systems, Proceedings of NATO ASI 354, Kluwer Academic Publishers, Dordrecht, pp. 123–142.

Mallet JL (2002) Geomodelling. Oxford University Press, New York, 599p.

Moka (2006) Modeleur de Cartes. http://www.sic.sp2mi.univ-poitiers.fr/moka/ (accessed on 26.2.2009).

Mueller W (2000) Skalierbare Datenstrukturen für große Mengen räumlicher 3D-Daten – am Beispiel von Dreiecksnetzen in einem GeoToolKit, Diploma Thesis, Institute of Computer Science III, University of Bonn, Germany.

Paterson J, Edlich S, Hörning H, Hörning R (2006) The Definitive Guide to db4o. Apress.

Rolfs C (2005) Konzeption und Implementierung eines Datenmodells zur Verwaltung von zeitabhängigen 3D-Modellen in geowissenschaftlichen Anwendungen, Diploma Thesis, University of Vechta, Germany.

Thomsen A, Breunig M, Butwilowski E, Broscheit B (2007) Modelling and managing topology in 3D geoinformation systems. Proceedings of 3D Geoinformation 2007, Delft, Lecture Notes in Geoinformation and Cartography, Springer, Heidelberg, pp. 229–246.

Trauner F-X, Boley C (2009) Application of geotechnical models for early warning systems to mitigate threats by landslides. Accepted for: 17th International Conference on Soil Mechanics and Geotechnical Engineering 2009, Alexandria, Egypt.

Zlatanova S, Holweg D, Stratakis M (2007) Framework for multi-risk emergency response. In: Tao CV,,Li J (eds.) Advances in Mobile Mapping Technology, Tay-lor&Francis, London, ISPRS Book Series, pp. 159–171.

Zlatanova S, Prosperi D (eds.) (2006) Large-scale 3D Data Integration – Challenges and Opportunities. CRCPress, Taylor & Francis group, London, pp. 88–116.

GIS Procedure to Forecast and Manage Woodland Fires

Antonio Condorelli and Giuseppe Mussumeci

Abstract

Over the last few years, the phenomenon of woodland fires has become ever more widespread, this is also due to changes in climatic conditions; unfortunately, experience has shown that the consequences for the environment and human life are very serious if the extinguishing operations are not timely. Starting from a brief analysis of the most important available models to simulate fire evolution, the aim of this work was to propose a dynamic prototypal GIS to support the forecast of woodland fire spread and the management of the available resources for extinguishing operations. Currently GIS technology offers many tools for the implementation of analytical models that are useful to evaluate the temporal and spatial evolution of natural phenomena. Specific functionalities for network and lifeline management are also available that make it possible to create a decisional support system (DSS) to obtain the real-time optimization of resource allocation.

In this chapter we introduce a model-based approach developed in the GIS environment, in which a well-known fire spread model has been implemented and a specific analysis of the road network has been carried out to supply real-time information about the optimal route to reach fires. The proposed methodology has been applied to two different case studies: to the territory of Mount Etna (Province of Catania), and to the Province of Enna, both in Sicily. The result is a DSS GIS environment by which it is possible to evaluate and to display real-time maps about: (1) the velocity and the preferential direction of a simulated fire; (2) the best route to reach the front of the woodland fire.

M. Konecny et al. (eds.), *Geographic Information and Cartography for Risk and Crisis Management*, Lecture Notes in Geoinformation and Cartography,
DOI 10.1007/978-3-642-03442-8_7,

1 Introduction

Woodland fires are part of the 'historical' problems that threaten our natural environment. The high number of events that each year is registered around the world and the consequent great loss of wooded areas make it necessary to pay great attention to the measures that can be adopted to protect one of the most precious resources of a territory. In this sense, if it is opportune to adopt preventive and repressive measures, both useful to reduce the number of the events, on the other hand, when an event has begun, it is necessary to provide an efficient and quick coordination of air and land extinguishing means to limit devastating effects.

Therefore, to contain damage, it is absolutely indispensable to use the most advanced technology that allows the elaboration of realistic forecast scenarios of the evolution of the fire event, which will then be of great importance for planning and managing the emergency resources and services. In this sense, GIS technology, by which it is possible to elaborate spread models to simulate the front of fire movements, is really an answer for the above described problems. Moreover, as we propose in this chapter, it is possible to integrate these evaluations in a more articulated procedure that allows the identification, in different contexts, of the best path for extinguishing means to reach the fire.

The developed GIS, which represents a *Decision Support System* (DSS), can be profitably used not only during emergencies, but even during the phase of planning because, using the simulations of multiple predictable scenarios, it allows the identification of critical points of road networks for linking the fire front to the territorial resources necessary for the organization of the interventions.

2 Fire spread models

There are many available fire spread models: most of them are based on equations of energy conservation opportunely adapted by many tests and experiments to simulate the fuel behaviour of different materials. Some of the models, such as the one proposed by Rothermel (1972) or another proposed by Albini (1976), are not very recent but they are still frequently used today. There are also more recent models such as FARSITE (Fire Area Simulator, Finney, 1998), which is a deterministic and mechanistic model of forest fire growth and spread based on existing fire behaviour models of surface fire spread (Albini 1976, Rothermel 1972), crown fire spread (Rothermel 1991, Van Wagner 1977, 1993), spotting (Albini 1979), point source fire acceleration (Forestry Canada Fire Danger Group 1992), and fuel moisture (Nelson 2000). Another important model is Prometheus (2006), which is a deterministic fire growth simulation model, coming from a national interagency project endorsed by the Canadian Interagency Forest Fire Centre (CIFFC) and its members. It uses spatial fire behaviour input data on topography (slope, aspect and elevation) and FBP fuel types, along with weather stream and FBP fuel type lookup table files. Another important model is Phoenix, which is one component of a bushfire risk management model, developed by Bushfire CRC, for southern Australia.

In this chapter the focus is not connected to the qualities or the differences between models, but the purpose is to verify the possibility of implementing models into the GIS environment, in order to relate the simulating results to the other typical spatial functionalities, such as network analysis or management. For this reason, the model we chose to realize this GIS application is the 'traditional' one proposed by Rothermel in 1972, a milestone in this sector: this model, in fact, is still today one of the most used and it is the basis of many others. Running the simulation model into GIS environment, we wanted to propose a wider integrated woodland fire emergency management application in which also resources for extinguishing are taken into account, as well as providing the best routes on the available road network.

In particular, the model of Rothermel can represent the 'behaviour' of an advancing fire that spreads in homogeneous and continuous fuel, carrying out the speed of the fire-head (that is the quickest part) so that it is possible to evaluate the potential path of the fire itself. In a more detailed way, Rothermel's equation has been developed into the hypothesis of superficial, uniform, continuous layers adjacent to soil, made of live and dead combustible material, of 2 m thickness, in stationary conditions and distant from heat source that started the fire. The model describes the way in which a fire moves forward in combustible material as a series of ignitions that progress like a 'contagion', and it expresses the speed that brings the fuel bands adjacent to the fire front to the ignition temperature. These characteristics are particularly useful in implementing the model in GIS environment. Without any other theoretical details, for which we suggest specialist studies, we draw your attention to Rothemel's equation:

$$R = \frac{I_{\mathrm{r}}\xi\left(1+\Phi_{\mathrm{W}}+\Phi_{\mathrm{S}}\right)}{\rho_{\mathrm{b}}\varepsilon Q_{\mathrm{ig}}} \qquad (1)$$

where R is the fire front propagation speed [m/min], I_r is the reaction intensity [Kcal/m^2 min], ξ is the propagating flux ratio, Φ_w is the wind coefficient, Φ_s is the slope coefficient, ρb is the ovendry bulk density [Kg/m^3], ε is the effective heating number, and Q_{ig} is the heat of pre-ignition [Kcal/Kg]. In the formula (1), the numerator represents the quantity of heat that combustible material has received; the denominator, instead, is the heat quantity necessary to bring it to ignition temperature. Because a detailed knowledge of these factors is very difficult, Rothermel reduced the different vegetal typologies to different combustible material 'models', useful for immediate application by the formula.

3 Implementation of the model and simulation of a fire event by GIS

This study was applied to two sample areas: Etna Park (Province of Catania), on the east side of the volcano, where a thick road network that drives into the wooded areas is present, and in another area in the Province of Enna, where many fires have burnt large tracts of woodlands over the last few years. The latter is very good because we have much information about woodland characteristics and data about past fires useful for comparison with the results of simulations. As concerns the Etna Park sample area, its territorial conformation is perfect to the GIS implementation we propose in this chapter, both for the model application and for network analysis. The model of Rothermel was implemented in GIS environment by a series of GRID themes, each one corresponding to one of the variables in the equation. The informative contents of some themes (Ig, ξ, ρ_b, ε, Q_{ig}) was mainly obtained not only from scientific literature data (combustible models) that referred to the characteristics of dominant vegetation species in studied wooded areas, but also from information (such as tree density, or the average concentration of combustible materials on the soil, etc…) deduced from documentation and thematic maps that are the property of the Province of Catania Administration and Etna Park Organization, or the property of the Province of Enna. The GRID theme linked to slope coefficient (Φ_s) was calculated by a *Digital Terrain Model* (DTM) of the studied area that resulted in a slope map. As concerns the wind coefficient (Φ_w), an index that represents the exposition and the scalar speed of wind, a uniform value on all cells was assumed.

For each of the realized GRID themes we adopted just one cell dimension, 10 × 10 m, because of the necessity of a continuous comparison with the vector themes of the road network. This dimensioning, that in relation to the general scale of the work (1:25000) could appear of excessive detailing, is absolutely necessary both for the physical elements that are involved in the modelling procedure, and to give reliability to the verification of the approaching path of extinguishing means, as we show in the next part of the chapter. Using the application of Rothermel's equation to the above described themes, we obtained a new GRID map where the value of each cell represents the potential spreading speed of the fire front expressed in meters/minutes (Fig.1).

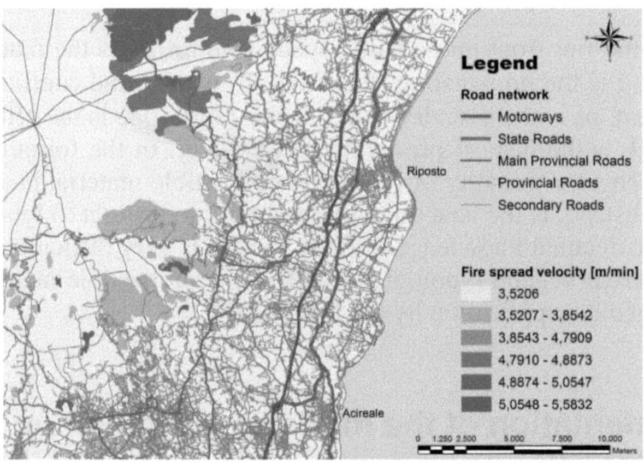

Fig. 1. Speed of fire front spreading evaluated by the Rothermel model (Etna)

At this point, because the first objective was to determine the spreading times and the distances covered by the fire front, a process of data elaboration was activated, finalized to use some proper GIS functions opportunely adapted to specific themes that we were working on.

A first elaboration was carried out to produce a new GRID theme in which each cell is associated to the inverse of the calculated spreading fire velocity. On this last GRID, after we made the hypothesis of a punctual origin of a fire, a spatial analysis using the function *Distance→ CostWeighted* was realized; in this way, we obtained a new GRID theme, in which the value of each cell represents the time (in minutes) necessary for the same cell to be reached by the fire from the supposed origin. The *CostWeighted* function was adapted to the modelling of the phenomenon, using as 'impedance' a referring GRID theme (to be more precise, the inverse of the fire spread speed, calculated thanks to Rothermel's model) and multiplying the values of each cell for the relative distances from the punctual hypothesized origin (in meters). In this way we obtained, cell by cell, a map of the necessary times for the fire ignition, based on the calculated fire spread speed and on the hypothesized origin (Figs. 2 and 3).

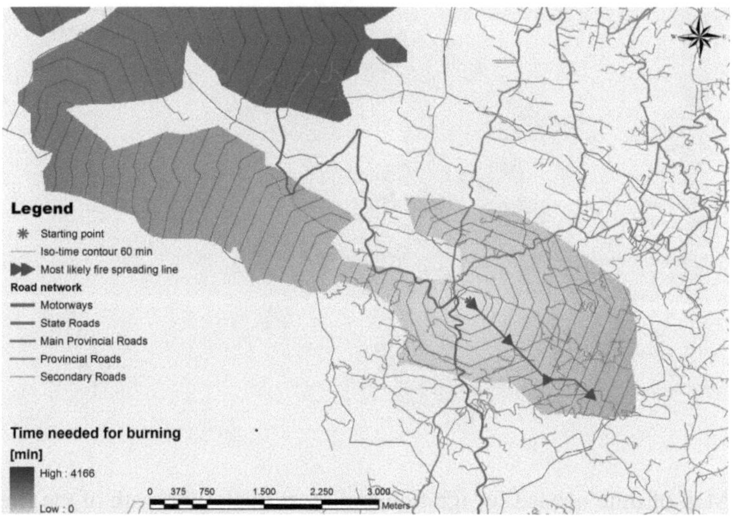

Fig. 2. Map of time needed for ignition and most likely line-path of the fire spread (Mount Etna, wind coming from North-West to South-East)

Thanks to this last elaboration, it was possible to elaborate another spatial analysis using the Distance→ ShortestPath function to identify the most probable approaching path of the fire. The ShortestPath function works on the GRID obtained by the Costweighted function, elaborating it again in function of a punctual destination that has to be assumed. The line that joins this destination point with the hypothesized other point for the origin of the fire defines the main wind direction that, in case of considerable air speed, is certainly the same direction of the most likely spreading of the fire front. Therefore, the ShortestPath function carries out, on the basis of the last evaluation about potential fire speed and the main wind direction, a path, made by 'minimum cost' cells: these are the cells where the fire spread is most likely, because the tendency to burn is higher. The ShortestPath function, of course, allows the identification of only a 'line of contiguous cells', which is not directly connected with the real width of the fire front; to represent in a more realistic way the phenomenon (in the hypothesis of constant speed and wind direction) it could be possible to use an average 'opening angle' on the origin point, using values inversely proportional to the wind velocity. At the end of the described elaborations it was possible to extrapolate iso-time contours on the GRID theme obtained by the CostWeighted function, which represent the place of the points (or, better, of the cells) that could be reached by the fire at the same time (Figs. 2 and 3). Verifying the intersections between the iso-time contours and the main spreading path of the fire front evaluated by the ShortestPath function, it was possible to make available a methodology to understand the advancing of the simulated fire in time and space.

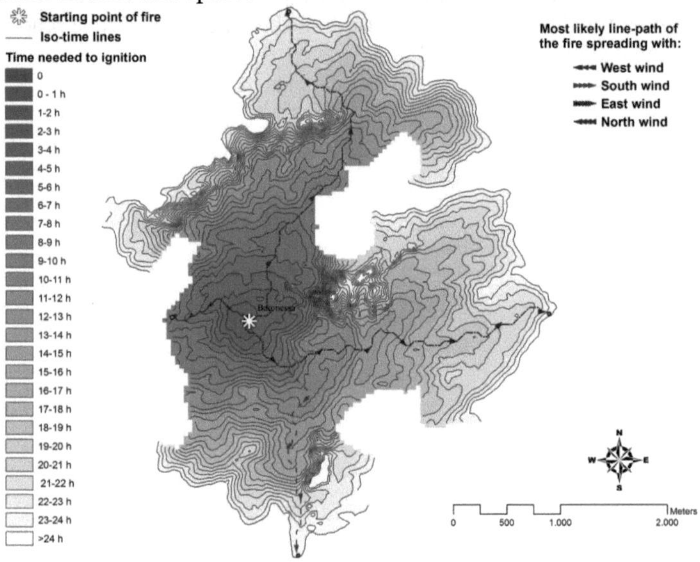

Fig. 3. Map of time needed for ignition and most likely line-path of the fire spread (Baronessa, in the Province of Enna, wind directions North, East, West, South)

4 Road network analysis to identify and update in real time the best approaching route for extinguishing means

The time-space simulation of woodland fires carried out by the proposed methodology can be profitably used to optimize the management of the emergency, based on the available territorial and infrastructural resources. Using a GIS network analyzer module, it is possible to identify the best operation center, (for example, a fire-brigade) that is able to make its land means reach the fire front first, and to choose the best approaching route. During the time needed to reach the 'destination' indentified when the alert started, it is possible that the fire front has gone beyond the destination itself or has destroyed some sections of the assigned path because of the variability of the phenomenon in relation to meteorological conditions. In this sense, a second quick verification of first selected path is needed and eventually its new definition, updating the destination and, if necessary, excluding road sections not useful for interventions.

The network analyzer, considering the characteristics of roads selected for the approaching path and the real traffic conditions can be also used to obtain real-time evaluations of travel time and to verify the position of means in relation to the contextual evolution of the fire front spread. To obtain these results, the RTK transmission of the position surveyed by GPS receivers installed on extinguishing means would allow the main operation center to evaluate, minute by minute, the movements and possible delays (for example due to traffic) of its squads and to elaborate again the proposed methodology to update the approaching route.

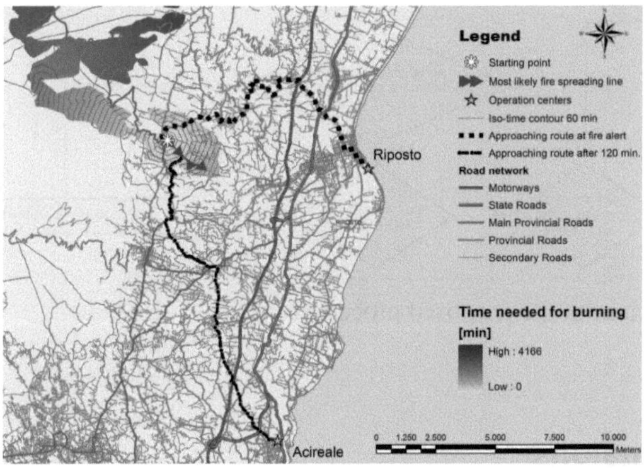

Fig. 4. Identification of the optimal approaching route in function of the most likely fire front position

For example, in Fig. 4 we show two different approaching routes: the one represented with dashed line has been evaluated by the network analyzer considering the position of the fire at the alert time; the one given with continues line refers to the most likely position of fire front after 120 min. It is useful to notice that the second solution (that also takes into account the travel time) consists not only of the evaluation of a new path, but also of the possibility of selecting a new starting point. This is very important to manage the emergency, because it is possible to identify the best operation center to be activated for the operations. The flow hart of Fig. 5 summarizes the different phases of the proposed procedure.

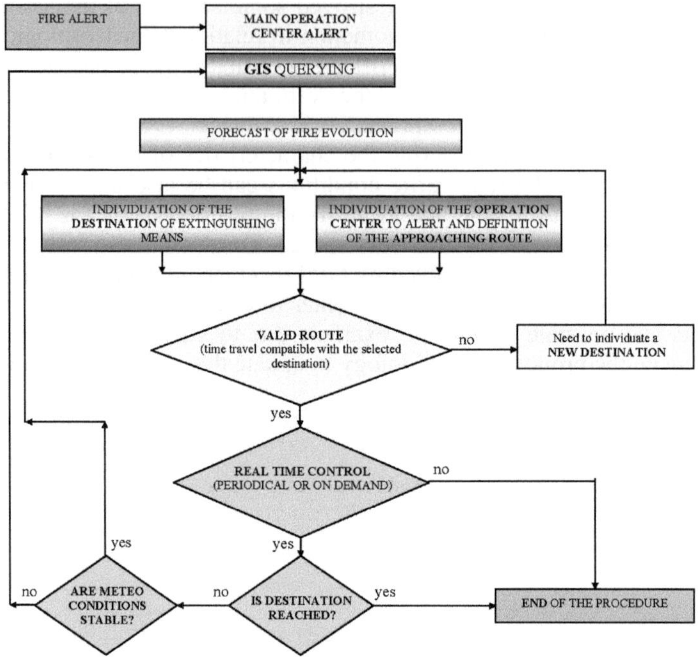

Fig. 5. Flow chart of the proposed procedure

5 Conclusions

The proposed procedure is based on GIS functions that can build a decisional support system useful both to manage fire emergencies and to provide precautionary plans, related to critical points that the system is able to highlight by the simulation of scenario events. After a short analysis of available fire simulation models, we chose Rothermel's model, which is one of the most used. Different tests were carried out on the mountainous area of Etna and in the territory of the Province of Enna, implementing the chosen model on wooded areas. The procedure is articulated in two different, independent but interconnected phases: a spatial analysis that allows the simulation of the fire spread by the implementation of Rothermel's model on specific GRID themes; a network analysis extended to the road network system, that defines and, eventually, updates the best origins and routes for extinguishing means, also evaluating travel time.

The emergency management would be helped by the use of 'real-time' GIS, using information that could be transferred to the main operation center by GPS receivers installed on the extinguishing means moving to the fire front, and by a special meteorological parameter monitoring system.

References

Albini FA (1976) Estimating wildfire behavior and effects. General Technical Report INT-30. USDA Forest Service, Intermountain Forest and Range Experiment Station, Ogden, UT, 92p.

Albini FA (1979) Spot fire distance from burning trees: A predictive model. General Technical Report INT-56. USDA Forest Service, Intermountain Forest and Range Experiment Station, Ogden, UT, 73p.

Finney MA (1998) FARSITE: Fire Area Simulator Model development and evaluation. Research Paper RMRS-RP-4. USDA Forest Service, Rocky Mountain Research Station, Ogden, UT, 47p.

Forestry Canada FireDangerGroup (1992) Development and structure of the Canadian Forest Fire Behavior Prediction System. Information Report ST-X-3.

Nelson RM Jr (2000) Prediction of diurnal change in 10-h fuel stick moisture content. Can J For Res 30:1071–1087.

Rothermel RC (1972) A mathematical model for predicting fire spread in wildland fuels, General Technical Report INT-115, USDA Forest Service, Intermountain Forest and Range Experiment Station.

Rothermel RC (1991) Predicting behavior and size of crown fires in the northern rocky mountains. Research Paper INT-438. U.S. Department of Agriculture, Forest Service, Intermountain Forest and Range Experiment Station, Ogden, UT, 46p.

Van Wagner CE (1977) Conditions for the start and spread of a crown fire. Can J For Res 71(3):23–34.

Van Wagner CE (1993) Prediction of crown fire behavior in two stands of jack pine. Can J For Res 23:442–449.

Cartographic Support of Fire Engine Navigation to Operation Site

Vít Voženílek and Lenka Zajíčková

Abstract

This chapter deals with traffic accessibility for fire engines within the city of Olomouc. The present way that drivers navigate (by hardcopy pre-prepared maps) is inefficient. Modern geoinformation technologies offer powerful capabilities to make navigation more convenient. The authors discuss the possibilities of GIS application based on network analysis over spatial databases fully equipped with attributes which are crucial for simulation of real-time fire operation. The attributes are mainly time of day, traffic density, road-works etc. Existing web-based map servers with a route planner module are tested as a technological environment for the application. Preliminary results show that various servers come up with different solutions. Inaccuracy is not allowed. Analytical aspects of the application reflect the authors' collaboration with Olomouc Fire Rescue Service. The main emphasis is put on cartographic aspects of navigation instructions fitted to drivers' abilities to read traffic situations.

1 Introduction

Many studies classify spatial information as one of the most valuable commodities of the current economic environment. They are part of a number of strategic decision-making processes in economy, politics, science and other areas. Cartography plays an important role in this. A map is a unique document that can give us a significant amount of spatial information in a quick and accurate way (Voženílek 2005). This is the reason why maps are commonly used for decision making and navigation in various practical fields (Charvát et al. 2008).

M. Konecny et al. (eds.), *Geographic Information and Cartography for Risk and Crisis Management*, Lecture Notes in Geoinformation and Cartography,
DOI 10.1007/978-3-642-03442-8_8, © Springer-Verlag Berlin Heidelberg 2010

The article presents a critical view of the use of cartographic works for the activity of the Fire Rescue Service e.g. in the Czech Republic. It is based on the fact that fire brigade operations can be considered typical spatial tasks. Solving such tasks can become much more efficient when using thematic maps. Together with spatial analyses in the GIS environment they make a strong tool for effective solution of tasks connected with fire brigade operation.

However, the reality is different. The Fire Rescue Service rarely uses maps, mostly because they are not aware of the advantages of acquiring spatial information from them (see Fig. 1). It is quite surprising that even though there has been considerable investment into highly developed and costly equipment for effective operations, the fire fighters have managed to get along without maps.

The authors are convinced that if maps (proper ones, i.e. thematic maps with emphasis on fire brigade needs) are used by the Fire Rescue Service, there will be considerable increase in the efficiency of their activity, especially shorter arrival time and better assessment of land-based activities, e.g. localization of operation sites (the most frequent location of fires, the most common arrival routes, the best/worst routes, etc.).

The aim of the article is to specify the role of cartography to support, ensure and improve fire brigade with navigation, using the example of the Olomouc Fire Rescue Service.

2 Current situation of using maps during fire brigade operations

One of the crucial tools for orientating in real space is the map as a product of cartographic visualization which is a set of map-related graphic procedures for analysis of geospatial data and information. Most activities in emergency management are spatially based and require tools with spatial concept (Crampton 2003). The relevant analytical and cartographic information referring to locational information about emergency operations from public maps provides Forrest who conducted research on maps for public information (Forrest 2007). Public maps tend to be very simple and generally have little information beyond target sites of fire brigade operations. Such maps with low accuracy of relevant information do not give confidence and should be replaced by scientifically completed thematic maps. These can be part of public participation in local policy decision making by web-mapping according to a concept presented by R. Kingston (2007).

Currently, maps are used rarely during fire brigade operations. An operation officer in the relevant operation centre is notified of an emergency. The officer enters information acquired from the caller into the Spojař system (Maršík and Uchytil 2007). The type of event and its extent, information about the caller and his or her contact number, as well as spatial information, especially the municipality, street or highway designation, type of building, building number, floor, are entered into the system.

The operation officer then adds the arrival route, which also includes possible water sources, and determines the operation technique. Each workplace of an operation centre is equipped with two monitors. One shows the operation information in the Spojař system, the other shows map materials in the GIS product "GISCln" by the company RDC Kladno s.r.o. The product is part of an integrated system "Výjezd". After entering information into the Spojař system, the GISCln monitor shows a 1 : 30 000 map segment localizing the operation site (Maršík and Uchytil 2007) (See Fig. 1).

Fig. 1. The present method of a driver's navigation to the operation site used by the Olomouc Fire Rescue Service: a textual file of list of streets in firefighter's hands

The GISCln system is interconnected with the Spojař system unilaterally. If the operation officer finds out that the stated address is not correct and wants to correct the operation site address in the GISCln system, the Spojař system does not react by changing the operation site address as well (Maršík and Uchytil 2007). Therefore, the address has to be corrected by a complicated process of entering all operation site parameters, not simply by "clicking" on the map. The drawing of the operation location is represented by one big red precinct on the map. The indication of the arrival route is absent altogether.

In addition to the above mentioned technology the Olomouc Fire Rescue Service uses the GIS Viewer administered by the GIS Unit of the Communication and Information Systems Department. The application's weakness is the out-of-date data. The operation centre also has the GIS data from the Call Centre of the emergency line 112 at its disposal, using the same principle of operation site localization.

Map materials from the above mentioned applications come from the internal data warehouse of the Fire Rescue Service of the Czech Republic situated in the town of Lázně Bohdaneč. They make use of various sources, e.g. the ZABAGED database (COSMC), StreetNET (CEDA), the T. G. Masaryk Water Research Institute, Czech Statistical Office, etc. (Mikulík et al. 2008). However, the data in the applications are not complete and are only seen by the operation officer in the office. The fire fighters' operation technology contains only a paper operation order which contains the operation site address and textual description of the arrival route.

2.1 Calculation of arrival route

The implementation of GIS-based network representations for transportation applications has increased dramatically over the past decade, with nearly ubiquitous availability of location services and address-based driving directions (Curtin 2007). Although network analysis in GIS has been largely limited to the simplest routing functions, the recent past has seen the development of object-oriented data structures, the introduction of dynamic networks (Sutton and Wynam 2000), the ability to generate multinodal networks and the use of simulation methods to generate solutions to network problems. Some network flow modelling functions have also been implemented, although there are substantial opportunities for additional theoretical advances and diversified application.

The arrival route is determined in a nonanalytical way. The operation officer uses the monitor to identify with the use of the GISCln system an approximate localization of the operation site (black precinct) and disposes of a textual file with a description of the arrival route. The file had been created in the following way: a selected worker used the map of the city of Olomouc to create a computer file containing a list of arrival streets for a particular operation site from the central office situated in the street Schweitzerova 222/91. Such a file with a list of arrival streets had been created for each street in the city of Olomouc. However, the arrival route is not calculated, it was decided as the worker saw fit. Since then the files have not been updated, only supplemented with files related to newly built streets. In the same way the descriptions of routes related to municipalities of the adjacent districts had been created. These contain two parts – a route to reach the municipality and navigation throughout the municipality. Both parts count on good knowledge of streets and excellent orientation in the municipalities. If the fire engine takes the wrong way or does not know how to reach the operation site, the operation officer assumes the responsibility for the navigation of the fire engine to the operation site. However, the navigation means that the operation officer uses a paper map to determine the fire engine's position and uses a walkie-talkie to lead them to the operation site. This means that the operation officer is also supposed to have good knowledge of the whole district, be aware of all newly constructed roads, road changes (e.g. construction of street refuges which could hinder the fire engine) or current traffic situation (e.g. traffic jams, narrowed passageway, etc.).

2.2 Cartographic representation of arrival route

Definition of the arrival route is provided in the form of a textual file of approximately. three lines. This means that an excellent orientation of the fire fighters in their district is supposed, as well as their perfect knowledge of the total of 669 streets in the city of Olomouc. This knowledge is ascertained by monthly drives when the fire fighters drive through the streets, revise the street network and monitor the road negotiability, quality, restrictions, etc. When new drivers are hired, they drive through streets and learn them by heart. The arrival route description is in fact a sequence of street names, together with the information whether to turn right or left at intersections. The size of the Olomouc Fire Rescue Service district almost corresponds to the administrative district of Olomouc.

2.3 Information about arrival route

The information about arrival route has the form of a textual "Operation order". It is a spatial task described as in a cookbook. However, compared to the cookbook pictures, i.e. a map containing the arrival route, are absent. Pictures can be read very quickly. Therefore, it is quite surprising that fire fighters in emergency situations use arrival route representation and description in the form of several lines written on a piece of paper.

3 Case study

When dealing with the analytical aspect of the task (calculation of the arrival route) we took into account the fastest route, both the day-time and night-time alternative, as well as the seasonal alternative, because all these conditions influence the arrival time (Lehmann et al. 2005). Traffic density, car parking situation and conflicts with pedestrians are different during the day and at night. For example at 3 p.m. the rush hour in the city of Olomouc can considerably influence the arrival route calculation.

The pilot case study contains three tasks which demonstrate deficiencies of the current solution, a proposed solution and their comparison. Two different web map route planners (Google Maps and Mapy.cz) were used to discover if currently used arrival routes by Olomouc Fire Rescue Service are calculated as the fastest routes to selected operation sites. The authors use the case study for demonstration of one typical situation in fire rescue services, not for showing general state worldwide.

3.1 Task 1 – Short route in the city

Operation site: Hynaisova 9a, Olomouc (the winter stadium in Olomouc)
Arrival route alternatives:

- Schweitzerova ⇒ Velkomoravská ⇒ Albertova ⇒ Štítného ⇒ Havlíčková ⇒ tř. Svobody ⇒ Legionářská ⇒ Hynaisova; arrival route length: 3.2 km; arrival time: approx. 7 min:

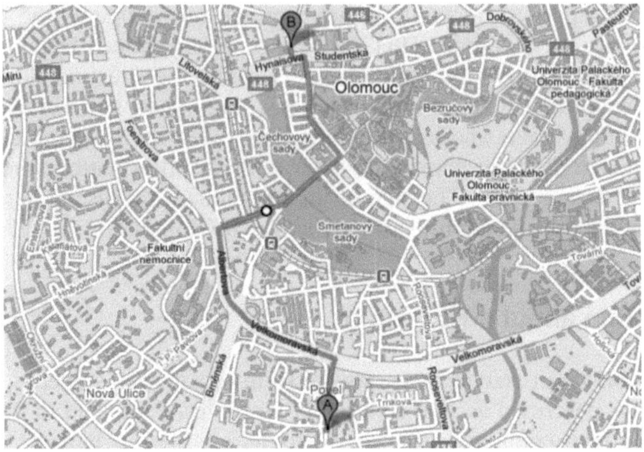

Fig. 2. First possible arrival route (*dark line*) to the winter stadium in Olomouc (automatic calculation at Google Maps)

- Schweitzerova ⇒ Velkomoravská ⇒ Albertova ⇒ Foersterova ⇒ Litovelská ⇒ Palackého ⇒ Hynaisova; arrival route length: 3.7 km; arrival time: approx. 7 min

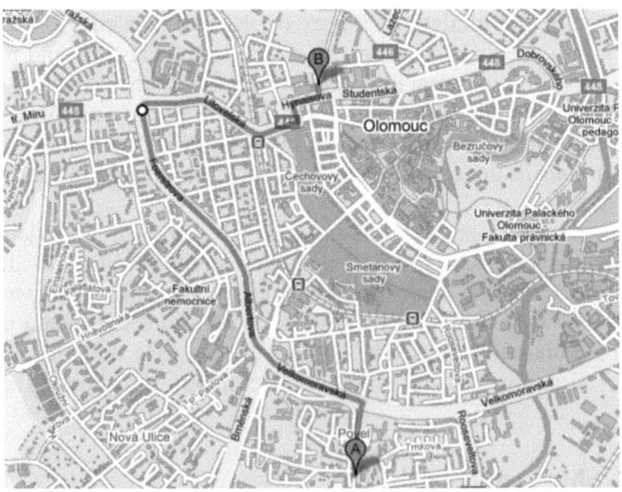

Fig. 3. Second possible arrival route (*dark line*) to the winter stadium in Olomouc (automatic calculation at Google Maps)

- Schweitzerova ⇒ Polská ⇒ tř. Svobody ⇒ Legionářská ⇒ Hynaisova; arrival route length: 3.0 km; arrival time: approx. 8 min; this is the route currently used by the fire engine drivers following a prescribed itinerary:

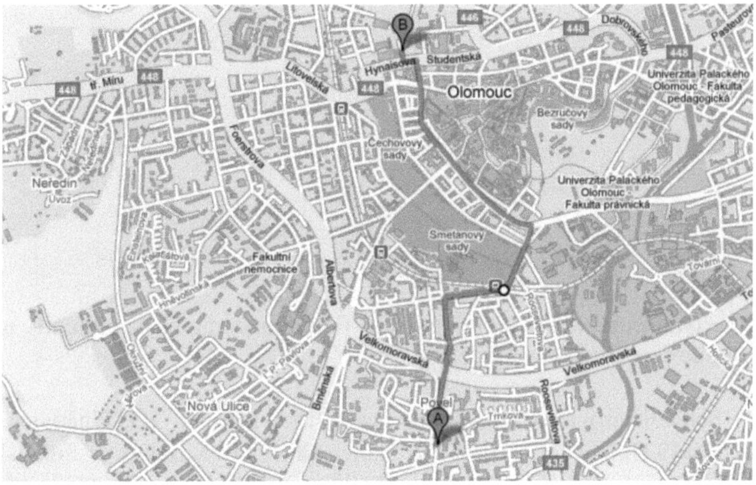

Fig. 4. Arrival route (*dark line*) to the winter stadium in Olomouc currently used by the Olomouc Fire Rescue Service

The first and second routes were created using the Google Maps route planner. Figures 2 and 3 represent the fastest route alternatives while the second route (Fig. 3) is longer than the first one (Fig. 2). Figure 4 represents the shortest but not the fastest route which is used in most cases by Fire Rescue Service. Only during the time when there are expositions on the Flora Olomouc exhibition grounds the route changes and the second alternative is used, so that the fire engine avoids the city centre and the Flora Olomouc exhibition grounds and saves time. It proves two findings: that the currently used arrival route differs from the fastest route and Google Maps finds different automatically calculated routes.

3.2 Task 2 - Exiting the city using minor outward-bound roads

Operation site: Nádražní, Senice na Hané (big hop-field)
Arrival route alternatives:

- Schweitzerova ⇒ Velkomoravská ⇒ Pražská ⇒ II/448 Ústín ⇒ III/44816 Vojnice ⇒ III/44815 Senice na Hané; arrival route length: 16.5 km; arrival time: approx. 28 min

Although this route goes off the highway (which is faster than lower class roads) it is the currently used arrival route by the Fire Rescue Service because the fire engine's velocity is restricted to 90 km/h. This is the reason why the drivers choose secondary roads, where the route is shorter (but not faster).

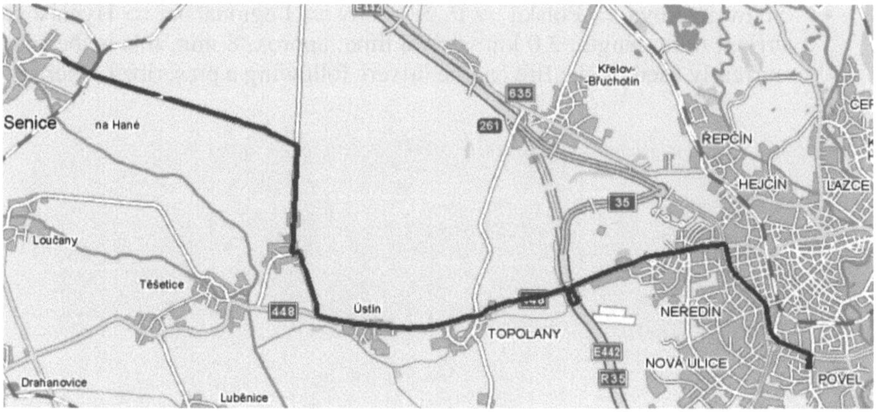

Fig. 5. Arrival route (*dark line*) to the hop-field Senice na Hané currently used by the Olomouc Fire Rescue Service

- Schweitzerova ⇒ Velkomoravská ⇒ třída Míru ⇒ I/35 Unčovice ⇒ II/449 Unčovice ⇒ III/44916 Senice na Hané ⇒ Žižkov ⇒ Sokolská ⇒ Nádražní; arrival route length: 20.9 km; arrival time: approx. 25 min

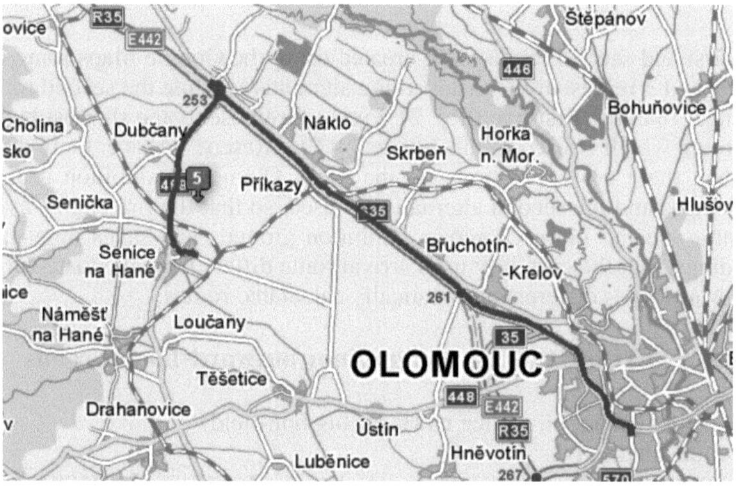

Fig. 6. The fastest route (*dark line*) to the hop-field Senice na Hané (automatic calculation at Mapy.cz)

Both above stated routes (Figs. 5 and 6) were derived using the web map route planner Mapy.cz. The first route (Fig. 5) is the shortest one, the second arrival route (Fig. 6) is the fastest one. The fire engine drivers use the first alternative strictly. If the operation site address is different or it is necessary to go through the Senice na Hané municipality to its other end, the drivers choose a route via the

R35 highway but arrival time is longer. The web Mapy.cz cannot bring into route calculation any additional requirements but is popular among engine drivers for their private activities.

3.3 Task 3 – Operation in a different district (a great distance from the city)

Operation site: Dykova 8, Prostějov (Palírna u Zeleného stromu Starorežná Prostějov distillery; one of the biggest distilleries in the CR)

Arrival route alternatives:

- Schweitzerova ⇒ Velkomoravská ⇒ Albertova ⇒ Brněnská ⇒ R46 Prostějov-sever ⇒ Konečná ⇒ Olomoucká ⇒ Vápenice ⇒ Blahoslavova ⇒ Plumlovská ⇒ Březinova ⇒ Dykova; arrival route length: 19.6 km; arrival time: approx. 21 min

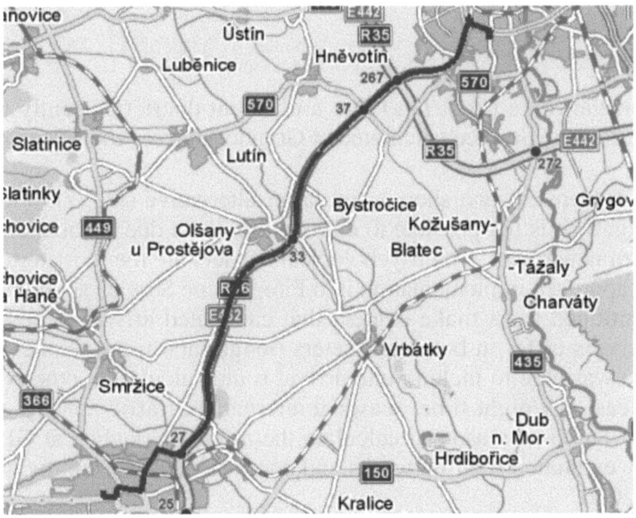

Fig. 7. The shortest route (*dark line*) to the distillery Palírna u Zeleného stromu (automatic calculation at Mapy.cz)

- Schweitzerova ⇒ Velkomoravská ⇒ Albertova ⇒ Brněnská ⇒ R46 Prostějov-centrum ⇒ Kralická ⇒ Dolní ⇒ Wolkerova ⇒ Palackého ⇒ Plumlovská ⇒ Březinova ⇒ Dykova; arrival route length: 20.6 km; arrival time: approx. 20 min; this is the route currently used by the fire engine drivers:

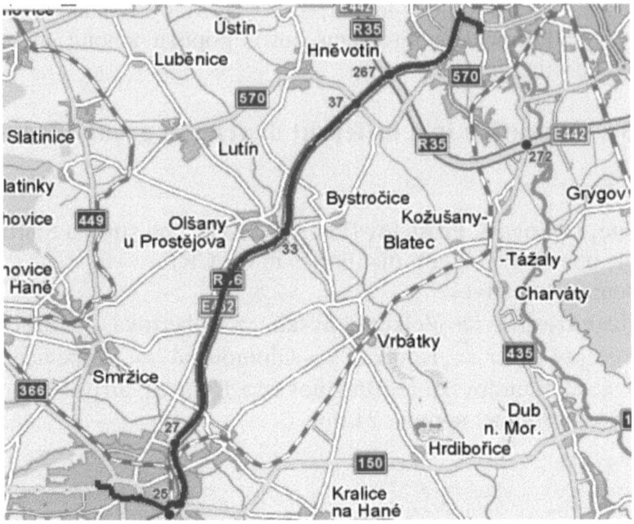

Fig. 8. The arrival route (*dark line*) into a different district currently used by the Olomouc Fire Rescue Service calculated at Google Maps as the shortest route

Even though there exists a shorter and faster alternative (Fig. 7) to the currently used route (Fig. 8) it is not possible to enforce this route due to the drivers' unwillingness to learn new routes. The psychological barriers are the most serious troubles in a new map approach implementation into Fire Rescue Service activities.

Above-mentioned tasks make evident that calculated arrival routes within different map servers differ in both parameters (length and time). None of web map route planners was able to include into arrival route calculation other valuable parameters (at least day/night time, seasonal alternative, traffic density or car parking situation). This is a reason to calculate the routes in analytical GIS packages with correctly evaluated road networks and sophisticated algorithms, for example Dantas et al. 2000.

4 Related problems and possible use of GIS

The current solutions of spatial tasks by the Olomouc Fire Rescue Service have a relatively high number of deficiencies. The issue of fire hydrants has become very important. The best solution would be to create a layer of fire hydrants, and thus provide information about available water supplies in the case of a more extensive fire. However, the low-quality management of hydrants provided by the water company Moravská vodárenská a.s. obliges the Fire Rescue Service to use natural water sources in most cases. Hydrants are often out of service, locked or lack high-enough pressure. In the case of a fire the minutes spent without using water have damaging consequences. The fire engine has to go kilometres away to fetch water. Therefore, up to four fire units are called to deal with one fire. Another reason for this is that the operation officer wants to take precautions for the possibility that the owners of the property may claim for damage caused by lack of water (one fire unit runs out of water and another one has not reached the site yet).

Another problematic issue is that of the accuracy of the operation site localization. The caller informing the fire brigade about a fire is often not able to localize the site or localizes it inaccurately. This should have been solved by a localization technology making use of incoming call phone numbers. Unfortunately, this is often hindered because of the strength of a transmitter signal and the position of the transmitter in relation to the caller. In case the transmitter is obscured, the accuracy can range in order of thousands of metres. Even localization of fixed lines is faulty because it is based on the address of the owner, not the line itself.

It is compulsory to give notification of burning brushwood in forests and burning waste. Some subjects perform this on a regular basis and always notify the relevant authorities. Such notification contains date and location, responsible person and his or her contact details (most often it is a mobile phone number). A possible alternative would be an online notification of regular or repeated burning using an electronic form; this record would be automatically assigned a cartographic symbol (e.g. an orange flame in the map) in the operation centre (Li et al. 2007, Kebair and Serin 2008). After reporting the end of the burning the symbol would disappear.

Another effective use of the GIS for the Fire Rescue Service is related to fires in newly built houses and industrial objects. According to the regulation no. 23/2008 Coll. such objects shall be equipped with an electronic fire alarm system. In the case of fire, a fire alarm would be activated and a symbol of fire would automatically appear on the location of the building on fire. After clicking the symbol an operative card would appear with prefilled contact details of responsible persons and further information needed by the fire brigade.

In the case of a car accident, it is quite difficult to ensure accurate localization. Every year a high number of deaths related to car accidents are due to late arrival of the first-aid and fire rescue services (who have to cut the injured person out of the car in more serious accidents). Either the people are in a critical state of health and are not able to call for help themselves or they are in shock and are not able to provide correct information after calling 112. Often the operation officer gets two different descriptions of the site from two or more callers. Such difficulties could be solved within the eCall project guaranteed by the European Union. It consists in wireless communication between the car and the 112 Call Centre operator. Immediately after the accident, due to impact detectors or data from the car's control unit, an automatic notification is sent to the staff of the integrated rescue system. The system accurately determines the number of passengers in the car, the car's velocity before crashing and the number of activated airbags. Using the GPS receiver the car's location is determined, with accuracy within metres. The accident location is shown at the operator's monitor, with immediate evaluation of arrival route length. The first such cars are planned to emerge in 2012. Originally, the system should have been part of cars' compulsory equipment but now it seems that the car manufacturers will offer it as optional. In the Czech Republic the system is being tested.

5 Implementation of map navigation into fire rescue service operation

Using map navigation as part of Fire Rescue Service operation seems more than logical. Primarily maps with contents thematically adapted for fire engine drivers should be used. However, it is necessary to keep in mind that incorrect use of such maps in real situations could be counter-productive, or even highly dangerous (Labadie 2005). The challenge for fire rescue service operations are ubiquitous maps which receive, analyse and present data to a user in a remote location using mobile and wireless technologies (Raper et al. 2007).

The danger lies in an incorrectly calculated arrival route in the sense of not taking into account some of the roads' parameters. Road works, speed bumps or street islands designed to regulate vehicle speed have to be taken into account. This has to be done in case large-volume equipment is used and it could be slowed down or even stopped altogether. The restriction in the form of bridges is also important. This can concern both height and weight of the vehicle, e.g. the emergency vehicle CAS 24-Scania 4x2 is 8 × 2.55 × 3.37 m big and 18 t heavy. The restrictions mentioned are not unexpected. They can be taken into account beforehand if the terrain is well mapped. However, the fire brigade does not take into account road closures and road works. These are temporary and because the communication between the fire brigade and the administrative authorities is defective the operation officer cannot work with such data. In practice the drivers' observation is relied upon – they communicate to each other all new road closures, road works or plans for repair and maintenance works.

The content of maps designed for fire engine drivers is important. New operation maps (Fig. 9) have been designed for texting. The topics of these maps are both related and diverse. However, all the maps have to provide legible and transparent cartographic representation. The speed of communicating spatial information needed to navigate to the operation site is crucial. Therefore, it is necessary to take into account the following requirements (Voženílek 2005), for both analogue and digital maps:

- the map's topographic base must be neither too detailed (thematic design would fade out), nor too vacant (orientation points would be missing);
- the cartographic symbol for the route has to be distinct enough not to hinder reading other map elements (e.g. road obstacles, power lines);
- the map's digital version must contain a number of relevant layers and the map's environment has to enable activating and deactivating them as needed;
- all data have to be updated regularly.

The selection of data, their quality and in what form they are handed over to the fire fighters in the fire engine, is essential. The technological aspect of the task represents an important problem – what medium to use to hand over the particular map to the drivers. An analogue map would not help the drivers. The driver cannot drive and at the same time read a paper map, not even if the co-driver read the map. In that case there would be hardly any changes. The driver either listens to a co-driver or operation officer or knows the route by heart. Therefore, it is necessary to hand the route over in a digital form. The question is whether to use GPS navigation because it is necessary to provide the fire engine driver with an arrival route created by the operation officer, with the use of GIS (Voženílek 2004, Laben 2002). Map sharing has to be wireless.

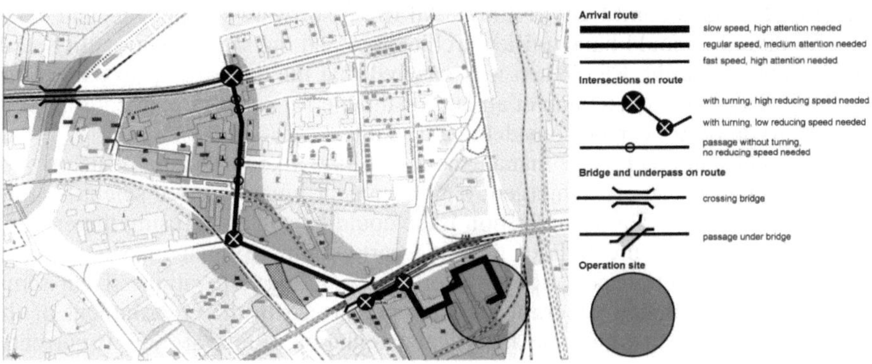

Fig. 9. Navigation map to operation site proposed for emergency vehicle driver

The visualization aspect of the task represents another problematic issue – what elements of the map content should be depicted using the GPS receiver and how should they be handed over. The base of the map content is the arrival route; therefore, it has to be represented by the most distinct symbol (Dymon 2003). It is still not clear whether or not the map's legibility will be hindered if we represent related topics that are seen by the operation officer because they often represent important links with the surroundings. GPS receivers use small monitors to depict the arrival route. Moreover, in this case we cannot count on activating and deactivating individual layers. This could be solved by sending the fire brigade a map that can be changed by the operation officer according to important circumstances. A similar solution was tested in the city of České Budějovice. A GPS receiver was used only for specifying the operation site in the municipality; the drivers had to use their memory to get to the municipality. Voice navigation disturbed the drivers and their habit and urge to use the known route was stronger (Voženílek 2005).

6 Conclusion

The speed of decision making is one of the key elements of an effective Fire Rescue Service operation. Among the different decision-making steps, from notification of fire to its extinguishment, there are many decisions of spatial character. Therefore, the effort to implement cartographic communication to this process is more than logical. It consists of three basic steps: analytical for calculating the fastest arrival route, visualization for its representation and technological for handing over the information to the fire engine drivers.

Analytical evaluation of the current situation is based on network analyses in vector GIS. They can be applied as part of the crisis management information system (ISKŘ) (Maršík and Uchytil 2007). All ISKŘ software components and existing programmes should have unified architecture and the same functionality at all end workplaces in the CR. When it is ensured that the data are up to date in the topological structure (Lehmann et al. 2005) suitable for network applications, generating arrival routes can be considered a trivial task.

Cartography in emergency management can play a key role as a decision support tool. Future maps for emergency management must be more scientific than contemporary maps. The visualization issue could be solved by using the existing and easily accessible methods of thematic cartography. Similar map outputs are produced in related transportation tasks or during military operations. Their semi-automated or automated generation can be performed in the environment of the same GIS products in which the analytical calculation of arrival route is currently performed (Voženílek 2004, Somers and Svara 2009).

The technological realization of the proposed solution represents the most demanding part of the innovation process. Handing over the generated map to the fire engine driver can make use of various commercial solutions. We must observe that all of them are easy to apply. Further they can join emergency management systems as responders and then improve time, communication and state-of-art public-related policy (Gibbs and Kish 2006).

It is very difficult to describe the general situation concerning map use in fire brigade operations worldwide. Maps are definitely used in many (probably most) national fire rescue services with various differences. There is no better assisting document for orientation in the area than a map. However the specific necessities concerning maps stemming from real rescue operations (mainly the fastest arrival route calculated with respect to valuable parameters) and user issues (driver's reservations) are still not resolved. The chapter has demonstrated that arrival route calculation cannot be processed within common web map route planners (they differ in results and do not give access to calculation equations) designed by non-professional (thematic cartography) workers.

The proposed approach based on cartographic visualization of the shortest route generated using network analyses in the GIS environment leads to considerable increase of efficiency of Fire Rescue Service operation. Three tasks simulating different approach to determining arrival routes for the Olomouc Fire Rescue Service enabled the authors to illustrate the advantages of the above described approach. Nevertheless, the authors are aware of a number of obstacles restricting the launch of the new approach. Among these are primarily financial demands, habitual constraints in humans, difficulties with quick and accurate updating of data, deficient cooperation with other elements of the Integrated Rescue System of the Czech Republic (especially the Police of the CR and the Municipal Police), the absence of national spatial infrastructure, etc.

References

Charvát K, Kubíček P, Talhofer V, Konečný M, Ježek J (2008) Spatial data infrastructure and geovisualization in emergency management. In: Pasman HJ, Kirillov IA (eds.) Resilience of Cities to Terrorist and Other Threats. Springer, Dordrecht, pp. 443–473.

Crampton JW (2003) Cartographic rationality and the politics of geosurveillance and security. Cartogr Geogr Inf Sci 30(2):135–148.

Curtin KM (2007) Network analysis in geographic information science: review, assessment, and projections. Cartogr Geogr Inf Sci 34(2):103–111.

Dantas A, Yamamoto K, Lamar MV, Yamashita Y (2000) Neural network for travel demand forecast using GIS and remote sensing. Proceedings of International Joint Conference on Neural Networks (IJCNN'00), 4.

Dymon UJ (2003) An analysis emergency map symbology. Int J Emerg Manag 1(3):227–237.

Gibbs W, Kish K (2006) MacDill builds on the initial efforts by PACAF and ESRI with emergency response tools. GIS in the Defence and Intelligence Communities 3:14–15.

Kebair F, Serin F (2008) Towards an intelligent system for risk prevention and management. Proceedings of the 5th International ISCRAM Conference, 526–535.

Kingston R (2007) Public participation in local policy decision-making: The role of webmapping. Cartogr J 44(2):138–144.

Labadie JR (2005) Problems in local emergency management. Environ Manage 8(6):489–494.

Laben C (2002) Integration of remote sensing data and geographic information systems technology for emergency managers and their applications at the Pacific Disaster Center. Opt Eng 41(09):2129–2136.

Lehmann KA et al. (2005) Network Analysis – Methodological Foundations. Springer, Berlin.

Li Z, Liu R, Liu N (2007) The discussion of crucial techniques in the emergency solution of special equipment security systems. Proceedings of the SPIE.

Maršík V, Uchytil J (2007) GIS Informačního Systému Krizového řízení – problematika datového skladu (in Czech). Proceeding of GIS Ostrava 2007.

Mikulík O, Voženílek V, Vaishar A et al. (2008) Studium rozvoje regionu založené na vizualizaci geoinformačních databází (in Czech). Univerzita Palackého v Olomouci, Olomouc.

Raper J, Gartner G, Karimi H, Rizos Ch (2007) A critical evaluation of location based services and their potential. J Loc Based Serv 1(1):5–45.

Somers S, Svara JH (2009) Assessing and managing environmental risk: Connecting local government management with emergency management. Public Adm Rev 69(2):181–193.

Voženílek V (2004) Atlas geo-production: a role of distributed GIS and spatial multimedia in atlas production. Proceedings of Geospatial World 2004, Miami, USA, 15–17 May 2004 [CR-ROM].

Voženílek V (2005) Cartography for GIS – geovisualization and map communication. Univerzita Palackého v Olomouci, Olomouc.

Strategies for the Automatic Extraction of Water Bodies from TerraSAR-X / TanDEM-X data

Thomas Hahmann, André Twele, Sandro Martinis and Manfred Buchroithner

Abstract

Medium-resolution SAR satellite data have been widely used for water and flood mapping in recent years. Since the beginning of 2008 high-resolution radar data with up to 1 m pixel spacing of the TerraSAR-X satellite are operationally available. Due to the improved resolution of the sensor more details of the Earth's surface become visible. A number of different appearances of water bodies in TerraSAR-X data are shown that are relevant for a general water mapping concept. Existing water body detection approaches that have been applied to medium-resolution SAR data are reviewed with regard to their applicability for TerraSAR-X data. As a complementary mission to TerraSAR-X the launch of TanDEM-X is planned for October 2009. The data of both satellites will be used to generate a global DEM (Digital Elevation Model) with an interferometric data acquisition concept. As a byproduct to the DEM data set a global water mask will be derived from the SAR data. The concept for this water detection process within the TanDEM-X project is introduced in this paper.

1 Introduction

Due to their cloud penetration capability, SAR (Synthetic Aperture Radar) satellites are almost independent from weather and daylight. Therefore they are more suitable than optical sensors to reliably and timely map land cover features. This is especially relevant for the detection of water bodies and inundated areas in flood situations, which often are accompanied with overcast sky conditions. Since the beginning of the 1990s operational space-borne systems like the European satel-

M. Konecny et al. (eds.), *Geographic Information and Cartography for Risk and Crisis Management*, Lecture Notes in Geoinformation and Cartography,
DOI 10.1007/978-3-642-03442-8_9,

lites ERS-1/2, Envisat ASAR and the Canadian Radarsat-1 have been used to map water bodies and flood situations at C-Band wavelength with medium resolution.

In recent years several high-resolution SAR satellites have been launched: Radarsat-2 (Canada), COSMO-SkyMed (Italy) and TerraSAR-X (Germany). The latter was successfully launched on June 15, 2007. With the end of the commissioning phase operational service started in January 2008. Main orbit characteristics of TerraSAR-X are sun-synchronous, near-polar dusk-dawn at a flying altitude of 514 km. The satellite features a nominal repetition rate of 11 days. By using different incidence angles an imaging frequency of 2–4 days can be reached. The versatile X-Band antenna of TerraSAR-X has the following imaging capabilities: in the High Resolution SpotLight (HS) and SpotLight (SL) modes, a spatial resolution of up to 1 m can be achieved. The size of the ground track is either 5 (HS) or 10 (SL) km in azimuth and 10 km in range. In the standard StripMap (SM) mode, a scene has a swath width of 30 km and a maximum length of 1500 km. Depending on the incidence angle, the ground resolution can be up to 3 m. In the ScanSAR mode (SC) a maximum image size of 100 km by 1500 km can be acquired with 16 m resolution. For each imaging mode a variety of acquisition and processing parameters can be defined (incidence angle, polarization, orbit accuracy, spatial resolution and geocoding).

The TanDEM-X mission (TerraSAR add-on for Digital Elevation Measurements) is a spaceborne SAR interferometry mission that is based on the TerraSAR-X satellite and a second identically constructed satellite called TanDEM-X. The launch of the latter is planned for October 2009. The main goal of the TanDEM-X mission is the generation of a global Digital Elevation Model (DEM) with a spatial resolution of about 12 m. As a by-product the generation of a global surface water body mask is planned.

The high resolution of the new class of SAR satellites offers enormous potential in the domain of water body and flood mapping. However, the improved spatial resolution of the SAR image data results in small-scaled image objects, which makes image analysis even more challenging. The aim of the current research activities is the development of (semi-) automatic algorithms, which allow near-realtime SAR data processing and reliable water body mapping.

In this contribution we focus on concepts of water body extraction from TerraSAR-X / TanDEM-X data. This chapter is organized as follows: In Sec. 2 the appearance of water bodies in TerraSAR-X data is presented. Section 3 contains a review of existing methods for flood extraction and their applicability to high resolution TerraSAR-X data. The water body detection design concept for the TanDEM-X mission is presented in Sec. 4. Section 5 shows a flood mapping example with TerraSAR-X data.

2 Appearance of water bodies in TerraSAR-X data

To understand the complexity of the task to produce a global water body data set in the frame of the TanDEM-X mission it is necessary to start with an overview about how different water bodies may appear in TerraSAR-X and respectively

TanDEM-X data. The ideal case of the appearance of water bodies in SAR data, especially TerraSAR-X data, is that the water surface is smooth with respect to the X-Band wavelength. Due to specular reflection almost the entire transmitted signal is reflected away from the sensor. A very low backscatter value is received from the instrument in contrast to higher backscatter values of other surface types such as e.g. soil or vegetation.

Comprehensive experience with TerraSAR-X data in the analysis of water bodies however shows that they are usually characterized by a varying degree of roughness. This means that these water bodies do not appear dark in the TerraSAR-X images, but show a certain structure, texture or pattern. The influence of wind leads to the origination of ripples and waves on water surfaces (Fig. 1a, b). The larger a water body the more susceptible it becomes for the formation of waves. Narrow rivers seldom show wind structures, while the sea surface along coastlines is often influenced by strong wind-induced waves, which are visible in TerraSAR-X data.

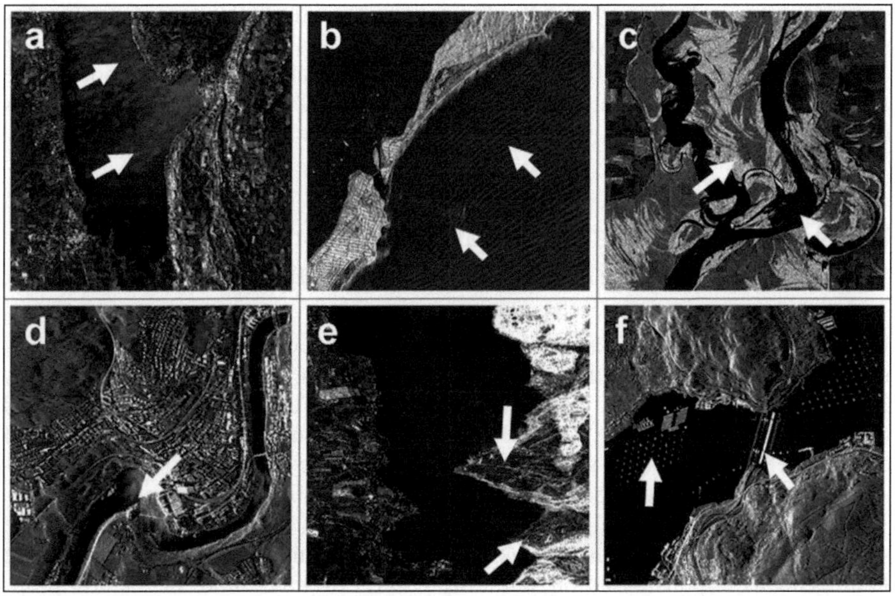

Fig. 1. Different forms of appearance of water bodies in TerraSAR-X data, © DLR (2007-2008) – (a) Wind pattern on Lake Ammersee (Germany), (b) Sea waves at Australian Coast, (c) Mississippi river flooding (USA), (d) Turbulent water surface at the Rhine Falls (Switzerland), (e) Layover effect at the Eastern slopes of Lake Traunsee (Austria), (f) Obstacles – e.g. a bridge with ghost effect and fish farming basins near Vigo (Spain)

In the general case of rough water bodies the visible structures do not show regular wave patterns. While regular patterns can mainly be observed at sea surfaces, irregular patterns occur more frequently at the surface of inland lakes. Increased roughness leads to increased backscatter values of the transmitted radar signal. Other roughness difference effects at water surfaces include for example wind shadow effects, oil spills (which can lead to smoothened water surfaces) and turbulence effects at rivers (Fig. 1d).

Depending on the type of vegetation and polarization, the double-bounce return can be clearly visible in TerraSAR-X images of flooded areas. The double-bounce effect leads to brighter returns from the flooded vegetation areas in contrast to the very low return due to specular reflection from smooth water surfaces (Fig. 1c). Lakes, rivers or the sea are in some cases associated with vegetation along the shoreline. The double-bounce effect along these lines can cause an underestimation of the water surface since the bright lines would be classified as non-water area.

For the detection of water bodies in mountainous terrain with steep slopes the side-looking radar imaging geometry effects have to be considered. These are radar shadow and layover. This can lead to dark radar shadow areas adjacent to dark water surfaces, which may not be distinguishable. At shallow incidence angles steep slopes can lead to layover, which may result in a mountain being imaged at the lake's surface. In Fig. 1e the eastern parts of the lake Traunsee are not visible as they are hidden by the mountain's western slopes. Obstacles like bridges, ships or fish farming basins may also occlude the open water surface (Fig. 1f).

3 A review of existing approaches for water body mapping in high-resolution SAR data

Pixel- and segmentation-based classification techniques can be distinguished as two main concepts for the detection of water bodies in SAR data (Heremans et al. 2003). Conventional classification approaches use pixels as smallest components of raster data. However, pixel-based classifiers hardly use context information and are not well suited to deal with heterogeneous land-cover units. Classification results may show a salt-and-pepper effect, which makes filtering necessary.

Some problems of pixel-based image analysis can be solved by using image segmentation. Segmentation means to partition an image into non-overlapping homogeneous regions based on similarity criteria of gray values or textural properties (Pal and Pal 1993) with the objective of generating segments, which resemble real objects of the Earth's surface. Especially for data of high-resolution SAR sensors segmentation methods appear promising. These images exhibit a very high spectral variance of the individual thematic classes due to the reduced mixed pixel phenomenon. In addition to spectral-related characteristics of the image objects further parameters like contextual information, texture and object geometry can be used for an improved classification.

Amplitude thresholding is one of the most frequently used techniques to distinguish water bodies from land surfaces in SAR imagery (e.g., Martinis et al. 2009, Herrera-Cruz and Koudogbo 2009, Townsend and Walsh 1998). Thereby, all elements of the SAR amplitude data with a backscatter value lower than a given threshold are assigned to the water class. This method is computationally very fast and most of the extent of a smooth water surface can be derived by this technique. Auxiliary information like digital elevation models can be used to improve mapping results, especially in significant topographic terrain. On the one hand misclassified areas in regions higher than the main expanse of water can be erased. These may be e.g. objects with a low radar backscatter similar to calm water like roads, airfields or radar shadow. On the other hand, terrain information can be used in combination with contextual information to integrate water bodies with backscatter intensities higher than the originally defined threshold, e.g., due to the effect of vegetation, into the water class. Thresholding generally works satisfactory for smooth water bodies that reflect radiation away from the SAR sensor, thus generating a very low backscatter. In contrast, the surrounding terrain exhibits higher backscatter values due to increased surface roughness. The value of the threshold depends on several disturbing factors like wind-induced waves, vegetation and the incidence angle of the sensor. Therefore, thresholds need to be set for every SAR data set individually.

Several approaches have been developed to improve water detection results. These include fairly speckle-resistant active contour models (Horritt 1999), which had been used by Ahtonen et al. (2004), Heremans et al. (2003) and Matgen et al. (2007) for flood boundary delineation in medium-resolution SAR imagery. A Bayesian segmentation technique along with energy minimization functions to separate land and sea regions in TerraSAR-X data is used by Ferreira and Bioucas-Dias (2008). The main limitation of this approach is that user interaction is necessary for the selection of training samples for land and sea. Fully automatic texture-based maximum likelihood methods, which are able to detect water bodies independent of the sensor incidence angle are applied by Ahtonen et al. (2004). Different multitemporal change detection approaches for the derivation of flood dynamics between SAR data have successfully been applied in the past. These include amplitude based (Heremans et al. 2003, Townsend and Walsh 1998) and coherence based techniques (Nico et al. 2000).

4 Design of the water body detection processor for TanDEM-X

As a contribution to the TanDEM-X project a water body data set with a global coverage will be derived. TanDEM-X is a SAR interferometry mission of German Aerospace Center (DLR), which shall generate a worldwide, consistent, high-precision Digital Elevation Model with an unprecedented accuracy corresponding to the HRTI-3 (High Resolution Terrain Information) specifications (12 m posting, 2 m relative height accuracy for flat terrain) (e.g. Krieger et al. 2007, Zink et al. 2008). The data acquisition will be performed in StripMap mode. The required minimum mapping unit for water bodies in the global water mask is rather conservative. Lakes with a diameter of more than 300 m shall be included in the water mask product. Rivers shall have a minimum width of 183 m. A module of the DEM Mosaicking and Calibration Processor (MCP) (Wessel et al. 2008) will be the water body detection processor. The design of this module that shall detect water bodies reliably in TanDEM-X data is described in this section (Fig. 2).

Amplitude and coherence values of the coregistered SAR data of the TerraSAR-X and TanDEM-X satellites are used as input to this module. In a preprocessing step it will be tested if the imaged area does contain potential water bodies. For this purpose a global land/water mask that consists of the SRTM (Shuttle Radar Topography Mission) water mask and the GSHHS (Self-consistent, Hierarchical, High-resolution Shoreline) Database (Wessel and Smith 1996) will be used. These reference data are used as they are available globally (GSHHS) and free of charge.

This mask will be built up before TanDEM-X data processing starts and is stored as a binary data set in the framework of DLR's DEM database in a tile structure. The value of each tile indicates whether water bodies are expected or not. If a TanDEM-X raw data area contains only tiles with a dry area flag no water detection has to be done.

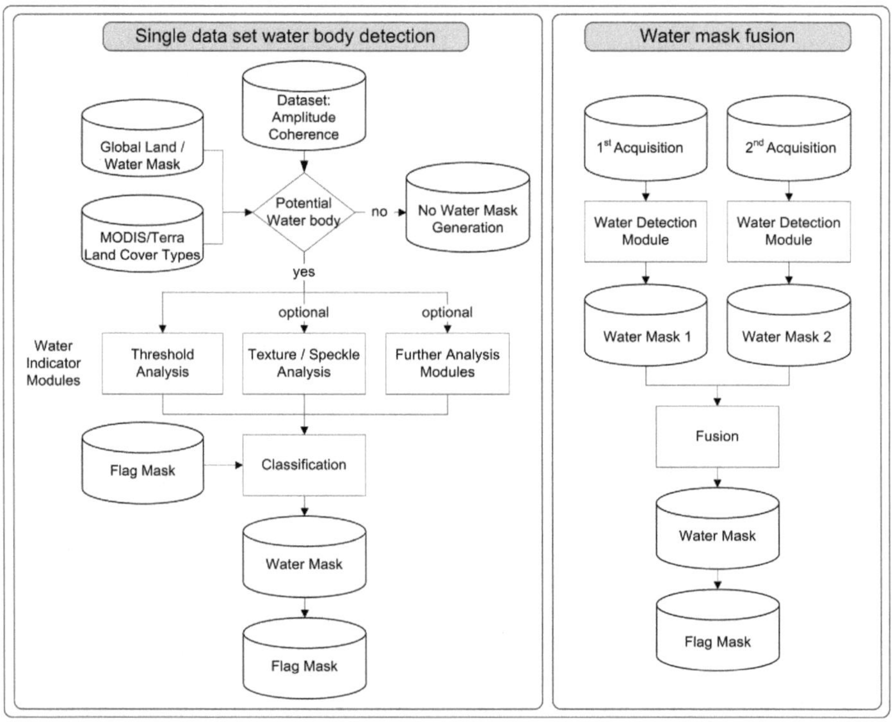

Fig. 2. Flowchart of the design concept for the water body detection processor for the TanDEM-X mission

The primary aim of this approach is to exclude desert regions from the water mask generation and thus save processing time and storage capacity. Furthermore the Polar Regions will be included in this data set due to big variances in the water /ice boundary, which makes it impossible to generate a consistent water mask for these regions with automated procedures. A further data set that will be used to exclude desert and Polar Regions is the MODIS/Terra Land Cover Type (Moderate Resolution Imaging Spectroradiometer) data set. It is available as global 1 km raster data set (MOD12Q1) or 0.05° data set (MOD12C1). These data sets include the land cover classes "Unvegetated", "Barren or Sparsely Vegetated" and "Snow and Ice." The minimum spatial extent of these land cover types was derived from data of the years 2001–2004 (Fig. 3).

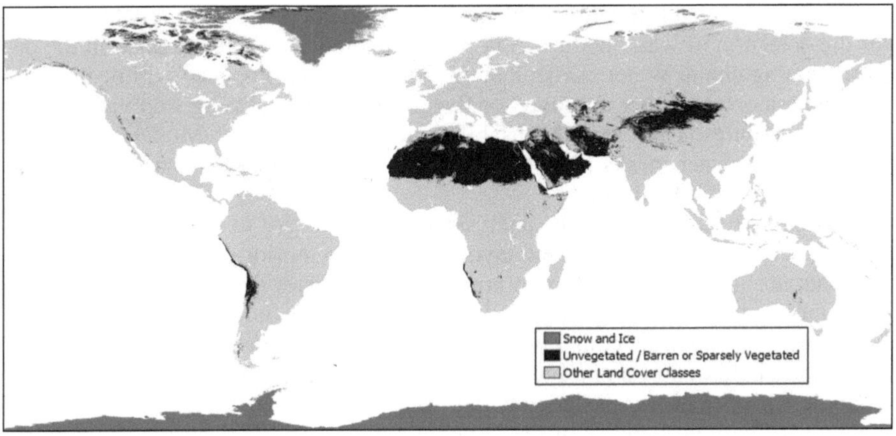

Fig. 3. World map showing snow/ice and unvegetated land cover classes based on "MODIS/Terra Land Cover Types—MOD12C1". These will be excluded from the water body detection process

If the potential water body check is positive a set of analysis methods, so-called Water Indicator Modules, will be applied to the data set. Amplitude thresholding will be conducted, which will detect most of the smooth water bodies. An individual threshold will be applied for every single scene.

Coherence threshold based methods are analyzed. For the repeat-pass-case (e.g. 11-day repeat interval for TerraSAR-X) they can be used in arid regions with negligible vegetation cover. It is however impossible to distinguish between vegetation areas and water bodies in data sets of the temperate zone as the coherence values are very low for both land cover types.

As described above TanDEM-X will be a satellite mission consisting of two satellites that fly very close together. The geometrical baseline between the two satellites will differ in a range of 200 m to several kilometers. That means that the temporal baseline of the two acquisitions will be in the order of milliseconds to seconds which is very small. In the so-called commissioning phase of the TanDEM-X mission in the first month after the launch of the satellite there will be a set of tests with different baselines and their effect to the coherence behaviour of rough and smooth water bodies and land areas. These small baselines can not be tested with TerraSAR-X data as there is only the normal 11-day repeat interval available. It is expected that the results will be completely different for the very small TanDEM-X temporal baseline compared to the 11-day interval. It is furthermore expected that waves on the water surface will have an influence on the coherence values so that coherence thresholding can contribute to detect rough water bodies.

Further methods that will be applied optionally include the use of local texture filters following Ahtonen et al. (2004) and speckle statistics approaches. These tools will be applied if the threshold approaches do not prove to be reliable enough for the detection of water bodies. An operator approach will be included in the data processing chain. Water masks that are automatically produced will be checked by an operator and either approved or flagged for a reprocessing step with further analysis modules. Currently the usability of Active Contour Models initialized from existing water masks (SRTM, GSHHS) for the TanDEM-X water body detection processor is analyzed.

The single result layers of the analysis tools are merged to one file and a classification is conducted. Shadow and layover data from a prior processing step of the raw data are used to remove these areas from the water mask. This is especially important in high mountainous terrain. The result will be saved to the so-called Flag Mask. This data set contains at least three so-called probability levels that describe the likelihood of a pixel to either belong to a water body or a dry area. For example a pixel that was assigned being water within every detection module will have a very high probability of really being a water area. If there are different results for a pixel within the several methods the water probability will be assigned a low value.

According to the TanDEM-X data acquisition concept at least two data sets for every continental area on Earth will be recorded with single polarization. These two coverages will be acquired subsequently in the first and second DEM acquisition year. For high mountainous terrain with steep slopes a third data acquisition will be performed with different sensor parameters (flight direction, incidence angle). This allows improving the result of the DEM generation in difficult terrain. It also helps to detect further water bodies in high mountain valleys. The water body detection will be executed individually for the first and second year coverage and if applicable also for the third year coverage. After the TanDEM-X data acquisitions have been conducted, all single water masks for a certain area are fused to classify the water area and to produce the final water mask product. This product may be used for further editing of the TanDEM-X DEM, e.g. flattening of water bodies.

Important information is the usage of a certain polarization for the data acquisition. VV polarization is better suited for the interferometric data processing for the DEM generation, whereas HH polarization seems to be more useful for the detection of water bodies (Henry et al. 2006). This result could be confirmed with TerraSAR-X data. Dual-polarization (HH + VV) data of lake surfaces and coastlines showed higher backscatter values in the VV polarization at water surfaces with increased roughness. So the main approach of amplitude thresholding would allow more reliable and accurate results when using HH polarization for water body detection. For standard TanDEM-X data acquisition however the VV polarization will be used as the DEM production is the primary aim of the mission. The TanDEM-X water mask described in this chapter may be seen as a by-product of the project.

5 A flood mapping example using TerraSAR-X data

In this section a flood mapping example using local texture features and a Digital Elevation Model is described. A TerraSAR-X data set of a flood situation along the White River in the state of Arkansas (USA) was analyzed. The available data set was acquired in StripMap mode (3.5 m spatial resolution) with horizontal polarization (Fig. 5a). The meandering river is clearly visible in the image as a dark line (smooth water surface). The river is surrounded by inundated forest areas. These flooded vegetation areas are characterized by very high backscatter values that are caused by the so-called double-bounce effect. This means that multiple reflections between the trunks and branches of trees and the water surface lead to increased values of the reflected signal. The inundated forest areas are visible as bright areas in the SAR image.

Figure 4 shows the flowchart of the water and flooded forest detection algorithm that is applied. Research results by Ahtonen et al. (2004) were used. The input datasets for this algorithm are the TerraSAR-X amplitude image and a Digital Elevation Model (DEM) of this area (SRTM C-Band data with 90 m resolution). From the DEM a slope data set is calculated.

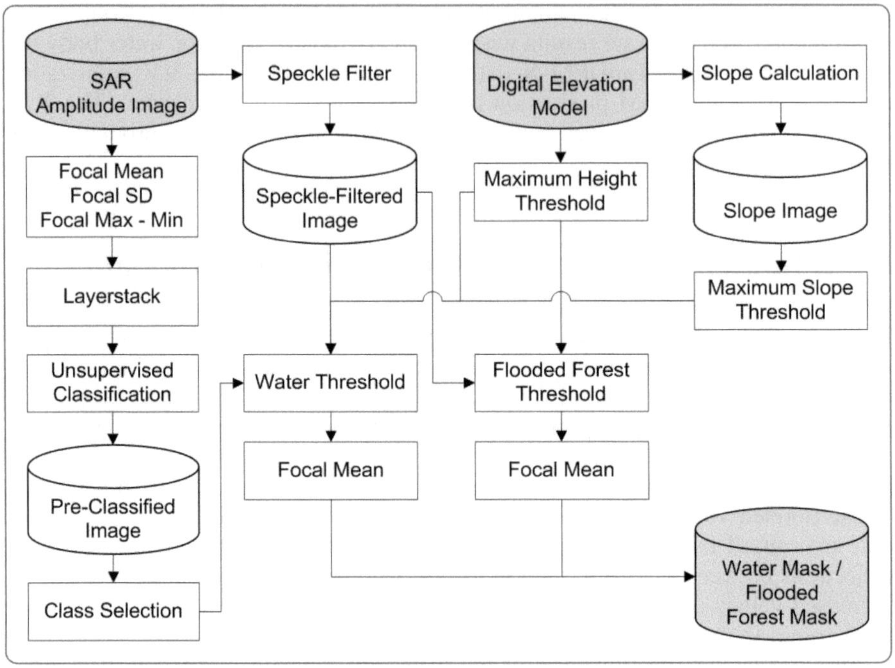

Fig. 4. Flowchart showing the water body detection algorithm for the Arkansas flood example using texture filters and a digital elevation model

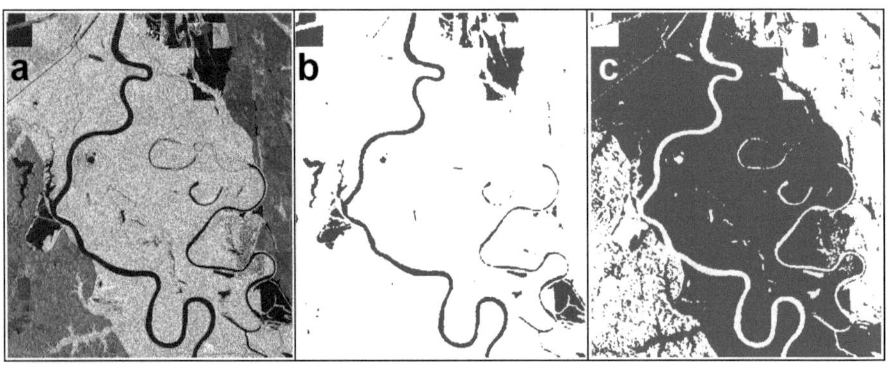

Fig. 5. Flooding of the White River South of Clarendon (Arkansas, USA) – (a) TerraSAR-X scene, StripMap mode, horizontal polarization, 27th March, 2008, © DLR (2008), (b) Water Mask, (c) Flooded Forest Mask

Both the DEM and the slope data set are used to exclude water bodies at steep slopes and in mountainous terrain from the resulting water mask. From the amplitude image a speckle-filtered image and three local texture features using a 5×5 estimation window are calculated. These are mean, standard deviation and data range (max-min). A logarithmic transform is conducted with the three result layers to improve the separability of water bodies and land areas. A layerstack of the three data channels is produced, followed by a subsequent unsupervised classification step, which leads to a preclassified image. Potential water classes and flooded forest classes are interactively selected.

By using thresholds for water bodies and flooded forests potential areas of these two classes are derived from the speckle-filtered image. The flooded forest threshold results from the knowledge that the backscatter values from these forests are significantly increased against non-flooded forests due to the strong double-bounce effects described above. The results of the supervised classification and the thresholding approach are finally fused and filtered with a 5×5 mean filter window to eliminate small areas from the result. The result of this procedure is a raster layer that contains both a water mask and a flooded forest mask (Fig. 5b, c). Some final manual editing for urban areas is necessary to delete settlement areas from the flooded forest mask. This results from the fact that buildings (due to the effect of corner reflection) show similar high backscatter values like inundated forest areas and are classified as flooded forest by the automatic thresholding approach.

Research by Ahtonen et al. (2004) and by the authors shows however that there exist restrictions of the applicability of these texture features in areas with increased backscatter value (due to wind/wave influence). For these areas additional methods have to be applied, e.g. Active Contour Models.

6 Discussion

This chapter shows that high-resolution SAR image data from the TerraSAR-X satellite have an enormous potential for accurate and reliable mapping water bodies as well as flood areas. The geometric resolution of up to 1 m makes a large amount of image objects visible. The well-known water mapping approach of amplitude thresholding works satisfactory for smooth water surfaces which is sufficient for most inland water bodies. In the scope of the TanDEM-X project a global water mask shall be derived. At coastlines and the surfaces of larger lakes rough water surfaces occur due to wind-induced waves. This necessitates the development of new algorithms to detect these water bodies reliably. Current work includes the analysis of the suitability of texture and speckle based methods. The texture filter approach was successfully used to derive flooded areas and flooded forest areas in a flood mapping example with high-resolution TerraSAR-X data. Active Contour Models along with the use of existing water masks are tested for the applicability in the TanDEM-X water body detection processor. Coherence threshold methods were tested with TerraSAR-X 11-day repeat pass scenes. This approach works well in arid areas with less vegetation influence. In temperate zones however it is impossible to distinguish between water bodies and vegetation areas as both land cover types feature very low coherence values. The applicability of coherence thresholding within the TanDEM-X mission with very small geometrical and temporal baselines will be tested in the commissioning phase after the launch of the satellite.

Acknowledgement

This work is mainly funded by the German Ministry of Education, Science, Research and Technology (BMBF) in the context of the RIMAX Research Programme (Risk Management of Extreme Flood Events), sub-project SAR-HQ (grant ref. nr.: 0330719). Thanks to German Aerospace Center (DLR) for providing TerraSAR-X data within the project nr. HYD0326. Finally, the authors wish to thank the anonymous reviewers for their helpful comments which greatly improved the paper.

References

Ahtonen P, Euro M, Hallikainen M, Solbø S, Johansen B, Solheim I (2004) SAR and optical based algorithms for estimation of water bodies. Technical report, FloodMan Project, http://projects.itek.norut.no/floodman.

Ferreira JPG, Bioucas-Dias JM (2008) Bayesian land and sea segmentation of SAR imagery. Proceedings of 3. TerraSAR-X Science Team Meeting, Oberpfaffenhofen, Germany, http://sss.terrasar-x.dlr.de/papers_sci_meet_3/paper/OCE0050_ferreira.pdf.

Henry JB, Chastanet P, Fellah K, Desnos YL (2006) Envisat multi-polarized ASAR data for flood mapping. Int J Remote Sens 27(9–10):1921–1929.

Heremans R, Willekens A, Borghys D, Verbeeck B, Valckenborg J, Acheroy M, Perneel C (2003) Automatic detection of flooded areas on ENVISAT/ASAR images using an object-oriented classification technique and an active contour algorithm. Proceedings of IEEE Conference on RAST 2003, Istanbul, Turkey, pp. 289–294.

Herrera-Cruz V, Koudogbo F (2009) TerraSAR-X rapid mapping for flood events. In: Proceedings of ISPRS Hannover Workshop High-Resolution Earth Imaging for Geospatial Information, Hannover, Germany, pp. 170–175.

Horritt MS (1999) A statistical active contour model for SAR image segmentation. Image Vision Comput 17:213–224.

Krieger G, Moreira A, Fiedler H, Hajnsek I, Werner M, Younis M, Zink M (2007) TanDEM-X: A satellite formation for high-resolution SAR interferometry. IEEE Trans Geosci Remote Sens 45(11):3317–3341.

Martinis S, Twele A, Voigt S (2009) Towards operational near real-time flood detection using a split-based automatic thresholding procedure on high resolution TerraSAR-X data. Nat Hazard Earth Sys Sci 9:303–314.

Matgen P, Schumann G, Henry JB, Hoffmann L, Pfister L (2007) Integration of SAR-derived inundation areas, high precision topographic data and a river flow model toward real-time flood management. J Appl Earth Observ and Geoinform 9:247–263.
MODIS/Terra Land Cover Type Yearly L3 Global 1km SIN Grid – MOD12Q1: http://edcdaac.usgs.gov/modis/mod12q1v4.asp, MODIS/Terra Land Cover Types Yearly L3 Global 0.05Deg CMG – MOD12C1: http://edcdaac.usgs.gov/modis/mod12c1v4.asp
Nico G, Pappalepore M, Pasquariello G, Refice A, Samarelli S (2000) Comparison of SAR amplitude vs. coherence flood detection methods – a GIS application. Int J Remote Sens 21(8):1619–1631.
Pal NK, Pal SK (1993) A review on image segmentation techniques. Pattern Recognit 26:1277–1294.
Townsend PA, Walsh SJ (1998) Modelling flood plain inundation using integrated GIS with radar and optical remote sensing. Geomorphology 21(98):295–312.
Wessel B, Marschalk U, Gruber A, Huber M, Hahmann T, Roth A, Habermeyer M (2008) Design of the DEM mosaicking and calibration processor for TanDEM-X. In: ITG/VDE (eds.) Proceedings of the 7th European Conference on Synthetic Aperture Radar (EUSAR), Friedrichshafen, Germany, vol. 4, pp. 111–114.
Wessel P, Smith WHF (1996) A global, self-consistent, hierarchical, high-resolution shoreline database. J Geophys Res – Solid Earth 101(B4):8741–8743.
Zink M, Krieger G, Fiedler H, Hajnsek I, Moreira A (2008) The TanDEM-X Mission Concept. In: ITG/VDE (eds.): Proceedings of the 7th European Conference on Synthetic Aperture Radar (EUSAR), Friedrichshafen, Germany, vol. 4, pp. 31–34.

GIS Techniques in the Evaluation of Pipeline Networks Seismic Hazard

Michele Maugeri, Ernesto Motta, Giuseppe Mussumeci and Erminia Raciti

Abstract

To evaluate seismic risk, it must be taken into account that modern towns depend daily on lifelines and utility systems, that become essential after natural disasters, but are often without any earthquake threat. To evaluate lifeline networks seismic vulnerability, we usually refer to damage models, requiring parameters dealing with pipe features, soil behavior, and seismic hazard of the studied area (peak ground acceleration or velocity, PGA, PGV, or permanent ground displacement, PGD). In this work, models evaluating seismic hazard in a studied area and expected seismic damage for pipeline networks will be applied. A model is shown to assess earthquake induced slope displacements. Some attenuation laws will be selected to evaluate PGA, PGV and PGD. Finally, Repair Rate will be calculated for pipes of an important Italian water network feeding 20 towns of Etnean area, referring to three seismic scenario events. Applications will be developed in a GIS environment.

1 Introduction

Lifelines seismic problems have recently attracted researchers for the great damage potential. Pipeline networks are vital systems in modern towns: they convey natural gas, oil, and water, or could contain communication and power cables, all of which are very important to maintain functional residential and industrial facilities. Pipelines are also required for economic and community recovery after natural disasters. Pipelines could be severely damaged by shaking, liquefaction, and landslides during earthquakes and the immediate assessment of pipeline facilities

M. Konecny et al. (eds.), *Geographic Information and Cartography for Risk and Crisis Management*, Lecture Notes in Geoinformation and Cartography,
DOI 10.1007/978-3-642-03442-8_10,

is critical to prevent fires, explosions, and pollution from broken gas or sewage lines. Therefore their vulnerability and expected damage assessment is very important to estimate seismic risk.

Lifeline networks' intrinsic features (their structure, the different types of nodes in each system) are complex. Moreover for the possible geographical discrepancy between resource and demand/consumption, their spatial distribution usually exceeds the urban area, and this implies a spatial variability of seismic motion (ground acceleration and velocity) and a higher probability of exposure to permanent ground displacement induced by fault offset, liquefaction phenomena, or landslides. The pipeline network digital data are generally necessary when the bureaus and disaster prevention organizations examine the restoration strategy and restoration period, but accessibility to data is often very difficult for many reasons, like the complexity, poor knowledge of buried sections (most of waterworks bureaus do not hold the detailed digital data of the whole supply area), the strategic importance and market competition, and the limited knowledge and documentation specifically in Europe on earthquake observations of damages. Many researches on pipeline network seismic damages have been found in technical literature. We will refer to damage models suggested by 'RISK-UE' (Monge et al., 2004) regarding seismic risk, where seismic damages to buried pipes is described by a 'repair rate', 'RR', combining breaks (complete fracture of the pipe) and leaks that require the water agency to perform a repair in the field. Damage models require some parameters dealing with pipe characteristics and soil behaviour. Moreover a synthetic parameter representing seismic hazard of the studied area is required.

In this work, models evaluating seismic hazard in a studied area (in terms of ground displacements or slope instability) and expected seismic damage for pipeline networks will be applied. A model is shown to assess earthquake-induced slope displacements according to Newmark approach. Attenuation laws will be selected to evaluate synthetic seismic parameters peak ground acceleration, PGA, peak ground velocity, PGV, or permanent ground displacement, PGD. Finally, Repair Rate will be calculated for pipes of an important Italian water network feeding 20 towns of Etnean area, referring to three seismic scenario events. The applications will be developed in a GIS environment, by 'Spatial Analysis' and 'Field Calculation'.

2 Damage algorithms for pipeline networks

Water supply facilities consist mainly of underground pipeline networks, vulnerable to seismic disturbance, especially when constructed in soft ground, and requiring much time and cost for implementing anti-earthquake measures.

Studies on seismic damage to water pipelines dates back to 1920s, for example in the Kanto 1923 earthquake. More recent earthquakes, like the 1971 San Fernando (California, USA) earthquake or the 1985 Mexican earthquake in Mexico, provided opportunities to study damage to water, sewage, and gas lines (Kawashima, 2006). In the 1995 Kobe earthquake in Japan, subway columns were damaged in the most extensive seismic ground motion ever seen in large engineering structures in soft ground.

On the other hand, the history of seismic design of underground structures is much shorter than that of above-ground structures. Studies of the seismic effects on underground structures were initiated in Japan and the United States in the 1960s, associated with the construction of underwater tunnels, necessitated by the soft and weak ground common at construction sites.

Measured data on the seismic response of underground structures has been accumulated, especially for pipelines. The behavior of an embedded pipeline intersected by a surface rupture was observed at Parkfield, CA in the United States (Isenberg et al. 1988). The seismic response of pipes with a diameter of 150 mm and 300 mm was measured in Sodegaura, Japan (Kawashima 2006).

In the last years, several methodologies were introduced for lifeline risk assessment in urban environment aiming to minimize losses, enhance the reliability of the systems, and improve mitigation policies. For water and waste-water system we can refer to Scawthorn et al. (1999), Reed and Cook (1999) and Seligson et al. (2003). Nagata et al. (2008) compared the performances of two analytical models based on past post-disaster restoration strategies, a detailed and a simplified one, which predict damage, post-earthquake restoration strategies, and its period of water supply pipeline network. Multihazard methodology tools were recently introduced to evaluate the vulnerability and the performance of lifelines under a variety of natural and technological hazards. HAZUS (NIBS 1999, 2004) is a typical example of an advanced GIS multihazard methodology.

Pitilakis et al. (2006) present the RISK-UE methodology for the seismic risk assessment of utility systems (potable water, waste-water, gas system, telecommunication, electric power) and transportation infrastructures (port, airport, road, and railway system). Weak points of urban systems are evaluated through detailed seismic hazard assessment including local soil conditions, and complete the inventory databases of all elements at risk. For selected seismic scenarios direct and indirect damages are calculated mainly for the building stock. Damage models suggested by 'RISK-UE' (Monge et al. 2004) expresses seismic damages to buried pipes by a 'repair rate' (RR) per unit length of pipe, that is the rate between the number of repairs and the length (km) of a pipe exposed to seismic hazard: this number is a function of pipe material, joint type, soil conditions, and diameter size, and of ground shaking level, in terms a synthetic seismic parameter. Referring to seismic wave propagation the following Eqs. (1) and (2) are proposed. As regards fault crossing, the presented model refers to permanent deformations (3):

$$\text{RR (repair/km)} = \text{C}_\text{p} \cdot \text{C}_\text{d} \cdot 0.00311 \cdot \left(\frac{\boldsymbol{PGV}}{100} - 15 \right)^{1,3} \quad \text{(Isoyama 1998)} \qquad (1)$$

$$\text{RR (repair/km)} = \frac{0.00187}{0.3048} \cdot \boldsymbol{K}_1 \cdot \left(\frac{\boldsymbol{PGV}}{0.0254} \right) \qquad \text{(ALA 2001)} \qquad (2)$$

$$\text{RR(repair/km)} = \frac{1.06}{0.3048} \cdot \boldsymbol{K}_2 \cdot \left(\frac{\boldsymbol{PGD}}{0.0254} \right)^{0,319} \qquad \text{(ALA 2001)} \qquad (3)$$

In the above equations RR is the number of repair per unit length of pipe [km]; C_p, C_d, K_1, and K_2 are coefficients according to various pipe material, joint type, soil type and conditions, and diameter size and can be found in specific tables (American Lifelines Alliance 2001). PGV and PGD are, respectively, Peak Ground Velocity and Permanent Ground Displacements. The numerical coefficients have been determined experimentally.

3 Evaluation of synthetic seismic parameters

Displacements potentially induced on natural slopes or earth structures in case of a seismic event can be evaluated applying Newmark (1965) displacement method. When seismic acceleration a exceeds a critical value a_c, the slope reaches a limit equilibrium condition and a potential landslide body starts sliding. It will stop only when seismic acceleration, changing its sign, will cancel relative velocity of the sliding mass. Afterwards the slope will not show any displacement until the critical acceleration value will be exceeded again. The critical acceleration of a slope is related to the examined instability mechanism geometry and to the slope's mechanical features. A generic approach could refer to an indefinite slope scheme in seismic conditions. In particular, assuming that shear strength of soils involved is only frictional and that soil is dry, the critical acceleration a_c, expressed as a gravity acceleration 'g' percentage, is

$$k_c = \frac{a_c}{g} = \frac{\cos\beta \cdot \tan\varphi' - \sin\beta}{\cos\beta^* + \tan\varphi' \cdot \sin\beta^*} \qquad (4)$$

where β is the slope angle, 'φ' is the soil shear strength angle, $\beta^*=\beta+\theta$ and θ is the slope of the resultant force from seismic action and weight in vertical direction. In this work, if ω is the angle between the seismic action direction and the horizontal direction we can assume $\omega=\omega_{cr}=\varphi-\beta$, to obtain the minimum value for critical acceleration k_c and then to perform a precautionary estimation of induced permanent displacements, so that
$\beta^*=\beta+\varphi-\beta=\varphi$. Replacing in (4):

$$k_c = \sin(\varphi' - \beta) \qquad (5)$$

The induced displacement entity could be estimated by a statistical correlation between its value and a seismic acceleration threshold value, function of geometrical and geotechnical soil parameters, like that proposed by Ambraseys and Menu (1988):

$$\log d = \log\left[\left(-\frac{k_c}{k_{max}}+1\right)^{2.53}\cdot\left(\frac{k_c}{k_{max}}\right)^{-1.09}\right]+0.90 \quad \left(0.1<\frac{k_c}{k_{max}}<0.9\right) \tag{6}$$

This equation has been obtained using a database of worldwide earthquake accelerometric records (M=5.5–7.5). d [cm] is the induced permanent displacement value.

The spatial distribution of seismic maximum acceleration values imposed at the ground level by a scenario earthquake could be evaluated by empirical attenuation laws that relate synthetic seismic parameters describing a seismic event (peak ground acceleration — PGA, peak ground velocity — PGV, or permanent ground displacement -PGD) a geometrical parameter describing the distance of a site from epicentre (epicentral distance 'R' [km]) and one or more parameters describing geotechnical and seismic soil features (S or S_A and S_S). The reliability of each attenuation law is deeply correlated with the mathematical model adopted for its development and with the features of the used seismic records database. In this work six attenuation laws suitable for the studied area have been selected among literature available models (Sabetta and Pugliese 1996, Eqs. (7) and (8); Troman and Bommer 2002, Eq. (9), Herrero Eq. (10), Sirovich Eq. (11), Langer Eq. (12) (see Faccioli and Rovelli 2004–2006), based on the seismic records database they are built on.

$$\log \mathrm{PGA} = -1.562+0.306\cdot M-\log\sqrt{R^2+5.8^2}+0.169\cdot S \tag{7}$$

$$\log \mathrm{PGV} = 0.710+0.455\cdot M-\log\sqrt{R^2+3.6^2}+0.133\cdot S \tag{8}$$

$$\log \mathrm{PGV} = -0.195+0.390\cdot M-1.074\cdot\log\sqrt{R^2+4.5^2}+0.142\cdot S_A+0.185\cdot S_S \tag{9}$$

$$\log \mathrm{PGD} = -4.68+1.08\cdot M-0.95\cdot\log_{10} R \tag{10}$$

$$\log \mathrm{PGD} = -1.8944\cdot\log(R+14)+0.4144\cdot M+1.1026 \tag{11}$$

$$\log \mathrm{PGD} = -1.4642-0.5512\cdot\log R \tag{12}$$

where, R[m] is the epicentral distance, M is earthquake magnitude and the parameters S, S_A, and S_S represent soil mechanical features (Sabetta and Pugliese 1996; Tromans and Bommer 2002).

4 Application

A Public Service manages a very important water supply system feeding 20 Etnean towns (Catania district, Italy), involving about 90,000 customers (about 400,000 people). The pipeline network develops from the western Etnean flank, with two main adduction pipes, to the southern-western and southern flank, were the distribution network feeds the customers. The adductor pipes are 'Maniace' Aqueduct, with a concrete main line about 46 km long (diameter varying from 300 to 450 mm) and three secondary lines, and 'Ciapparazzo' Aqueduct, completed in 1975, with a main line about 34 km long and 11 secondary lines, all in cast iron (diameter varying from 400 to 800 mm). These two main lines, after a nearby parallel route, along which some bounding bridges can be found, feeds the distribution network. The information stored in the water agency three-dimensional geodata base is quite good for the two main pipes, while is fragmentary and sometimes inadequate for the distribution network. The whole 1327 km of pipes have been surveyed with a traditional topographic approach, using a total station, with a centimetric accuracy. Among the detected pipes 271,39 km are adductor pipes, 871,27 km have distribution function, 53,57 km have relaunch function, and the remaining part is not defined. The material the pipes are made of is unknown for 29.54% of them; 23.22% of pipes are cast iron, 18.94% steel, and other small percentages are made of other materials. Over 90% of pipes cross noncorrosive soil, mainly volcanic. Only a small percentage crosses colluvial or clayey outcropping, which easily retains moisture and tend to be corrosive. Diameter is known for over 90% of detected pipes. For the remaining 10% a small diameter, typical of distribution pipes, has been hypothesized. As for all pipes the most influent factor for pipe seismic performance, that is the joint type, is unknown; the worst condition of rigid joints have been hypothesized.

All the models to calculate water pipes seismic vulnerability in terms of damage require a seismic scenario to be chosen, to evaluate PGA, PGV, or PGD. Analysing Eastern Sicily and more specifically Etnean area (Azzaro et al. 2000) seismic hazard, for first-level seismic scenario Catania, 11 01 1693 event (M= 7.3; I_{max}= X MCS; TR = 250–500 years) and for second level, Acicatena (south-eastern Etnean flank) 20 02 1818 (M = 7.2; I_{max}= IX MCS; TR = 250–500 years). As moderate scenario event Etnean earthquake of Bronte, 31 10 1832 (M = 3.4) has been chosen, as its magnitude is undoubtedly over the average, and for the geographical allocation of its epicentres, quite near the first part of Maniace Aqueduct, one of the main pipe feeding the distribution network.

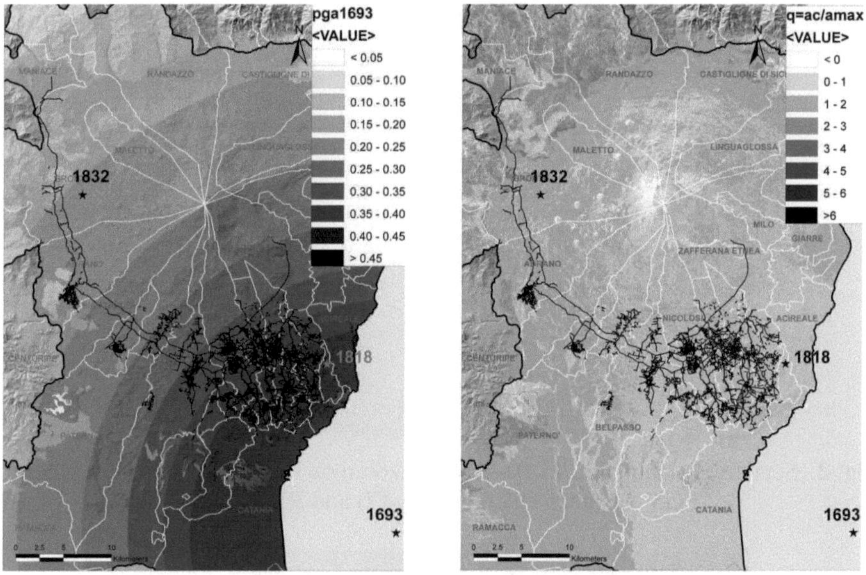

Fig. 1. Scenario earthquake: 11-01-1693. (**a**) Map of PGA (**b**) Map of unstable zones ($k_c / k_{\max} < 1$).

The calculation models have been applied to Catania district area and implemented in a GIS environment by Spatial Analysis techniques, using ArcGIS (ESRI) Model Builder. As cartographic background it has been used: the Regional Technical Cartography - CTR (1:10,000 scale; *.dxf - 612100 and 612140 sections); a Digital Terrain Model (DTM) of Catania District, 20×20 settlement, derived from isolines 1:25,000 scale (by Italian Geographic Military Institute – IGMI); a Lithological Map of Sicily, derived from a simplification of Italian Geological Map 1:100,000 scale. Moreover the pipeline network geodatabase, surveyed by total stations, with a centimetric accuracy, and the road network of Catania District, extracted from a vectorial commercial data set, with a centimetric accuracy, and the scenario earthquakes (1693, 1818, 1832) epicentres coordinates shape files were available. From DTM, by 'Slope' interpolation algorithm (ArcGIS 'Spatial Analyst'), the slope map has been obtained, with an accuracy between 84 and 86%. To assign soil mechanical features a vectorial map of the district area surfacial lithology has been used: basing on literature data, values of φ' (Lambe and Withman 1969), S, S_A, and S_S (D.M. 14/01/2008) have been assigned for each lithotype. The corresponding grid themes have been created.

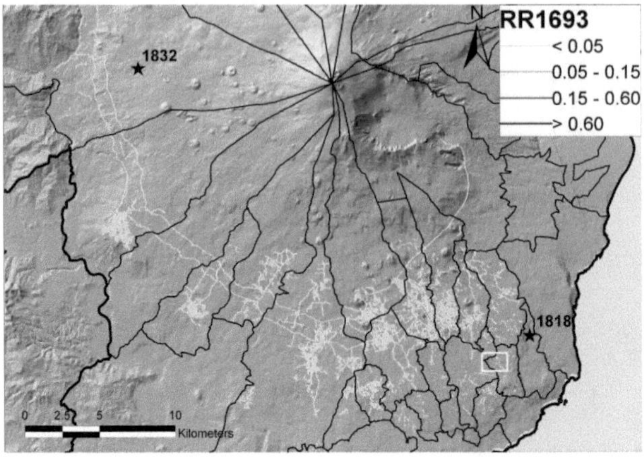

Fig. 2. Scenario earthquake: 11-01-1693. A zoom on map of repair rates, where PGV has been calculated by (7) and RR by (2)

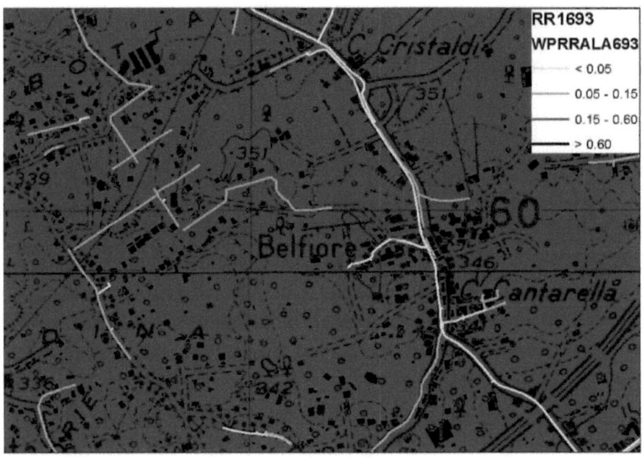

Fig. 3. Scenario earthquake: 11-01-1693. A greater zoom on map of repair rates, where PGV has been calculated by (7) and RR by (2)

To calculate Newmark displacement method critical acceleration (5) can be used, implementing it by Map Algebra techniques, with φ' and β GRID as input. A new GRID of critical accelerations k_c will be obtained. For the next calculations cells where $k_c < 0$ have been omitted, as they correspond to a static safety value numerically smaller than one, but not necessarily identifying a slope static instability context, as this value could be due to the unavoidable approximation in the estimation of φ' average value. For every input earthquake (1693, 1818, 1832), applying attenuation laws (7, 8, 9, 10, 11, 12) the grids of PGA, PGV, and PGD have been calculated and the concerning thematic maps have been created. Figure 1a is the map of PGA evaluated by (7).

According to the infinite slope scheme, permanent displacements can take place when $k_c / k_{max} < 1$: the cells where $k_c / k_{max} < 1$ are highlighted in Fig. 1b. For the unstable cells, for every scenario earthquake, the value of the induced permanent displacements d have been calculated by (6). The raster maps of PGV and PGD have been converted in shape files. The information about seismic hazard and about pipes has been overlaid, calculating 'RR' by different approaches. For example, in Fig. 2 the thematic map of 'repair rates' evaluated by (2) with a PGV evaluated by (8) is drawn, and in Fig. 3 a zoom of the previous map on a densely urbanized area is plotted.

5 Main results and conclusions

GIS environment clearly appears as a very efficient and highly productive tool for large-scale calculation and representation of complex numerical models. This experience shows that it is possible to calculate the parameters of interest and produce georeferred thematic maps of a very large territory in a short time: this is very important for preventive risk analysis and for emergency management.

Observing the obtained thematic maps it can be noticed that, as expected, the parameters PGA, PGV, and PGD attenuates with distance from the epicentre of the scenario earthquake chosen (it goes from dark to tan colors). The seismic scenarios of 1693 and of 1818 show significant ground accelerations k_{max} (even over 0.45 g). However, the interested areas are also characterized by high critical acceleration k_c values, due to small slope angles β or to the big shear strength angle values φ' assumed. Consequently the potentially unstable areas, characterized by significant permanent displacement arising, are not wide and most of them are isolated regarding the two main adductor pipes. Moreover, the displacement analysis carried out regarding these zones highlighted small displacement values, generally lower than 4 cm.

All the more so, for 1832 seismic scenario, k_{max} values being lower, for the same critical acceleration values k_c, the potentially unstable areas are less than the previous cases and focus near the epicentral area. However this epicenter is very close to the two main aqueduct fonts, so that the potentially induced displacements by such an earthquake would involve the first part of Maniace aqueduct. Since it is a big cement pipe with quite rigid joints, it could be damaged with consequent water losses. Comparing the calculated displacement value with the potential induced permanent ground displacements and an admissible value for the examined structures, a judgement about a certain seismic scenario event effect on a water network can be expressed.

Table 1. Serviceability of the analysed water network [% of length] versus repair rate RR (ALA 2001) (from Raciti 2008)

Repair Rate (repair/km) (ALA 2001)	Damage States	**% length of pipeline network**					
	Scenario earthquake	**1693**		**1818**		**1832**	
	Attenuation law	**Bommer et al.**	**Sabetta and Pugliese**	**Bommer et al.**	**Sabetta and Pugliese**	**Bommer et al.**	**Sabetta and Pugliese**
< 0.05	Minor	96.02	90.06	84.99	82.46	95.47	87.66
0.05–0.15	Moderate	3.98	9.94	15.01	15.30	2.20	4.11
0.15–0.60	Extensive	0.00	0.00	0.00	2.34	2.33	8.23
> 0.60	Complete	0.00	0.00	0.00	0.00	0.00	0.00

The definition of an admissible threshold for displacements must consider the displacement effects on partial or total pipes serviceability losses. Moreover it is important to take into account how much money and how long it will take to recondition the network and the importance the temporary unserviceability of this pipe has on socioeconomic life of the interested region. This is a very complex problem that is affected by the unavoidable subjectivity of the opinion.

Both national and international laws regarding pipeline seismic damage are extremely poor. Alaska Geotechnical Evaluation Criteria Committee establishes five damage classes: minor ($d < 3$ cm); moderate ($d < 15$ cm); very high ($d < 30$ cm); extensive ($d < 90$ cm); catastrophic ($d < 300$ cm). Based on this classification, we found that the scenario earthquakes adopted in this work would cause minor or moderate damages to the water network.

In Table 1 seismic behavior of the studied water network for the scenario earthquakes '1693', '1818' and '1832' (1° vulnerability level) is represented by the synthetic parameter 'RR'. Comparing the values written in the tables, a kind of similarity between the results obtained by Tromans and Bommer (2002) equation and those obtained by Sabetta and Pugliese (1996) equation can be observed. The obtained results are quite comforting as the obtained repair rate 'RR' total values are low: a very high percentage of the network would undergo minor damages. The information provided estimating RR parameter can be profitably used to develop statistical investigations about the importance of the essential actions to mitigate the network risk, to plan appropriate procedures of emergency management, and to plan any improvement works in the network nodes or pipes where a greater vulnerability has been found.

However, it must be considered that the estimation could suffer of a significant uncertainty if detailed information about pipelines and above all about joint types are missing.

References

Ambraseys NN, Menu JM (1988) Earthquake-induced ground displacements. Earthquake Eng Struct Dyn 16:985–1006.

American Lifelines Alliance (2001) Seismic fragility formulations for water systems. guideline: ASCE-FEMA, Part 1, p. 104 and Appendices: ASCE-FEMA, Part 2, p. 239.

Azzaro R, Barbano MS, Antichi B, Rigano R (2000) Macroseismic Catalogue of Mt. Etna earthquakes from 1832 to 1998. Acta Vulcanologica 12(1–2):3–36 (CD-ROM).

D.M. 14/01/2008, New Italian Technical Seismic Regulation.

Faccioli E, Rovelli A (INGV Roma) (2004–2006) Project S5 – Seismic input in terms of expected spectral displacements. A collection of the research units final scientific forms.

Idriss I (1985). Evaluating seismic risk in engineering practice, Proceeding Eleventh International Conference on Soil Mechanics and Foundation Engineering, vol. 1. A.A. Balkema, San Fransisco, CA, pp. 255–320.

Isenberg J, Richardson E, O'Rourke TD (1988) Buried pipelines across San Andreas fault. Proceedings of 9th World Conference Earthquake Engineering. Tokyo/Kyoto.

Isoyama R, Ishida E, Yune K, Shirozu T (1998) Seismic damage estimation procedure for water supply pipelines. Proceedings of Water & Earthquake '98 Tokyo, IWSA International workshop, Anti-Seismic Measures on Water Supply, International Water Services Association and Japan Water Works Association, Tokyo Japan.

Lambe TW, Whitman RV (1969) Soil Mechanics. John Wiley & Sons.

Kawashima K (2006). Seismic analysis of underground structures. J Disaster Res 1(3):378–389.

Monge O, Alexoudi M, Argyroudis, Martin C, Pitilakis K (2004) RISK-UE. An advanced approach to earthquake risk scenarios with applications to different European towns. Vulnerability assessment of lifelines and essential facilities (WP06): Basic methodological handbook. Report n◦GTR-RSK 0101-152 av7, 71pp.

Nagata S, Kageyama K, Yamamoto K (2008) An emergency restoration model for water supply network damage due to earthquakes. J Disaster Res 3(6):2008, pp.

National Institute of Building Sciences (1999, 2004) Direct physical damage to lifelines-transportation systems-utility systems. Earthquake loss estimation methodology. HAZUS Technical manual, Vol. 2. Chap. 7, 8. Federal Emergency Management Agency, Washington, D.C.

Newmark NM (1965) Effect of earthquakes on dam and embankment, The Rankine Lecture, Geotèchnique 15(2).

Pitilakis K, Alexoudi A, Argyroudis S, Monge O, Martin C (2006) Earthquake Risk assessment of lifelines. Bull Earthquake Eng 4:365–390.

Raciti E (2008). Lifelines seismic risk: evaluations in a GIS environment. Master Thesis in Analysis, Monitoring and Mitigation of Environmental Risk; (In Italian).

Reed D, Cook C (1999) Multi-hazard analysis of utility lifeline systems. In: Elliot M, McDonough P (eds.) Proceedings of 5th US Conference on Lifeline Earthquake Engineering, TCLEE/ASCE, Monograph No.16, 940–949 390 Bull Earthquake Eng (2006) 4:365–390.

Sabetta F, Pugliese A (1996). Estimation of response spectra and simulation of nonstationarity earthquake ground-motion. Bull Seismolog Soc Am 86:337–352.

Seligson H, Ballantyne D, Huyck Ch, Eguchi R, Bucknam St, Bortugno E (2003) URAMP (Utilities Regional Assessment of Mitigation Priorities)—a benefit–cost analysis tool for water, wastewater and drainage utilities: methodology development. In: Beavers JE (ed.) Proceedings of 6th US Conference and Workshop on Lifeline Earthquake Engineering, TCLEE/ASCE, Monograph No.25, pp. 494–503.

Scawthorn Ch, Ballantyne DB, Eguchi R, Khater M (1999) Multi-hazard risk assessment for lifelines – Part 1 – overview and approach. In: Elliot M, McDonough P (eds.) Proceedings of 5th US Conference on Lifeline Earthquake Engineering, TCLEE/ASCE, Monograph No. 16, pp. 950–959.

Tromans IJ, Bommer JJ (2002). The attenuation of strong-motion peaks in Europe. Proceedings of 12th European Conference on Earthquake Engineering, p. 394.

Towards a Quick Damage Detection System Based on Grid Computing

Tuong-Thuy Vu and Masashi Matsuoka

Abstract

This chapter develops a quick damage detection system prototype, which produces reasonably detailed information of damaged buildings and accommodates high spatial resolution data. Parallel processing based on grid-computing is chosen to reduce computation time as well as to provide end-users with cost-effective access. The clusters of the GEO (Global Earth Observation) Grid system (www.geogrid.org) at the National Institute of Advanced Industrial Science and Technology (AIST) in Japan will serve as the main platform. The automated damage detection runs through a scale-space analysis. It is developed as a context-based approach integrating texture, shape, size, and spectral reflectance. QuickBird imagery acquired over Yingxiu town which was heavily damaged due to the 2008 Sichuan Earthquake is used to demonstrate the performance of damage detection algorithms. Damage information at building block level is successfully mapped. At the current stage of development, damage detection focuses on producing quick orientation damage maps. The data grid component of the GEO Grid with data federation capability will be employed to connect to various satellite image and field-survey databases in the next stage of development. This is to enhance the mapping capability for more accurate maps. In addition, future work will practically perform the processing of large-scale damage detection on the GEO Grid, and test the system performance with remote sensing data acquired in various catastrophes.

M. Konecny et al. (eds.), *Geographic Information and Cartography for Risk and Crisis Management*, Lecture Notes in Geoinformation and Cartography,
DOI 10.1007/978-3-642-03442-8_11,

1 Introduction

Remote sensing technologies have been playing an important role in disaster management, especially to provide the necessary data at an early stage after a catastrophe. As time is critical at this stage, an automated approach to both data acquisition and data processing is vital. The former has been solved to some extent through the establishment of the International Charter on 'Space and Major Disaster'. Data capture now efficiently operates to provide redundant data of disaster areas shortly after the event. Regarding data processing to derive the damage information, there have been numerous efforts in developing automated approaches for various remote sensing data types from optical to radar imagery (Adams et al. 2004, Estrada et al. 2000, Matsuoka and Yamazaki 1999, Sato et al. 2004, Singhroy and Mattar 2000, Stramondo et al. 2006, Vu et al. 2005a). Those approaches showed certain success in damage detection but no reliable approach exists yet. Quick damage detection still relies on a manual approach which produces an overview of damage extent. The accurate damage information in detail will only be produced long after a rigorous processing of remote sensing data in combination with field survey data.

Advanced remote sensing technologies provide the capability to have higher spatial resolution, higher temporal resolution and higher spectral resolution data. For instance, very high spatial resolution QuickBird and IKONOS satellite images are favourably used by disaster management practitioners and damage assessment analysts. As more aspects of damage information can be captured, a more advanced image processing approach is required. Following the current trend in development of an advanced processing approach, damage detection has recently employed object-oriented processing (Gusella et al. 2005, Kouchi and Yamazaki 2005, Vu et al. 2005b). Object-oriented analysis enables the integration of rich content of information in processing to produce an easy-to-understand form for further analysis. As the process becomes more complicated, computation time drastically increases. It is, hence, inapplicable in the context of damage mapping for emergency response. However, choosing simple processing like the system proposed by Gamba and Casciati (1998) to obtain a quick outcome cannot well accommodate new high-resolution data nowadays.

Our previous research efforts (Vu et al. 2005b, Vu et al. 2007a) were the step-by-step development of a reliable automated damage detection method. We then introduced the context-based damage detection method (Vu et al. 2007b) for high-resolution satellite images. This is a 2-stage processing approach in which the first stage is to extract the debris areas using edge texture analysis and the second stage is to delineate the standing buildings based on a scale-space analysis. Subsequently, the crosscheck of two results classifies the damage level. Integrating texture analysis into scale-space analysis is a further development which is expected to better describe the context for analysis. The whole idea of context-based damage detection with latest improvements is presented in Sect. 2. Its performance is then demonstrated (Sect. 3) using a QuickBird image acquired on June 03, 2008 which captured the damaged situation of YingXiu town, Sichuan Province,

China. This town was heavily devastated due to the main shock of the 2008 Sichuan earthquake on May 12, 2008 and continuing powerful aftershocks months later.

By sharing the heavy and complex damage mapping tasks among a number of operators/experts, the process can be sped up. Moreover, various expertises would contribute knowledge and resources to derive better damage maps. It is feasible to have such a 'collaborative mapping' approach with current Information and Communication Technologies (ICT). Maiyo et al. (2009) concluded that GeoWeb services provide a better solution to meet the post-disaster mapping challenge. It led to a proposed Collaborative Disaster Management System (CDMS) allowing feedback from near-real-time field survey information. This promises to obtain highly accurate damage maps (Maiyo et al. 2009). CDMS is mainly based on field-survey information. Other daily-use online map services such as Google Map and Microsoft Virtual Earth can act as the platform for collaborative mapping. Recently, Virtual Disaster Viewer (VDV) (www.virtualdisasterviewer.com) has been tested in response to the Sichuan earthquake. VDV developed on MS Virtual Earth acts as a 'social networking tool' for earthquake damage assessment. VDV relies on high-resolution satellite imagery for damage detection and requires participating experts to produce the damage maps in their own ways. It should be noted that the current spatial resolution satellite imagery could only help to detect the heavily damaged and collapsed buildings (Rathje and Adams 2007). In addition, top-view imagery is unable to capture vertical damage like pancake collapse. Detection from satellite images is also subject to problems with changes not due to the event between the pre- and post-scenes (Adams et al. 2004) and different acquisition angles (Chesnel et al. 2007). However, as satellite images play a unique role in acquiring damage information over large areas and difficult-to-access areas, it is necessary to utilize them to complement field-survey information. The solution proposed here aims at integrating heterogenous data sources from users, on the one hand, while providing a powerful satellite image analysis tool on the other. To implement such an approach, grid computing and web services are employed.

The main objective of this study is to design an applicable damage detection system prototype focusing on buildings in the context of emergency response. The aforementioned context-based damage detection approach is deployed as the core processing. The foundation of the proposed system will be the GEO (Global Earth Observation) Grid system as described in Sect. 4. The damage detection system prototype is then presented in Sect. 5. The first development step discussed in this chapter focuses on exploiting the computing grid, i.e., mounting the damage detection algorithms onto GEO Grid clusters. Data federation will be exploited in future when an efficient communication manner can be established with satellite data providers. This technology will be also utilized to integrate with field survey data for a better damage map. It is one of the foreseen further developments to be discussed prior to the concluding remarks in Sect. 6.

2 Context-based damage detection

Recently, the most reliable way to detect damaged buildings from a remotely sensed image has been by visual inspection. A damage analyst (or expert) starts locating the debris areas which are the cues of nearby damaged buildings. Subsequently, the still-standing buildings are of concern. In comparison with the status obtained from a pre-event scene or existing building inventory database, the operator can separate the damage into categories of heavy damage (totally collapsed buildings), moderate damage, light or no damage. Thus, there are two important indicators, i.e., debris area and still-standing buildings, which help to delineate the damage. Our developed context-based approach mimics the visual inspection by automatically extracting those two indicators (Vu et al. 2007b). Briefly, scale-space analysis based on area morphology theory (Vincent 1992) categorizes the objects according to their spectral information, size, and morphology and then extracts the objects of interest. It enables the extraction of still-standing buildings. On the other hand, the entropy of edge intensity, computed by the Prewitt operator, (T_e) representing uniformity of the edge structure in a local neighbouring area of a pixel, was used to characterize the debris areas. Since collapsed buildings show strong non-uniformity, the highest range of T_e indicates the debris areas. The ratio of T_e to a chosen unit area then was applied to accommodate the building sizes. This infers that better results can be expected if the texture analysis is integrated into scale-space analysis. The improved context-based damage detection algorithms are described in the following paragraphs.

Figure 1 depicts the general processing flowchart. Each rectangular box illustrated in this figure is implemented as an IDL (Interactive Data Language) script, which is efficient in quickly implementing image-processing algorithms. The prototype will work with IDL whereas real implementation on the GEO Grid platform will use Python or Perl instead. This is to maximize the number of idle CPUs used and hence, to speed up the processing. First, the entropy of edge intensity (T_e) is computed and assigned weights. The higher T_e value is assigned higher weight. Meanwhile, K-mean clustering is employed to assign a spectral index for each pixel. The operator has to input the meaning of all the spectral indices. Since the extraction focuses on building and debris, all the irrelevant indices such as the ones standing for vegetation, water or shadow are removed. Additionally, similar classes can be merged to obtain a smaller number of spectral classes.

Subsequently, the morphological scale-space is generated for both spectral and texture clusters. The classical morphological filters including dilation, erosion and their combination like opening and closing have been effectively used in image processing. Their extensions such as greyscale opening (and/or closing) by reconstruction and area opening-closing (Vincent 1992) enhance the capability and hence, expand the employment of morphological operators in various image processing tasks (Soille and Pesaresi 2002). Iteratively applying a combination of these morphological operations on an image with increasing disc size enables the generation of a non-linear scale space. The small details disappear when upscaling to a coarser scale and the image is smoothed or simplified by flattening.

In object detection here, we use a flat disc with greyscale opening-closing by reconstruction to generate the scale-space. The size parameter is determined through granulometry analysis (Serra 1982).

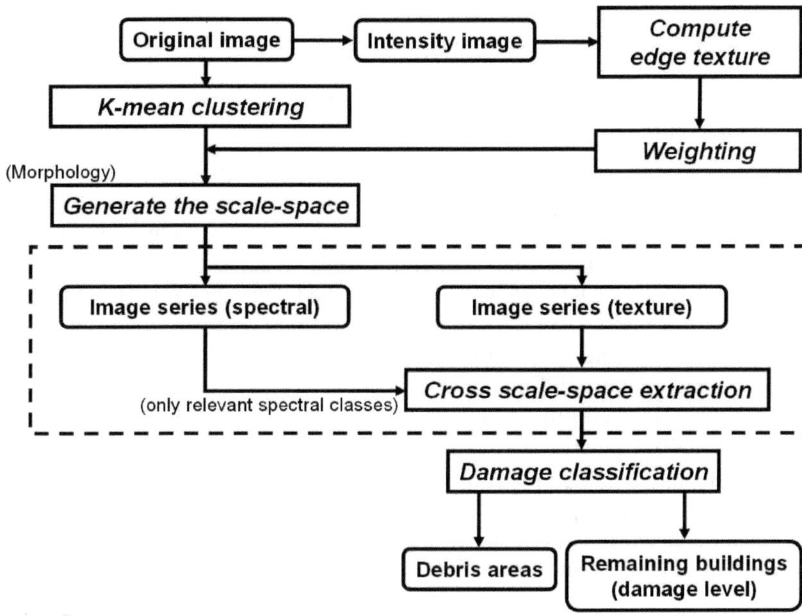

Fig. 1. Processing flowchart of context-based damage detection

Scale-space analysis is the second processing step as illustrated inside the dashed-line box in Fig. 1. Spectral information and edge texture are separately processed here but in a similar manner. The adjacent pixels of similar spectral signatures are grouped into one object with an assigned *ID* to prepare for the cross-scale-space analysis. Likewise, *TEXTID* is assigned to the texture-based blobs formed by pixels of same texture weight. For the sake of clarity, in the spatial domain a spectral-based cluster is called an object whereas a texture-based cluster is called a blob.

Depending on its size and shape, an object vanishes on a scale when down-scaling from the finest to coarsest scales. A pixel, therefore, belongs to one object at this scale but it might belong to another one at a coarser scale. Cross-scale-space analysis tracks the pixels' behaviour, and groups the ones which have similar behaviours together. Consequently, objects can be formed and extracted. In implementation of search and extraction, hybrid processing on scaled images and a relational database is developed. A database is prepared with the following attributes: *ID* (object id), *SCALE* (the scale where it exists), *SPE* (the spectral index), *AREA* (its size) and 2 other shape indices named *SHAPEplex* and *SHAPEcom*. The former is computed based on the complexity of the object's skeleton. A building object generally has a small *SHAPEplex* value. Alternatively, *SHAPEcom* represents the compactness of an object. It is simply the ratio of the object's area to the area of a bounding box. An object with a more compact shape has a high *SHAPEcom* value. In classification of texture blobs used to map the debris areas, area and the complex shape are not taken into account. Therefore, only *SHAPEcom* is computed for texture blobs. A blob, hence, possesses the following attributes *TEXTID*, *TEXTSCALE*, *TEXTURE* (texture weight) and *TEXTSHAPEcom*.

Three indices, *SHAPEplex*, *SHAPEcom* and *AREA*, are reclassified into *low* and *high* classes. The crosscheck between spectral-based objects and texture blobs is carried out as follows:

- First, big objects of complex and less compact shapes, i.e. *high SHAPEplex*, *high AREA* and *low SHAPEcom* values, are removed since they represent the ground or open land objects. The remnant is masked as potential buildings.
- Second, while working on the texture blobs to pick up the debris areas, the blobs with *high TEXTURE* values are masked. The less compact shape blobs, which are normally the edges of objects, are then removed using the *low* value of *SHAPEcom*. The remaining blobs can now be of two types: some small components of the buildings or debris areas. By overlaying those texture blobs on the potential buildings to eliminate the small buildings, the debris areas can then be finally located.
- Third, the above named potential buildings are further investigated. Since building objects are normally not very large and in less complex as well as very compact shapes, it is possible to pick up the *low* AREA, *low SHAPEplex* and *high SHAPEcom* as building objects.

The above extraction scheme is recorded as the *TYPE* attributes of spectral objects and texture blobs. Consequently, the extraction produces the final results in form of a 7-band image including *ID*, *scale*, *spectral*, *shape complex*, *shape compact*, *area* and *type* bands for spectral objects and a 5-band image including *ID*, *scale*, *texture*, *shape compact* and *type* bands for texture blobs. Alternatively, to enable further analysis in damage assessment, the first band (*ID*) can be used for conversion to ESRI shapefile format. Other attributes can also be added into the shapefile database.

3 Detection of damage due to the 2008 Sichuan earthquake

A QuickBird image (Fig. 2) acquired on June 03, 2008 which captured the damaged situation of YingXiu town, Sichuan Province, China is used to demonstrate how the context-based damage detection performs. Figure 2a illustrates the damage scene in intensity value, which was used to compute edge texture and its entropy. It showed that building objects have diverse spectral signatures and generally have various sizes in the small range in comparison with other object types such as vegetation or water body. To simplify the extraction task, we created a river mask to mask out the water since the water waves obviously appear as bright as debris.

Based on the result of granulometry analysis, the following sizes 4, 25, 64, 100, 225 and 400 were used for generation of the scale space. Figure 2b and 2c illustrate spectral values clustered in the scale space from fine scale to coarser scale.

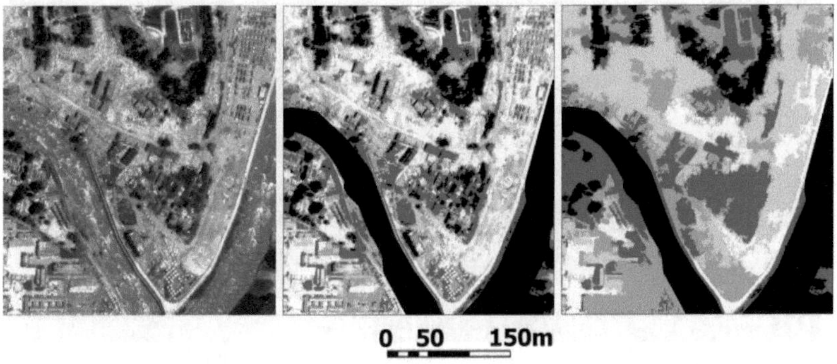

(a) (b) (c)

Fig. 2. QuickBird image June 03, 2008 of YingXiu town: (**a**) intensity image and scaled images generated from the spectrally clustered image with size (**b**) 4 and (**c**) 100

The extracted results are illustrated in Fig. 3 in which grey codes represent the object types, i.e. brighter means more likely to be an object of interest. As aforementioned, the results from texture information are called blobs (Fig. 3a) whereas the ones from spectral information are named objects (Fig. 3b). Following the hierarchical classification described in Sect. 2, the final classification produced the extracted debris areas and still-standing buildings as shown in Fig. 4. It shows that the developed context-based approach quite successfully extracts the buildings of various sizes and diverse spectral signatures.

The damage detection result posted on VDV (www.virtualdisasterviewer.com) was used as reference data for comparison. On VDV, damaged buildings were reported as points classified to *indistinguishable*, *slight damage*, *extensive damage* and *collapse*. The last two classes were picked up for checking with the detected debris areas. To be comparable, a buffer of 5 m was created for each reference point. The comparison showed that 4 of 81 reference points found no match with debris areas. Apart from an identical group to *indistinguishable* points, detected debris areas also introduced some commission errors in the areas of new temporary houses/tents. Detected still-standing buildings were checked with visually detected ones since no reliable reference data was available. Performing the visual detection confirmed that it is really difficult to delineate the standing buildings solely from satellite images, i.e. spectral signatures from the top view. Figure 4b presents the number of false detections due to objects of compact shape existing of in debris areas.

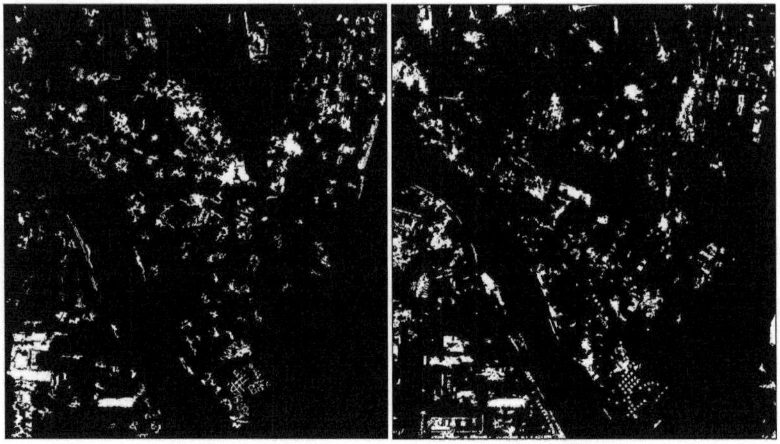

Fig. 3. Extracted results: (**a**) texture-based blobs and (**b**) spectral-based objects

There were two obstacles to obtain a better accuracy in extraction and mapping. First, numerous building roofs showed very similar spectral reflectance as that of the ground, i.e. concrete material. In addition, the collapsed scene shows unclear distinction of a building from the ground. As a result, those buildings barely formed into objects extruded from the ground. Second, plenty of small temporary houses or tents were built in the area. They showed similar behaviour to that of the debris areas presented in texture analysis. A group in the bottom of the image could be successfully clarified as building objects due to their spectral homogeneity whereas others in the top left could not. When tuning the size parameters to pick up those small houses, small homogeneous debris objects will also be included. Field survey data would help to overcome these problems. The requirement of satellite image acquisition time is another limitation of this approach. Since it uses debris as the main indicator, imagery should be acquired shortly after the event when the ground has not been cleaned up. In terms of quick damage mapping, the above debris and still-standing maps produced from a single post-event scene could provide necessary information at an early stage. Our context-based damage detection approach will be the core processing of the damage detection system as described in the following sections.

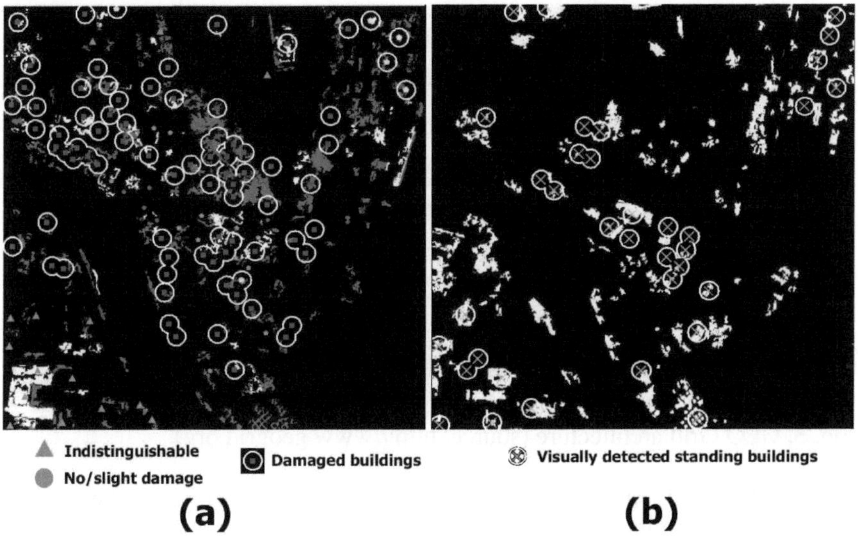

Fig. 4. (**a**) Extracted debris and (**b**) extracted still-standing buildings in comparison with other visually detected results

4 Grid computing and GEO (global earth observation) grid system

To accommodate the critical time requirement for emergency response and diverse demands on detailed levels of information from end-users, we propose a damage detection system based on powerful grid computing. Grid computing is the exploitation of numerous computers to solve a large-scale problem which requires a great number of computer processing cycles and/or access to large databases. The size of a grid computing system varies and can be in the form of network-distributed parallel processing or large-scale cluster computing. The remote sensing society has recently turned its attention to deploying high-performance computing, especially grid computing in remotely sensed data archiving and processing. Employing the grid-computing approach brings many advantages such as cost savings through resource-sharing, scalability, fast computing, increased flexibility and resilient operational infrastructures, enabling wide collaboration, increased productivity through standardization, etc. (Gasster et al. 2008).

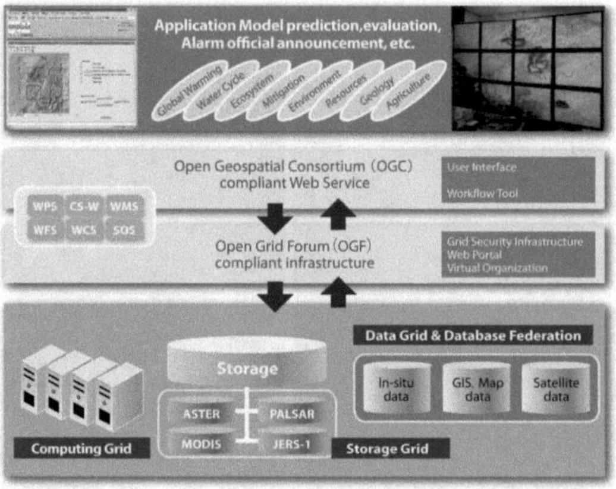

Fig. 5. GEO Grid architecture (source: http://www.geogrid.org)

The GEO Grid aims at providing an e-Science infrastructure for the worldwide earth sciences community. Its concept is the marriage of grid technology to global earth observation. It provides securely and rapidly large archives of different attributes of earth observation satellite data (ASTER, MODIS, etc.) and integrated services with various observation databases and geoscientific information. The principal design of the system is the open and standard-protocol-based architecture including OGC (Open Geospatial Consortium) and Open Grid standards (Sekiguchi et al. 2008). An overview of GEO Grid infrastructure is shown in Fig. 5.

The GEO Grid system implements an IT infrastructure for flexible, secure and coordinated sharing of satellite images, other geospatial databases and computing power. Virtual Organization (VO) designed by the GEO Grid development team is to ensure security and accommodate the diverse data sources and end users. There are four different roles for the usage model of GEO Grid: service providers, VO managers, end users and GEO Grid administrators. To enable the integration of various distributed data sources, the GEO Grid provides database functionalities based on a middleware called Open Grid Services Architecture Data Access and Integration (OGSA-DAI) (Sekiguchi et al. 2008). Data analysis takes advantage of fast computing of grid technology in the form of parallel processing. With numerous loosely coupled computers on the GEO Grid system, solving large-scale damage mapping becomes feasible at the emergency response stage.

5 Damage detection system prototype

In the context of post-disaster and emergency response, time is the most critical factor. It is also the main reason why remote sensing could play an important role at this stage of the disaster management cycle. Our ultimate goal is to develop a damage detection system which can minimize both acquisition time, i.e. excluding the capture and down-link time of space segments, and computation time. The foundation of our proposed damage detection system is GEO Grid. Figure 6 sketches the system diagram in which our damage detection tool is emphasized. Numerous other tools have been currently developed by the GEO Grid team and other research partners. They will all be provided as web services to end users.

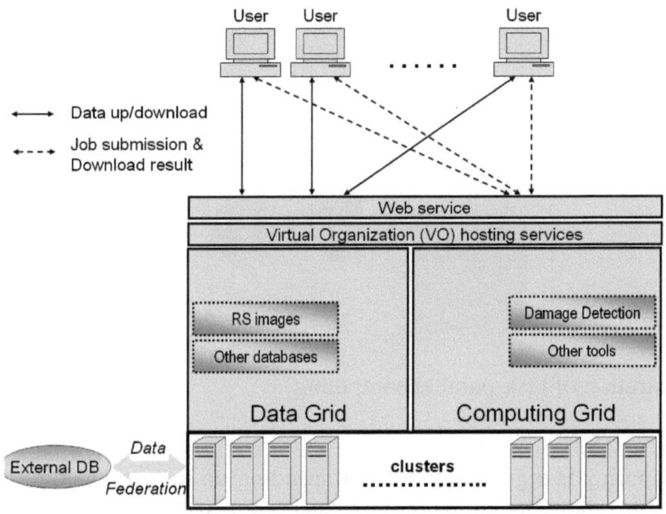

Fig. 6. Damage detection as a service provided by the GEO Grid

At this step of development, the prototype exploits only the computing grid component of GEO Grid. Satellite images are directly accessed in GEO Grid system. By design employing data federation, the end users can virtually access any connected satellite image database on GEO Grid; even this database is managed by another organization or end user not GEO Grid. There are three currently activated and managed VOs (Sect. 4), alternatively called working groups, on the GEO Grid. An end user must be a member of one working group to be able to access and work with the provided tools. The developed damage detection tool is provided as a service of the "Geological Hazards and Disaster Mitigation Working Group (or VO)". Since data and processing tools in the form of IDL scripts are already on the system, an end user starts damage detection work by submitting the tasks including the parameters of each step and the names of data sets to work with.

Relying on the GEO Grid system, damage detection algorithms are designed as task-parallel computing as illustrated in Fig. 7. Each processing box in Fig. 1 is defined as a main task comprised of several independently running sub-tasks. For example, the main task of scale-space generation is divided into the tasks of running an open-and-close morphological filtering by reconstruction of size s on band B of an image I. Let's assume that an operator is working with two 3-band images and he/she wants to run on 6 scales. Thus, he/she should prepare and submit $2 \times 3 \times 6 = 36$ sub-tasks. The task scheduler on the GEO Grid system will then allocate those 36 tasks to the idle CPUs, 36 CPUs for instance. It is logical that the processing could be speeded up 36 times.

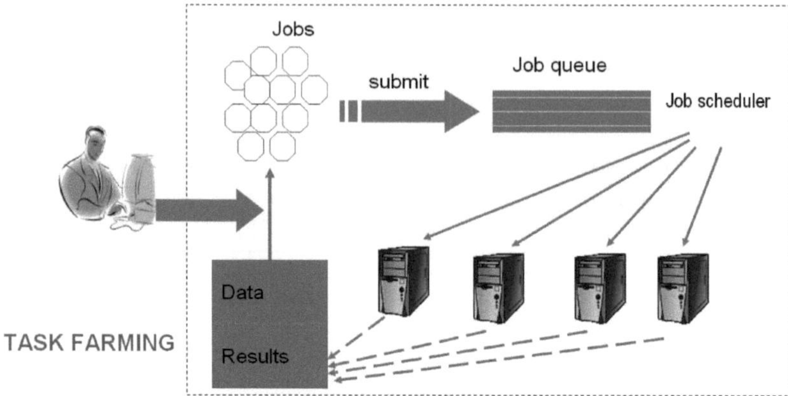

Fig. 7. Illustration of task-parallel computing

Grid computing is a good solution for speeding up time-consuming processing steps. It enables the detailed and complicated context-based analysis of the large scenes in which diverse aspects of acquired information are integrated. It should be noted that the currently developed damage detection prototype works on a single post-event image. The outcome is the extracted debris and still-standing objects as described in Sect. 2. Extraction of buildings from a pre-event image can exploit this context-based image analysis. The comparison of post-event and pre-event extraction results to further classify the damage levels will be developed and added in further research. In addition, the GEO Grid with data federation capability can help to drastically reduce the time and increase the efficiency of data distribution. Acquired satellite images which are quickly down-linked to the ground station of a space agency or company are virtually accessible by the damage analyst through the GEO Grid. End users with a standard PC and access to the Internet can easily request the performance of large-scale problem solving in disaster mitigation or post-disaster/emergency response.

6 Conclusions

Aimed at quickly providing reliable damage information over a disaster-affected area, this study has presented a damage detection system prototype based on the GEO Grid system. Only the computing grid is the focus at this stage of development. The computation for damage extraction employs a newly developed context-based image analysis technique. Its good performance was demonstrated by using a QuickBird image of Yingxiu, Sichuan, China. The extracted results here show that detected debris areas were matched with the nearby damaged points detected by VDV contributing experts. More improvement should be carried out in building detection to make it applicable as a context-based tool for image classification/segmentation and feature extraction from images of various spatial resolutions. Exploiting the great capability of grid-computing clusters could accommodate the critical-time requirement and rich detail in satellite images. Moreover, the proposed system can reduce the monetary investment in data and processing systems by end users. Thus, it provides a feasible solution to developing countries, which have faced numerous large-scale disasters recently. Future work will add one more function to integrate with field-based information, to compare the post- and pre-information, to compute the building-block-based damage index for further damage assessment, and concentrate on optimization of parallel computation on the GEO Grid. One recommendation is to compare the computation time of the entire process including grid computation and visual assessment to the time spent for conventional visual inspection. Additionally, a portal site will be developed as the interface and hence damage detection service will be made available to end-users.

Acknowledgements

This research is a part of an international collaboration project funded by New Energy and Industrial Technology Development Organization (NEDO), Japan. The authors greatly appreciate the support of the GEO Grid Research Group in idea development and further implementation of the proposed system. Further acknowledgement is given to the "Sichuan Earthquake Data Consortium" that purchased QuickBird imagery of the earthquake on behalf of the participants of the Consortium.

References

Adams BJ, Huyck CK, Mansouri B, Eguchi RT, Shinozuka M (2004) Application of high-resolution optical satellite imagery for post-earthquake damage assessment: The 2003 Boumerdeas (Algeria) and Bam (Iran) earthquakes. Multidisciplinary Center for Earthquake Engineering Research Progress and Accomplishments 2003–2004, University of Buffalo. http://mceer.buffalo.edu/publications/resaccom/04-SP01/12_Eguchi.pdf. Accessed 20 Dec. 2007.

Chesnel A-L, Binet R, Wald L (2007) Damage assessment on buildings using very high resolution multimodal images and GIS. In: Proceedings of 5th International Workshop on Remote Sensing, George Washington University, Washington DC, CD-ROM.

Estrada M, Yamazaki F, Matsuoka M (2000) Use of Landsat images for the identification of damage due to the 1999 Kocaeli, Turkey earthquake. In: Proceeding of the 21st Asian Conference on Remote Sensing. Singapore, 1185–1190.

Gamba P, Casciati F (1998) GIS and image understanding for near real-time earthquake damage assessment. Photogramm Eng Rem Sens 64(10):987–994.

Gasster SD, Lee CA, Palko JW (2008) Remote sensing grids: Architecture and implementation. In: Plaza AJ, Chang C-I (eds.) High Performance Computing in Remote Sensing. Chapman & Hall /CRC, Floria, USA, pp. 203–235.

Gusella L, Adams BJ, Bitelli G, Huyck CK, Mognol A (2005) Object-oriented image understanding and post-earthquake damage assessment for the 2003 Bam, Iran, Earthquake. Earthquake Spectra 21(S1):S225–S238.

Kouchi K, Yamazaki F (2005) Damage detection based on object-based segmentation and classification from high-resolution satellite images for the 2003 Boumerdes, Algeria earthquake. In: Proceeding ACRS 2005. 7–11 Nov 2005, Hanoi, Vietnam, CD-ROM.

Maiyo L, Kobben B, Kerle N (2009) Collaborative post-disaster damage mapping via GEO web services. In: Proceeding Cartography and Geoinformatics for early warning and emergency mapping. 19–22 Jan. 2009, Prague, Czech Republic.

Matsuoka M, Yamazaki F (1999) Characteristics of satellite images of damaged areas due to the 1995 Kobe earthquake. In: Proceedings of the 2nd Conference on the Applications of Remote Sensing and GIS for Disaster Management. The George Washington University, CD-ROM.

Rathje E.M, Adams BJ (2007) An evaluation of earthquake damage scale used in remote sensing studies of the 2003 Bam, Iran Earthquake. In: Proceeding of the 5th International Workshop on Remote Sensing Applications to Natural Hazards. 10–11 Sept. 2007, Washington DC, USA.

Sato K, Spence RJS, Going C, Markus M, (2004) Using high-resolution satellite images for post-earthquake building damage assessment: A study following the 26 January 2001 Gujarat earthquake. Earthquake Spectra 20:1–25.

Sekiguchi S, Tanaka Y, Kojima Isao, Yamamoto N, Yokoyama S, Tanimura Y, Nakamura R, Iwao K, Tsuchida S (2008) Design principles and IT overviews of the GEO Grid. IEEE Syst J 2(3):374–389.

Serra J (1982) Image Analysis and Mathematical Morphology. London, Academic Press.

Singhroy V, Mattar KE (2000) SAR image techniques for mapping areas of landslides. In: Proceedings of ISPRS Congress. Amsterdam, The Netherlands, pp. 1395–1402.

Soille P, Pesaresi M (2002) Advances in Mathematical Morphology Applied to Geoscience and Remote Sensing. IEEE Trans Geosci Rem Sens 40(9):2042–2055.

Stramondo S, Bignami C, Chini M, Pierdicca N, Tertulliani A (2006) Satellite radar and optical remote sensing for earthquake damage detection: Results from different case studies. Int J Rem Sens 27:4433–4447.

Vincent L (1992) Morphological area opening and closings for greyscale images. In: Proceedings NATO Shape in Picture workshop. Driebergen, The Netherlands, pp. 197–208.

Vu TT, Matsuoka M, Yamazaki F (2005a) Detection and animation of damage using very high-resolution satellite data following the 2003 Bam, Iran, earthquake. Earthquake Spectra 21(S1):S319–S327.

Vu TT, Matsuoka M, Yamazaki F (2005b) Towards object-based damage detection. In: Proceeding of ISPRS workshop DIMAI'2005. Bangkok, Thailand, CDROM.

Vu TT, Matsuoka M, Yamazaki F (2007a) Dual-scale approach for detection of tsunami-affected areas using optical satellite images. Int J Rem Sens 28:2995–3011.

Vu TT, Yamazaki F, Matsuoka M (2007b) Context-based detection of post-disaster damaged buildings in urban areas from satellite images. Urban Remote Sensing Joint Event 2007. Paris, France, 5pp.

GMES Services for Conflict Prevention and Mitigation: Supporting the DG RELEX in Mission Planning

Olaf Kranz, Stefan Lang, Dirk Tiede, Gunter Zeug, Thomas Kemper, Mathilde Caspard and Stephen Clandillon

Abstract

Following a request from the DG RELEX in April 2008 for support in the planning of the EU mission to Darfur, the GMES-LIMES project partners in the Humanitarian Relief & Reconstruction cluster combined their expertise to contribute their support through the provision of satellite imagery based maps and information. Detailed damage assessment maps were produced for three villages that had been burnt down by militias early in 2008. Analyses were also carried out on the evolution of two large IDP camps, focussing on the population growth and densities within both the camps and the surrounding villages. In addition, rapid mapping techniques adapted to the complex humanitarian situation were applied in analysing the effects of militia raids on the three villages in West Darfur and for monitoring the evolution of the IDP camps between 2002 and 2008. The impact of the camps on the surrounding environment was also analysed using medium-resolution MODIS time series data (2000—2008).

M. Konecny et al. (eds.), *Geographic Information and Cartography for Risk and Crisis Management*, Lecture Notes in Geoinformation and Cartography,
DOI 10.1007/978-3-642-03442-8_12,

1 Introduction

1.1 Support to mission planning

The Darfur conflict has led to major human rights violations including systematic murder, rape and forced displacement. More than 2 million civilians have been forced to leave their homes and hundreds of thousands of civilians have died as a result of countless attacks. Although there are now considerably fewer deaths than during the period of intensive fighting in 2003 and 2004 the number of confrontations has multiplied and violence is again increasing at a time when international peacekeeping is not yet effective and access for humanitarian agencies is limited (International Crisis Group, 2007). By the end of 2007 a joint African Union (AU)/United Nations (UN) hybrid operation (UNAMID) had begun operations in the area.

In April 2008, the Directorate General for External Relations of the European Commission (DG RELEX) made a request to the *Humanitarian Relief & Reconstruction* cluster of the EC-funded GMES[1] LIMES (*Land and Sea Integrated Monitoring for European Security*) project to assist in improving the delivery of EU aid through NGOs and UN agencies in the field. The GMES-LIMES project aims to define and develop prototype information services to support security management of the EC (see www.fp6-limes.eu), and the cluster partners[2] jointly elaborated a detailed work plan to provide the requested information.

QuickBird satellite imagery over five areas in western Darfur (Fig. 1) was analysed to provide an overview of developments on the ground. Rapid mapping techniques adapted to the complex humanitarian situation were applied to analyse the effects of raids on three villages in West Darfur that had occurred in February 2008. The resulting damage assessment maps highlighted intact buildings as well as destroyed buildings and infrastructure, also indicating how many people might have been affected. Furthermore, the detection of temporary dwellings such as tents was helpful in estimating the magnitude of displacement. The evolution of two IDP (internally displaced persons) camps between 2004 and 2008 was also monitored in parallel to the analyses on the villages. The analyses utilised automated object-based image analysis methods combining segmentation, class modelling and knowledge representation techniques. Satellite-based

[1] GMES: Global Monitoring for Environment and Security

[2] SERTIT, France; Joanneum Research, Austria; European Union Satellite Centre (EUSC), Spain; Centre for Geoinformatics (Z_GIS), Austria; European Commission Joint Research Centre (JRC), Italy; GAF AG, Germany; German Aerospace Center (DLR), Germany

extraction of dwellings provides a guide to the number of people present at the time of data acquisition, but the exact number can only be estimated. Different scenarios were therefore generated for the estimation of population figures, based on the number of extracted dwellings, various figures from published literature and oral communications with Médecins Sans Frontières (MSF).

IDP or refugee camps themselves have an impact on the surrounding environment due to the pressure on scarce local resources such as water, grazing areas and firewood. Estimating the magnitude of such impact is of considerable importance with respect to conflict prevention. Analysis of Moderate Resolution Imaging Spectroradiometer (MODIS) time-series data from 2002 to 2008 using the Enhanced Vegetation Index (EVI) helped in detecting spatial trend patterns in areas surrounding the camps.

1.2 Situation analysis

An attack by government-backed militia in northern West Darfur on February 11th, 2008 destroyed much of the two towns of Abu Sorouj and Sirba, located 55 km north of El-Geneina, the capital of West Darfur. Both towns hosted thousands of IDPs and large parts were burned to the ground. The United Nations High Commissioner for Refugees (UNHCR) reported that about 12,000 people fled into neighbouring Chad, 150 people died and thousands were left without food or shelter as a result of the attacks. Abu Sorouj and Sirba had previously (in 2004) hosted about 40,000 IDPs who had fled from the insecurity of surrounding villages. No detailed information is available on militia activities in Dagrase.

Since December 2007, the northern corridor of West Darfur has seen repeated clashes between Sudanese government forces and the Justice and Equality Movement (JEM), which took control of key strategic villages north of El-Geneina. According to Human Rights Watch (HRW) the attacks in 2008 cut off about 160,000 civilians from aid. Amnesty International described the attacks as a major test for the AU-UN mission, which took over from the AU Mission in Sudan on December 31, 2007 but is still woefully under-resourced, and called on the mission to ensure the safety of all civilians in Sirba and Abu Sorouj (Reuters).

In April 2008 the DG RELEX requested more detailed information concerning the situation in 3 villages and 2 IDP camps in Western Darfur, to support their activities in the region (Fig. 1).

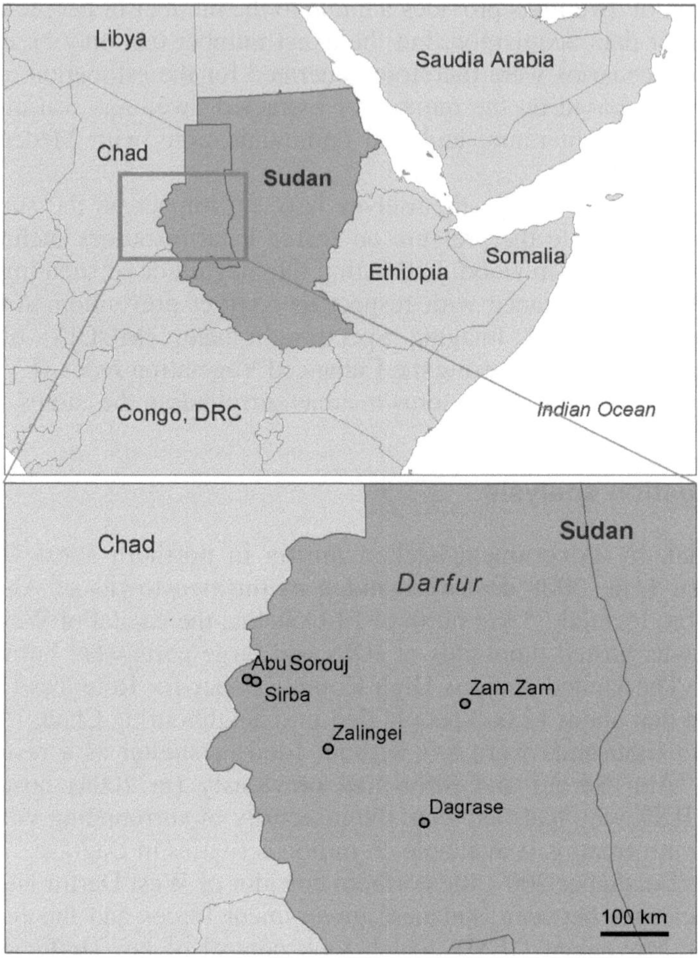

Fig. 1. Overview of the locations for the DG RELEX request

Damage assessments were requested for the Sirba, Abu Sorouj and Dagrase villages while a retrospective monitoring (time series analysis) was required for the Zalingei and Zam Zam IDP camps. The overall objective was to generate additional information for a better and more detailed view of the situation in the specified villages and IDP camps.

2 Methods

2.1 Damage assessment mapping

The objective was to provide damage maps reflecting the situation on the ground following reported attacks on villages in February 2008. User requirements were for A1-sized damage maps in pdf format that would provide answers to the following two questions: first, "What buildings, infrastructure and other facilities have been affected?" second, "What can now be seen on the ground?"

To accomplish this task very high resolution (VHR) satellite data acquired in late 2007 was used, together with data acquired in the spring of 2008.

An approach that combined automatic detection with computer-assisted photointerpretation, adapted to the complex humanitarian situation in Darfur, was used to generate the required set of damage assessment maps. It was decided to use QuickBird data, which was the highest resolution readily available colour imagery, since the characteristics to be analysed on the ground were very small with many huts being less than 5 m in diameter. Very recently archived QuickBird data dating from late 2007 and early 2008 was available over the three villages and was ordered, together with new (tasked) acquisitions, in a coordinated effort between the involved partner organisations (Table 1).

Table 1. Very high spatial resolution optical satellite data acquired (archived and tasked) for West Darfur

Location	Date	Sensor
Dagrase	08/12/2007	QuickBird (QB) panchromatic + multispectral bundle
	31/05/2008 (tasked)	QB panchromatic + multispectral bundle
Abu Sorouj	02/03/2006	QB pansharpened
	28/02/2008	QB pansharpened
Sirba	04/11/2006	QB pansharpened
	23/05/2008 (tasked)	QB pansharpened
Zam Zam	18/06/2002	QB pansharpened
	20/12/2004	QB pansharpened
	08/05/2008 (tasked)	QB pansharpened
Zalingei	14/09/2004	Ikonos pansharpened
	06/07/2008 (tasked)	QB pansharpened

The first step in the damage assessment was to clarify the appearance of dwellings in this particular region. The common dwelling hut has a circular outline with a diameter of about 3.5–4.5 m, and a few rectangular huts also exist. The spectral contrast between the huts and the ground is high (80–100%) and the outlines are therefore very sharp. Intact huts can thus be easily identified and the fact that huts are often surrounded by fences or walls facilitates their identification. In contrast, the appearance and detection of burnt-out huts is more complex since although their spectral signature is similar to that of intact huts the transition to the surrounding ground is more gradual, and the edges of the burnt-out huts are therefore more or less blurred.

The automatic analysis for Abu Sorouj and Sirba was conducted using object-based image analysis (OBIA), i.e. semi-automated information extraction based on segmentation, class modelling and rule sets (Lang, 2008). Because of the very similar shapes and spectral signatures of huts and single trees, the near-infrared (NIR) channel was used to distinguish between these two types of objects and thus avoid overestimating the number of dwellings. The results of the object-based image analysis were validated and optimised by visual inspection, particularly for the hut remains within the totally burnt-out village centres.

Although an initial analysis of Dagrase found that the urban landscape was not as structured as in the other two sites, the following dynamic types were identified for the period between December 2007 and the end of May 2008: (a) a significant area of forest that had been burned, close to the village, (b) huts and other urban features that had been rebuilt, (c) areas of undisturbed settlement, (d) a number of huts and other urban features that had disappeared and (e) a compound that had been established.

In a rapid mapping context this type of work is generally carried out by photointerpretation and this method was therefore utilised. The urban and damage dynamics are reported in the legend of the A1 maps produced.

2.2 Camp structure and dwelling density mapping

During the course of meeting the request from the DG RELEX, two refugee/IDP camps (Zalingei and Zam Zam) were also analysed with a focus on monitoring the evolution of these large camps. The main objectives of these satellite imagery analyses were (1) the mapping of the refugee/IDP camps, (2) the monitoring of population growth in the camps and (3) the monitoring of the impact of these camps on surrounding areas. As both IDP camps have come into existence since 2004, the minimum time frame for the investigation needed to cover the period between 2004 and 2008. Two archived QuickBird scenes (from 2002 and 2004) were therefore ordered for Zam Zam together with a current QuickBird image from 2008. For Zalingei an already available Ikonos scene from 2004 as well as a newly acquired QuickBird image from 2008 were used.

After pre-processing of the images (including pansharpening and ortho-rectification), two aspects of the evolution of the camps were analysed:

1. The growth of the camps from 2002 (Zam Zam) and 2004 (Zalingei) until 2008, including the structure of the entire populated area in terms of camp area, traditional settlements and dwelling densities.
2. The number of dwellings in the camps in order to derive an estimate of the number of people present, differentiating between two types of dwelling for Zam Zam (darker, traditional dwellings and lighter dwellings mainly representing tents or tarpaulins), while focussing on the extraction of only light dwelling structures in the Zalingei area.

Analyses were conducted using OBIA techniques. The rule set was developed for the 2004 Zam Zam imagery and optimised in such a way that it was transferable to both the 2002 and 2008 scenes as well as, with some modifications, to the Zalingei area. The rule set was adapted to work on the Ikonos imagery; only minor additional adaptations were required for the QuickBird scenes. The extraction of dark dwellings was limited by the presence of other dark structures in the images (mainly fences). The analysis in the Zalingei area focussed on the extraction of light dwelling structures only as they proved to be a sufficient indicator of the number of IDPs since, in contrast to the Zam Zam camp, the IDPs in Zalingei inhabited light tents and a significant number of traditional structures could not be identified within the main IDP areas of Zalingei. The camp areas were delineated automatically according to a certain threshold in dwelling density.

2.3 Estimation of population in Zam Zam and Zalingei

Satellite-based extraction of dwellings does not provide rigorous evidence of the actual number of people present at the time of data capture. Scenarios used for the estimation of population figures were based on published figures (Giada et al. 2003, Bush 1988, UNHCR 2006) and on oral communications with MSF. Population estimates were calculated for three scenarios with different occupation rates, based on the number of dark and light dwellings (for Zam Zam) or light dwellings only (for Zalingei). For Zam Zam, Scenario I assumes low occupancy rates of 3 inhabitants per dark dwelling structure and 4 inhabitants per light dwelling structure, Scenario II (moderate occupancy rate) assumes 3 inhabitants per dark dwelling structure and 5 inhabitants per light dwelling structure, and Scenario III (high occupancy rate) assumes 4 inhabitants per dark dwelling structure and 6 inhabitants per light dwelling structure. Similarly, for Zalingei Scenario I assumes 4 inhabitants, Scenario II 5 inhabitants and Scenario III 6 inhabitants per light dwelling structure.

2.4 Environmental impact assessment

IDP/refugee camps have a marked impact on the surrounding environment due to the depletion of local resources such as water, grazing areas and firewood. This may even lead to conflict between IDP/refugees and the host communities (Martin 2005, UNHCR 2005). There is growing concern about the environmental impact of Darfur's conflict, in particular the impact on Darfur's forest resources which were already being depleted at an estimated rate of 1% per annum before the conflict (UNEP 2008). In situations such as Darfur this depletion also has an impact on the security of the IDPs because the depletion of resources forces them to collect firewood further away from the camps, increasing the risk of attack (ibid).

The detection and, where possible, the quantification of this impact may provide important information for the management of these camps. However, this is often difficult since the situation prior to the arrival of IDPs/refugees is unknown and ongoing conflicts prevent detailed field assessments. In such cases satellite-based remote sensing may be the only instrument available for the derivation of such information. A standard application is the detection of changes between two (or more) dates, but such comparisons may be influenced by different phenological stages, especially in extremely fragile arid environments such as that in Darfur. Alternatively, archives of satellite data with coarse geometric but high temporal resolution may provide more reliable results (Cracknell 1997, Justice et al. 1998). Such an analysis aims to detect spatial trend patterns, which are an important factor in long-term environmental studies. A trend is characterised by its functional form, direction and magnitude and should be interpreted with respect to its statistical significance (Widmann and Schär, 1997).

In this study a time-series analysis of MODIS data from January 2002 to July 2008 was used. The time-series analysis is based on the Enhanced Vegetation Index (EVI) of 16-day maximum value composites with a spatial resolution of 250 m, which is believed to be superior to the commonly used Normalised Difference Vegetation Index or NDVI (Huete et al. 2002). The aim of the time-series analysis was to detect spatial trend patterns in the vicinity of the camps. These can be identified using parametric trend models such as regression analysis, but these models rely on specific assumptions that influence the derived trend function and when these assumptions are not met the standard error for the estimate of the regression coefficients becomes suspect, and significance tests could be erroneously interpreted (de Beurs and Henebry 2004). Nonparametric methods are, however, independent of such assumptions. The majority of nonparametric statistical significance tests for trend analysis are derived from the Mann-Kendall (MK) test, which is suitable for monotonic trend detection independent of the functional nature (e.g. linear, quadratic) of the trend. The MK test assumes that the error terms are identically distributed, which may lead to problems if a seasonal and/or serial correlation exists in the data, in which case the seasonal Kendall test should be used instead. The seasonal Kendall test is a generalization of the Mann-Kendall test that accounts for a seasonal component in the data by deriving the test statistics individually for each season (Hirsch and Slack 1984).

3 Results and discussion

3.1 Damage assessment mapping

3.1.1 Abu Sorouj and Sirba

Counting revealed approximately 4,350 huts in Abu Sorouj in February 2008, of which more than 1,000 had been constructed in the 2 years from March 2006 to February 2008. The main areas of growth were in the north-western and western parts of the village (Fig. 2). In contrast, the centre of Abu Sorouj was completely burnt to the ground and had not been reconstructed by the end of February 2008. Some larger buildings located close to the centre of the village might indicate a central market, or storage buildings.

Approximately 2,400 dwellings were identified in Sirba, of which 760 were burnt down by militias in February 2008. Once again there were some large buildings located in the village centre, possibly representing a central market or storage buildings, and some more solid buildings could be seen to the west and south-west of the village centre.

While many temporary dwellings (approximately 320) were established in Sirba to compensate for the loss of houses, only a few tents were detected in Abu Sorouj, posing the question of where those people that used to live in the totally destroyed centre of the village are now living.

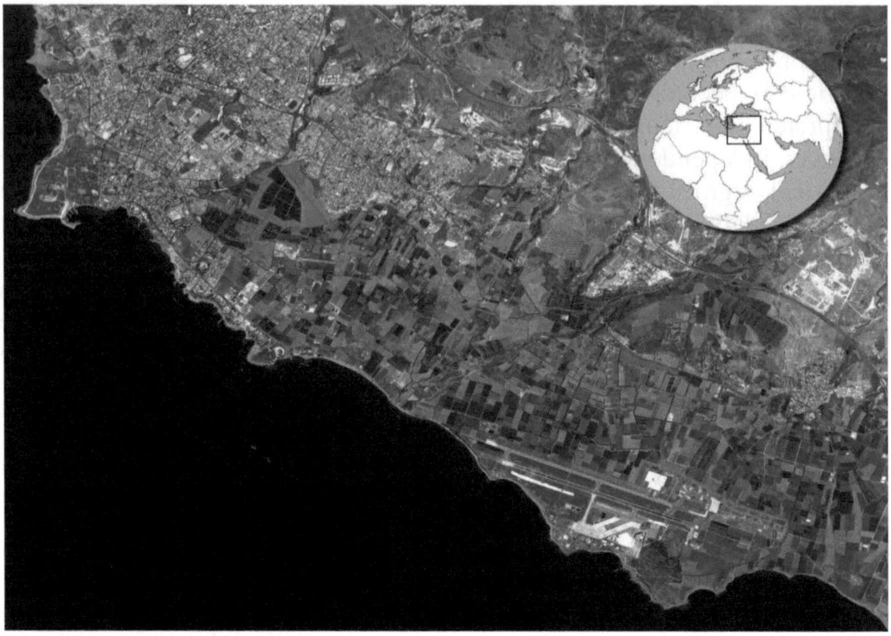

Fig. 2. Damage assessment map of Abu Sorouj. *Light dots*: destroyed dwellings, *dark dots*: intact dwellings

3.1.2 Dagrase

Once the urban dynamics and damage assessment had been evaluated 3 maps were produced covering the Dagrase village, all of which contained damage assessment type data for the period between December 2007 and May 2008 (Fig. 3). The situation in Dagrase appeared to be very complex: the village itself was very difficult to define in the December 2007 imagery since there appeared to be many shells of huts but very little urban structure.

Despite a destructive event occurring within this period many new dwellings (including tents) and other buildings were also erected between December 2007 and May 2008 in and around Dagrase. Many of the damaged huts, however, remained shells and do not appear to have been rebuilt. Although fire did affect a relatively large area during this period damaging many huts, there was an overall net gain in the number of intact buildings within the area. As a result of the unstructured nature of the settlement in Dagrase, automatic or semiautomatic processing has, however, so far been unable to yield satisfactory results in mapping the individual buildings, building (settlement) densities or settlement limits. In another R&D test, change detection procedures were applied to circular buffer areas around the points representing the huts in order to investigate whether similarities could be seen between photointerpretative results and change detection results. The results did not indicate the type of change or its cause, but could nevertheless provide an aid to rapid mapping photo-interpretation.

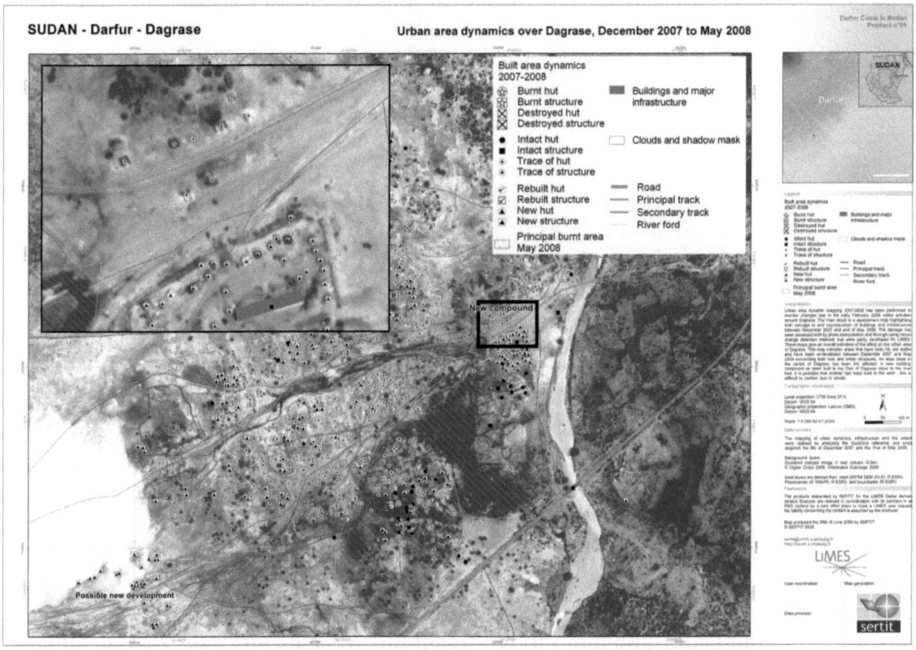

Fig. 3. Dagrase urban area dynamics map, December 2007—May 2008

Overall, the detailed analysis produced unexpected and complex results given the destruction already evident in 2007 and the complication of subsequent rebuilding having been interrupted by the fire and other destruction in early 2008. Taking into account the complexity of the situation and the detailed nature of the analysis, the processes involved require too much time to justify their use for a preliminary product within a classical rapid mapping context but might be utilised for a second-stage product that could be delivered within a few days of receiving a pre- and post-event pair of very high resolution satellite images.

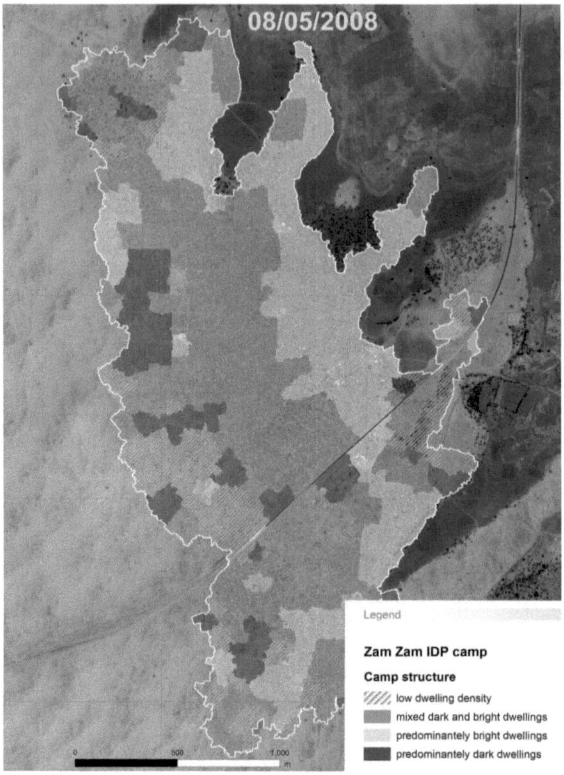

Fig. 4. Zam Zam camp structure (2008), based on automatically extracted dwelling structures

3.2 IDP camp structure and dwelling densities

With respect to the development of the Zam Zam camp site as a whole, areas were differentiated with (i) predominantly traditional dwelling structures or huts from those with predominantly new camp/tent structures), (ii) mixed structures and (iii) areas inside the camp with low dwelling densities (Fig. 4). Monitoring of developments between 2002 and 2008 reveals a major increase in the number of dwellings between 2004 and 2008; the number of light dwellings (tents/tarpaulins) in particular increased significantly during this period.

Further analysis encompassed kernel density calculations to create dwelling density maps (dwellings per square km), based on the semi-automatically extracted dwellings (Fig. 5).

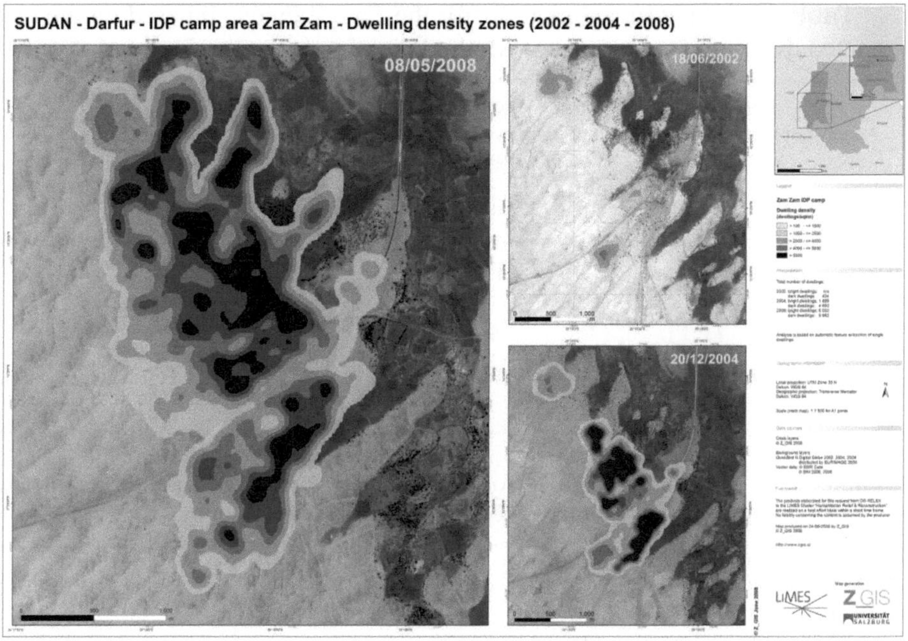

Fig. 5. Dwelling density zones for the Zam Zam IDP camp (2002–2004–2008)

3.3 Camp population estimates

Figures for the three different scenarios were compared with official population data from the Humanitarian Needs Profile (HNP) of UN OCHA (United Nations Office for the Coordination of Humanitarian Affairs), Humanitarian Information Centre (HIC) and the Spanish Red Cross (SpRC) (OCHA 2007). Depending on the particular scenario used for the dwelling extraction from the QuickBird images, these figures show good agreement (Tables 2 and 3). The dwelling numbers extracted from the Ikonos scene from 2004 for Zalingei, showed the widest variation between population estimations and official figures. This underestimation could be a result of the coarser resolution of the Ikonos image (1 m) compared to the higher resolution (0.6 m) QuickBird images, which complicates the automated dwelling extraction.

Table 2. Population estimations for the Zam Zam camp based on the extracted dwelling structures (see Methods section for scenario descriptions)

Semi-automatically extracted dwelling structures:					Population estimation:		
Acquisition date - satellite imagery	# of dark dwellings ('traditional')	# of light dwellings ('tents/ tarpaulins')	Overall	Population figures (from HNP/OCHA /SpRC)	Scenario I	Scenario II	Scenario III
June 18, 2002	434	–	434	–	1,302	1,302	1,736
December 20, 2004	4,690	1,899	6,589	18,190	21,666	23,565	30,154
May 8, 2008	6,982	6,032	13,014	> 50,000	45,074	51,106	64,120

Table 3. Population estimations for the Zalingei camp based on the extracted dwelling structures (see Methods section for scenario descriptions)

Semi-automatically extracted dwelling structures:			Population estimation:		
Acquisition date - satellite imagery	# of light dwellings (<35m²–'tents/ huts / tarpaulins')	Population figures ('affected population') from HIC (2004) and OCHA (2007/2008)	Scenario I	Scenario II	Scenario III
September 14, 2004	11,569	82,960 (Res: 20,000; IDP: 62,960)	46,276	57845	69,414
July 6, 2008	31,502	144,688 (Res: 45,257; IDP: 99,431)	126,008	157,510	189,012

3.4 Environmental impact assessment

The impact analysis using the seasonal Mann-Kendall test for the Zam Zam camp is illustrated in Fig 6. The significantly degraded areas are all on the western side, within 5—10 km of the camp. Areas closest to camps are generally more secure and are hence the most exploited. IDPs from a camp at Abu Shouk in Al Fashir (approximately 20 km away) recently reported that it takes 7 days to collect one cartload of wood (UNEP 2008). The land degradation in the vicinity of the Zam Zam camp becomes evident when comparing the VHR data from 2002 with that from 2008 (Fig. 7). While the area was sparsely covered with trees in 2002 and still being used for dry farming, by 2008 the camp had already extended into the eastern part of the image and the area was basically deprived of shrubs and trees.

Since the time-series analysis was performed for an entire MODIS scene covering an area of 1,200 by 1,200 km, the information is available for the whole of Darfur and allows the analysis of all IDP/refugee camps. The analysis might also be extended to include the deliberate destruction of resources during the conflict. Despite the coarse resolution, the MODIS satellite data provides valuable information for evaluating the impact of IDP camps due to its ability to identify even subtle changes and the availability of long time-series over large areas. The quantitative information may be used to enhance the qualitative information (e.g. UNEP, 2008) available from higher resolution data. Figure 7 shows a more detailed comparison between fine-scaled QuickBird data from 2002 and that from 2008.

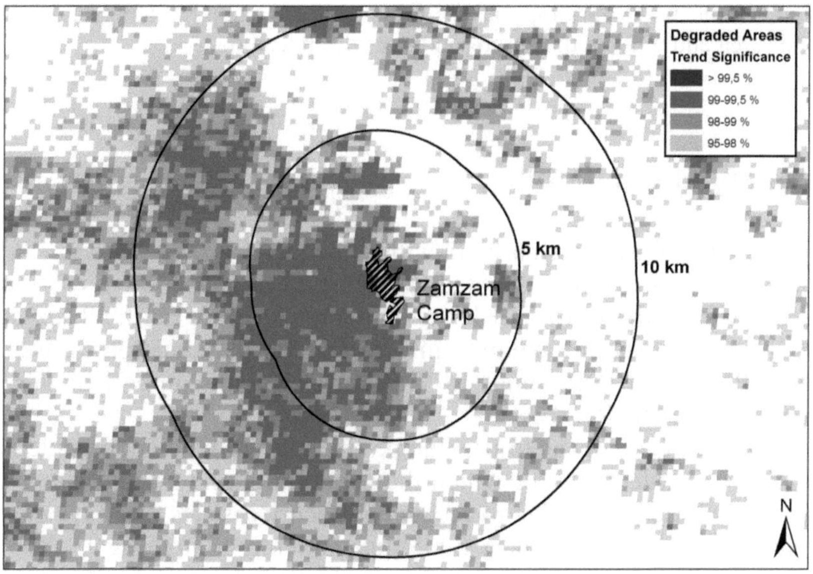

Fig. 6. Trend analysis results for Zam Zam camp overlaid on SPOT data, with the daily walking distance for firewood collection assumed to be between 5 and 10 km

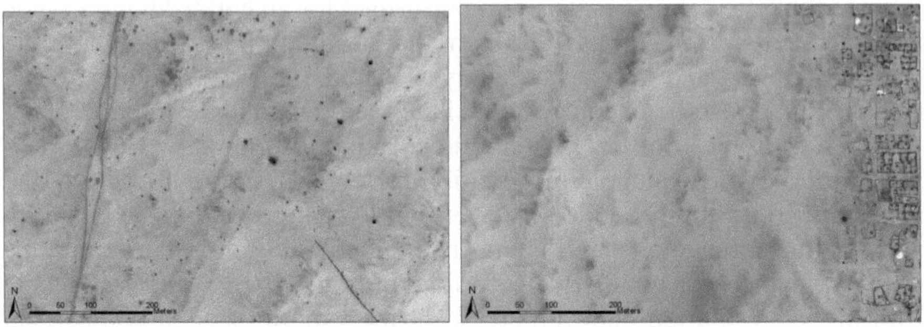

Fig. 7. Comparison of Quickbird data for 2002 (*left*) and 2008 (*right*). The dark spots, mainly visible in the 2002 image, are trees and shrubs and the fine linear structures indicate dry farming. In the 2008 image the camp borders had extended into this area and most of the shrubs and trees disappeared

Summary and outlook

In spring 2008 a group of partners within the LIMES *Humanitarian Relief and Reconstruction* cluster pooled their resources and know-how to provide a service to the DG RELEX, using mainly satellite Earth Observation (EO) imagery to acquire detailed information on the situation in user selected villages and IDP camps in Western Darfur, in support of mission planning by the DG RELEX.

The main objective was to provide pertinent up-to-date geographical information on settlements for which the situation on the ground was either unknown or little known due to access and security issues that have existed since armed attacks in February 2008. In an attempt to investigate the various options available both tried and trusted rapid mapping techniques and relatively untested R&D methods were used to demonstrate the potential of EO-based services developed within the LIMES project. To back up the standard photointerpretative thematic extraction methods, new automatic and semi-automatic procedures were tested for mapping the settled areas, individual buildings and infrastructure as well as both settlement densities and settlement compositions. Population estimates, population densities and their environmental impacts were also studied. Although the automatic detection of dwellings, damaged dwellings and other infrastructure may still have limitations and the full extraction of camp structures for population estimates remains challenging, several promising approaches were identified to automatically generate relevant information about conflict areas or at least support the visual interpretation of very high resolution satellite data.

It is important to note that the major objective of the analyses described in this chapter was the provision of preoperational services rather than to exclusively carry out research-oriented activities. To this end the information requested by the DG RELEX was prepared in close cooperation with the user to ensure that the final maps and interpretations provided the maximum possible benefit and support for the planning of the EU mission to Darfur. A questionnaire completed by the DG RELEX (to provide user validation on the information and map products provided) summarised the groups' work as follows: 'Overall very good quality work, which is unusual for a research project. The analysis shows great attention and use of additional information than geo-information sources.' (DG RELEX)

The cooperation between a number of partner organisations in these joint activities has demonstrated the advantages of bringing together their different capacities and areas of expertise, both operational and R&D, to provide rapid support to mission planning within the context of a complex crisis situation. Following this demonstration it was considered that our capacity to provide such services had been proven and the partners continued to work on other operationally orientated service provisions within LIMES. Further requests of a similar nature by the DG RELEX, and perhaps by other users, are already anticipated.

References

de Beurs KM, Henebry GM (2004) Trend analysis of the Pathfinder AVHRR Land (PAL) NDVI data for the deserts of Central Asia. IEEE Geosci Remote Sens. Lett. 1:282–286.

Bush R (1988) Hunger in Sudan: The Case of Darfur. African Affairs 87(346), (Jan., 1988):5–23.

Cracknell AP (1997) The Advanced Very High Resolution Radiometer (AVHRR). Taylor & Francis, London, 534 pp.

Giada S, De Groeve T, Ehrlich D, Soille P (2003) Information extraction from very high resolution satellite imagery over the Lukole refugee camp, Tanzania. Int J Rem Sens 24:4251–4266.

HIC (2005) Camp coordination in West Darfur. [http://www.humanitarianinfo.org/darfur/uploads/maps/IDP/West%20Darfur_IDP_Camp_Coordination_Map_A3_30Sept05.pdf] (20 August 2008).

Hirsch R, Slack J (1984) A nonparametric trend test for seasonal data with serial dependence. Water Resour Res 20:727–732.

Huete A, Didan K, Miura T, Rodriguez E (2002) Overview of the radiometric and biophysical performance of the MODIS vegetation indices. (Special Issue) Rem Sens Environ 83:195–213.

International Crisis Group (2007) Darfur's new security reality. Africa Report N°134, 26 November 2007.

Justice DH, Salomonson V, Privette J, Riggs G, Strahler A, Lucht R, Myneni R, Knjazihhin Y, Running S, Nemani R, Vermote E, Townshend J, Defries R, Roy D, Wan Z, Huete A, van Leeuwen R, Wolfe R, Giglio L, Muller JP, Lewis P, Barnsley M (1998) The Moderate Resolution Imaging Spectroradiometer (MODIS): Land remote sensing for global change research. IEEE Trans Geosci Rem Sens 36:1228–1249.

Lang S (2008) Object-based image analysis for remote sensing applications: Modeling reality – dealing with complexity. In: Blaschke T, Lang S, Hay G (eds.) Object-Based Image Analysis – Spatial concepts for knowledge-driven remote sensing applications. Springer, New York, pp. 3–28.

Martin A (2005) Environmental conflict between refugee and host communities. J Peace Res 42(3):329–346.

OCHA (2007): Camp data. [http://www.unsudanig.org/library/mapcatalogue/darfur/data/camps/], (20 August 2008).

UNEP (2008) Destitution, Distortion and Deforestation: The Impact of Conflict on the Timber and Woodfuel trade in Darfur. Geneva, United Nations Environment Program.

UNHCR (2005) Environmental Guidelines. Geneva: United Nations High Commissioner for Refugees.

UNHCR (2006): Specifications and list of tests for future reference Light-Weight Emergency Tent (UNHCR LWET). [http://www.unhcr.org/admin/ADMIN/44c48eca2.pdf], (20 August 2008).

Widmann M, Schär C (1997) A principal component and long-term trend analysis of daily precipitation in Switzerland. Int J Climatol 17:1333–1356.

Part II
Geo-information Services

Mobilizing Multi-source Geospatial Information for EW and EM: Maximize Sharing, Enhance Flexibility, and Minimize Costs

Mark Reichardt and Carl Reed

Abstract

It is difficult to think of any Early Warning (EW) and Emergency Management (EM) activity that does not involve geospatial information, that is, information about things and events that have location in time and space. It follows that EW and EM decision makers can make better decisions if their information systems provide geospatial information in ways that fully accommodate well-defined user requirements for information discovery, access, and use. Geospatial information for EW and EM is almost always created and maintained by multiple organizations, agencies, and companies, and the data almost always resides on different systems on one or more networks. Unfortunately, "non-interoperability" between these systems and their data has often prevented their discovery, access, and use. Sharing geospatial resources is now much easier, thanks to a well-developed, though still evolving, set of standards. EW and EM managers can maximize sharing, enhance flexibility, and minimize costs by making full use of these standards and contributing to their evolution.

1 Introduction

Every activity happens somewhere and "somewhen". This is particularly important to consider in regard to Early Warning (EW) and Emergency Management (EM) events. For decades, providers of information and communication technologies (ICT) have invested considerable effort and money in developing ways to effectively and efficiently use information about location and time. Different

M. Konecny et al. (eds.), *Geographic Information and Cartography for Risk and Crisis Management*, Lecture Notes in Geoinformation and Cartography,
DOI 10.1007/978-3-642-03442-8_13,

applications—GIS, Earth imaging, Cartography, Command and Control, Facilities Management, Location Services, Navigation, etc.—required different technologies and different approaches, and different vendors developed products that were innovative but not designed to interoperate with other vendors' products. Because users have a strong need to share geospatial information and to maximize the value of their software investments over time, they demanded standards that would bridge the gaps between different types and brands of geoprocessing systems. These user needs led to the creation in 1994 of the Open Geospatial Consortium, Inc, (OGC)®.

The OGC' Mission is to serve as a global forum for the development, promotion, and harmonization of open and freely available geospatial standards. The OGC is a voluntary consensus standards organization with over 370 members—companies, government agencies, NGOs, universities, and research institutions. The membership has developed and adopted more than 30 interface and encoding standards that have been implemented in hundreds of products in the marketplace (OGC, 2009). EM and EW organizations have been an important source of requirements for OGC standards, and there is broad use of OGC standards-implementing and OGC-conformant products worldwide.

2 What do we mean by interoperability?

Interoperability is the "capability to communicate, execute programs, or transfer data among various functional units in a manner that requires the user to have little or no knowledge of the unique characteristics of those units" (OGC, 2002). You know you *don't* have interoperability when your organization finds that it is burdened with

- Time-consuming custom integration
- Difficulty in rapidly mobilizing new capabilities
- Missed opportunities to collaborate
- Unnecessary duplication of effort

There are different kinds, or layers, of interoperability. The OGC has traditionally focused on "technical interoperability." Technical interoperability refers to different technology products being able to communicate and "work together" and is achieved through standard interfaces and encodings. Communication means exchanging information "through a *common system* of symbols, signs or behavior." (Merriam Webster Online) Standardization is about "agreeing on a *common system.*" The Internet, the World Wide Web, e-Commerce, and the wireless revolution have produced great wealth because of "network effects": That is, a node on the network (a product or service, for example) increases in value with the size of the network (such as the number of potential users of the product or service). As the daily use of the Web for millions of secure financial transactions demonstrates, secure applications can run on networks based on open standards. Open standards are a tremendous bargain: cheap to produce, free to use, and conducive to wide use, which results in large—and thus valuable—networks of cooperating systems.

"Information interoperability" is about semantics. Geospatial information involves the naming of geospatial features and phenomena. Non-interoperability results when the same short road, for example, is labeled as a "dead end" in one data set and a "cul-de-sac" in another data set. The problem can be solved by organizations "agreeing on a common system" of names and naming conventions, and it can also be solved by software that is able, after being properly configured, to "interpret" between different naming schemas. This capability is greatly enhanced when both communities use the same content model and encoding language, such as the OGC Geography Markup Language (GML) Encoding Standard.

Organizational interoperability and cultural interoperability are the high-level goals that require and drive technical interoperability and information interoperability. Leaders and managers working toward improved communication and cooperation want information systems that help them meet their goals, and this is the reason organizations increasingly support and participate in consensus standards organizations.

3 OGC standards

The OGC standards framework supports a variety of system architecture patterns, such as Service Oriented Architectures (SOA), with an emphasis on architectures that use the "standards ecosystem" of the Internet and Web. These standards enable communities of users to publish, discover, access, fuse, and apply geospatial information from multiple sources across the Web. Most geospatial data and services related to activities such as Critical Infrastructure, Emergency Management, Weather, and Decision Support can be managed through OGC web services.

OGC standards are defined, discussed, tested, and approved by members in a formal process. Approved OGC standards are available free of charge at http://www.opengeospatial.org. Though many OGC standards provide critical capabilities in EW and EM, we list only a few here to indicate the kinds of interoperability problems that OGC standards solve:

- Web Map Service – The OGC WMS Interface standard provides a simple HTTP interface for requesting geo-registered map images from one or more distributed geospatial databases. From an EM perspective, using WMS-enabled servers allows a client to access distributed and often disparate geospatial repositories and create a fused situational awareness or common operating picture map on the client (OGC Web Map Service Implementation Specification, 2006).
- Catalogue Services - supports the ability to publish and search collections of descriptive information (metadata) about geospatial data, services, and related resources. Providers of resources, such as maps and services, can register metadata in catalogues that conform to the provider's choice of an information model; such models include descriptions of spatial references and thematic information. Catalogs implementing this standard enable inventorying and discovery of distributed resources critical in EM and EW, such as map layers, emergency symbology, and plume models (OGC Catalogue Service Implementation Specification, 2007).

- The OGC Sensor Observation Service (SOS) supports interoperable web services based aggregation of feeds from both in situ and remote sensors and sensor networks. The ability to rapidly access, fuse, and use information from an array of sensors in a location and temporal context has been an important EW/EM user community goal (OGC, 2007).
- Geography Markup Language (GML) – The ability to easily share geospatial content between organizations, jurisdictions, and systems is critical, especially in a crisis situation when responders need the right data at the right time and in the right place. (GML) is an XML grammar for expressing geographical features. GML serves as a modeling language for geographic systems as well as an open interchange format for geographic transactions on the Internet. Many EM applications are already using GML or plan on using GML application schemas (OGC Geography Markup Language Encoding Standard, 2007).

4 Alliances

Because geospatial interoperability depends on interoperability standards in the broader Internet and Web computing infrastructure, the OGC maintains alliances with many other standards development and professional organizations (SDOs). These partnerships are critical to the success of both the OGC and its Alliance partner organizations. The partnerships enable the OGC and its partner organizations to work directly on standards and interoperability challenges of mutual interest. From the EW/EM community perspective, it is important that the OGC is currently collaborating with the following standards organizations (OGC, 2009):

- buildingSmart International and the buildingSmart Alliance – This is important to the EM community because first responders benefit from ready access to information about the buildings they enter. Such information is created by a wide variety of building stakeholders and is potentially made available in Industry Foundation Class (IFC) standards based Building Information Models (BIM). BIM standards are the focus of these two organizations (BuildingSmart International, 2009).
- OASIS (the Organization for the Advancement of Structured Information Standards) – The OGC and OASIS have a formal memorandum of understanding. OGC staff have been actively working on OASIS standards being developed and maintained by the Emergency Management Technical Committee. Such collaboration insures that OGC standards, such as GML, are compatible and work with OASIS standards such as EDXL.
- IETF (Internet Engineering Task Force) – for almost five years, OGC Staff have been working closely with IETF GeoPriv Working Group. This WG is focused on location payloads for use in a variety of applications, especially EM, and how to encode privacy rules. Currently, the WG has formalized mandatory guidance that all coordinate location payloads be encoded using a GML application schema known as the *GML PIDF-LO Geometry Shape Application Schema for*

use in the IETF (IETF, 2007). This work and payload encoding is now being incorporated in numerous other Internet standards.

- NENA (National Emergency Numbering Authority) – NENA is authorized to define the architecture, operating procedures, and standards for the IP-based Next Generation 9-1-1 system (NG9-1-1). OGC staff participate in NG 9-1-1 Working Groups to ensure common and consistent encoding of geospatial content and payloads. Here, GML is a key OGC standard.

These alliance partnership arrangements enable the OGC to address increasingly complex, cross community interoperability issues that could not be solved by any one organization in isolation. Furthermore, these relationships allow the OGC to work with broader IT standards organizations to ensure that location is represented consistently across the IT standards "stack".

5 Some examples of systems based on OGC standards

5.1 Tsunami information portal

Tsunami Information Portal (http://mapsherpa.com/tsunami/) is a rapid response application developed to support rescue and recovery activities following the Indian Ocean Tsunami of 2004 (Fig. 1). developed by DM Solutions (Canada) in concert with the Asian Institute of Technology, Chulalongkorn University, and Laboratory of Applied Geomatics. It uses the OpenGIS WMS Interface Standard.

5.2 Debris flow disaster information system

The Debris Flow Disaster Information System developed by the GIS Center at Feng Chia University, Taiwan, for the Taiwan Committee of Agriculture's Soil and Water Conservation Bureau supports debris flow disaster management (SAO/NASA, 2008).

The main purpose of this project, which began in 1998, is to address the four phases of disaster management in situations involving debris flow from river floods. Goals include:

- Avoid casualties caused by debris flow
- Record information during the lifecycle of a disaster
- Provide information to decision makers
- Manage recovery projects
- Capture data to support debris flow research
- Interoperate with heterogenous information systems to meet disaster management requirements.

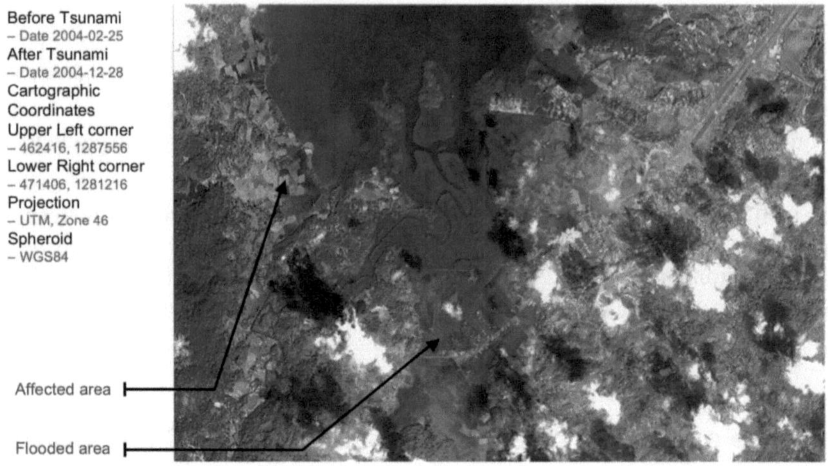

Fig. 1. "False color" earth image showing areas affected by the December 2004 Indian Ocean tsunami (original image is from Spot Image, Inc.)

Seventeen fixed real-time observation stations are deployed at high-potential debris flow locations and 3 mobile observation stations are available to be deployed as needed. All stations transmit real-time data (precipitation, water level, earth vibration, images, etc.) to the database. Several models provide rapid analyses to support decision making. Once a disaster happens, an application sends dispatches and reports on resources involved in rescues. In the recovery phase, another system provides support for engineering projects. The system is designed to facilitate collection and use of "lessons learned." The application architecture involves a portal that provides real-time information for common/entry level users. Web applications are provided for officers on duty. The architecture integrates commercial software (ESRI ArcIMS, ESRI ArcSDE, Oracle, and SQL Server) and custom software. Standards Implemented include EDXL, CAP, OGC Web Map Server Interface Standard, OGC Web Feature Service Interface Standard, OASIS WS-security, some of the OGC SWE standards, and potentially the OGC Web Processing Service standard.

5.3 SLEWS – A Sensor-Based Landslide Early Warning System

SLEWS (Sensor-based Landslide Early Warning System) (http://www.slews.de/) is being developed in Germany by the Department of Engineering Geology and Hydrogeology of the RWTH Aachen University, ScatterWeb GmbH, the Department of Geodesy and Geoinformatics of the Rostock University, and the Federal Institution for Geosciences and Natural Resources. The SLEWS project investigates the complete Early Warning System information chain, starting from data gathering using wireless sensor networks via information processing and analysis to information retrieval. The prototype addresses problems of landslides and mass land movements, but the overall objective is to develop principles and tools that can be applied in other EW applications.

SLEWS uses an innovative, open, service orientated information structure in which OGC standards play a key role. The prototype of a Spatial Data Infrastructure (SDI) in the context of an alarm- and early warning system will be orchestrated by a number of OGC SWE services such as Sensor Observation Service (SOS), Sensor Alert Service (SAS), Sensor Model Language (SensorML) and Observations & Measurements Schema (O&M), as well as other services: Web Map Service (WMS), Catalog Services – Web (CSW), Web Feature Service (WFS) and Web Coverage Service (WCS). Implementation of analytic processes and calculations involving business logic will be represented in self-contained web-based services that implement the OGC Web Processing Service (WPS).

5.4 Katrina map server interface

Soon after Hurricane Katrina hit New Orleans, a small group of individuals working in government, academia, and industry implemented an initial emergency mapping portal—the Katrina Map Server Interface—based on open standards and open source software. The portal provided access to a diverse and distributed collection of data provided by many agencies, companies and organizations (Fig. 2). The American Red Cross, the National Institute of Urban Search and Rescue (NIUSR, www.niusr.org) and other relief agencies used the data portal and map viewing application in their efforts. Many who fled the storm have used this Web service to see the condition of their homes before returning. This geocoding capability was quickly put to use by the US Navy, who used it to geographically locate military personnel as well as civilian employees and their extended family who were located in the disaster area.

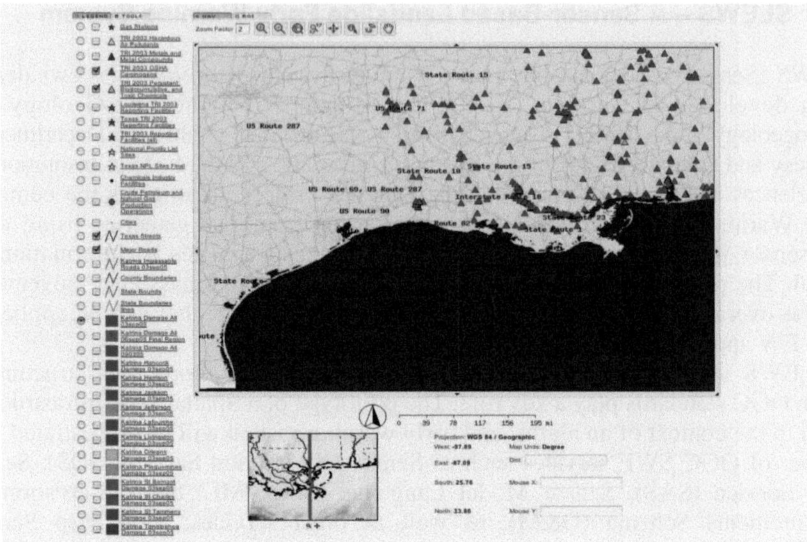

Fig. 2. Map provided by the Telascience Katrina Portal showing a carcinogen sites map layer from the US National Institute of Health's National Institute of Environmental Health Sciences (NIEHS)

6 Sensor Web Enablement

Building on a project that began in NASA, the OGC membership developed a vendor-neutral interoperability framework for Web-based discovery, access, control, integration, and visualization of online sensors. The "Sensor Web Enablement" (SWE) initiative was undertaken in the OGC for two reasons. The first reason was that every sensor—from a simple fixed rain gage to a complex imaging device on an orbiting platform—has location, and every observation occurs at some time, and these are often key parameters in using online sensors. The second reason was that the OGC members had a strong need to enable the rapid integration of real time feeds from various fixed and mobile sensors in order to improve their decision making capability. OGC members, familiar with a variety of different sensor applications, then leveraged the OGC consensus process and existing OGC standards to mature a complementary SWE framework. OGC formed an Alliance partnership with the IEEE technical committee working on IEEE 1451 "smart sensor" standards to make sure that a complementary and versatile set of sensor standards evolved to meet the broad requirements of the global community of users.

In the SWE framework, all sensors report position; all are connected (at least sometimes) to the Web; all may have metadata registered in a catalog for easy discovery; all are readable remotely; and some are controllable remotely. OGC Sensor Web Enablement (SWE) Standards enable discovery, access, and application of real time sensor observations and stored sensor observations.

7 Sensor Web Enablement applications

7.1 Ocean observation

The ocean observation community is one of many communities of interest using OGC standards to help them share and apply geospatial data. This community includes organizations with coastal and maritime EM and EW responsibilities.

The EU's InterRisk project (Interoperable GMES Services for Environmental Risk Management in Marine and Coastal Areas of Europe) (http://interrisk.nersc.no/), for example, is developing a pilot system for interoperable monitoring and forecasting services for environmental management in marine and coastal areas. (GMES is the EU Global Monitoring for Environment and Security program (http://www.gmes.info). InterRisk data and service providers will deliver their products using servers that implement the OGC Web Services standards. Catalogs and registries that implement the OGC Catalogue Services for the Web (CSW) will ensure the delivery of ISO 19139 metadata. These services will be accessed through the ESA Service Support Environment (SSE) portal (http://services.eoportal.org/).

SeaDataNet, a Pan-European project to provide Infrastructure for Ocean & Marine Data Management (http://www.seadatanet.org/) and the UK Met (Meteorology) Office's DEWS (Delivering Environmental Web Services) project (http://www.dews.org.uk) are also using OGC standards. The EUCC (EU Coastal Union) (http://www.eucc.nl/) promotes standards-based interoperability and facilitates stakeholder involvement in MOTIIVE (the EU Marine Overlays on Topography for Annex II Valuation and Exploitation) (http://www.motiive.net).

MOTIIVE and the Australian Oceans Portal project (http://www.aodc.gov.au/index.php?id=34) are collaborating on a registry to deliver OGC standards driven query models, presentation resources and processing chains.

The US Integrated Ocean Observing System (IOOS) program (http://ioos.noaa.gov) is embracing open standards including geospatial and SWE standards of the OGC as part of its technical architecture to improve sharing across federal to local organizations involved in ocean observation (NOAA, 2009). Cooperating organizations have also advanced an Open IOOS project (http://www.openioos.org/) that leverages OGC geospatial and SWE standards.

In January 2007, OGC members launched the "Ocean Science Interoperability Experiment" (OceansIE) (http://www.opengeospatial.org/projects/initiatives/oceansie) to study implementations of OGC Web Service (OWS) standards being used by the ocean-observing community. A number of cooperating ocean-observing organizations participated in the Interoperability Experiment (IE) to develop an initial set of best practice recommendations, available in the Oceans IE Phase I report (www.oostethys.org/outreach/working_folder/ogcreport/ogc-oie-20080822.pdf/view). A second phase of Oceans IE is now underway. Members of the Oceans Science IE are also involved in the OOTHethsys (http://www.oostethys.org/) effort to develop cookbooks, reference materials, and software to enable a "system-of-systems" of linked data providers in the marine domain based on OGC standards, with a focus on sensors.

7.2 Defense and intelligence

OGC standards are being applied in a range of Defense and Intelligence systems and enterprise activities to improve situational awareness, analysis and decision making. For instance, Northrop Grumman Corp. (NGC) has been using the SWE standards in an internal research and development (IRAD) project called Persistent Universal Layered Sensor Exploitation Network (PULSENet) and SAIC has been using the standards in the SensorWeb system they are developing for the Defense Intelligence Agency. A number of other projects are under way. Overall, the objectives are to enable users to:

- Quickly discover sensors (secure or public), send commands to them, and access their observations in ways that meet user needs
- Obtain sensor descriptions in a standard encoding that is understandable by a user and the user's software
- Subscribe to and receive alerts when a sensor measures a particular phenomenon.

7.3 German Indonesian Tsunami Early Warning System

The German Indonesian Tsunami Early Warning System (GITEWS) (http://www.gitews.org) is a 35-million Euro project of the German aerospace agency, DLR (http://www.dlr.de/en), and the GeoForschungsZentrum Potsdam (GFZ) (http://www.gfz-potsdam.de/index-en.html), Germany's National Research Centre for Geosciences. GITEWS uses SWE services as a front-end for sharing tsunami-related information. GITEWS uses real time sensors, simulation models, and other data sources. All those data sources are integrated into a single system through the use of SWE standards. The German organization 52North (http://www.52north.org) provides a complete set of SWE services under GPL license (http://www.gnu.org/copyleft/gpl.html). This open source software is being used in GITEWS and a number of other real-world systems.

7.4 NASA application of SensorWeb

The US National Aeronautical and Space Administration (NASA) has a comprehensive "SensorWeb" vision that involves the development of a wide range of capabilities, such as fire detection, emergency management support, discovery of available sensor assets over the Internet, "wizards" that assemble possible workflows, and workflow engines that control the creation of multi-sensor products for delivery to the user desktop. Satellite operations are part of the vision according to Dan Mandl of the NASA Goddard Space Flight Center, who summarized the benefits of OGC standards in late 2008:

> We continue to upgrade the ground and flight systems on EO-1 with SOA-based web service implementations that allow us to continue flying the satellite on less available operations funding while increasing our functionality, reliability, and performance. We have increased the flexibility of the systems while decreasing the turn-around and

> delivery times for new target-of-opportunity acquisitions important to the disaster management community through use of Sensor Web Enabled services we developed in concert with OGC-sponsored pilots and testbeds. We are implementing more new product generation capabilities and improving the ease of integration of these new products in our processing stream using the SOA-web services approach and OGC standards. (Mandl, 2009))

Many OGC members representing user organizations leverage the OGC process as an opportunity to reduce their technology risk. Through the OGC process they encourage broad industry development and acceptance of open standards, then apply these standards to their programs to reduce cost, improve flexibility, simplify integration, and promote reuse of services.

8 The OGC interoperability program

User requirements for Early Warning, Emergency Response and Crisis Management have been major drivers for OGC standards development, testing, and validation. Typically these requirements are codified in use cases that are assembled into application scenarios that provide the agenda for OGC testbeds and interoperability experiments. OGC pilot projects are designed to validate the practicality of adopted OGC standards in an operational context. In all "Interoperability Initiatives"—testbeds, interoperability experiments, and pilot projects—sponsors define the functional requirements for new standards and share the costs of each initiative. Technology developer participants are recruited to do the technical work in fast-paced technology prototyping and testing activities. OGC Interoperability Initiatives are based on "rapid prototyping" and "spiral engineering" principles.

8.1 OGC GALEON Interoperability Experiment

The OGC "Geo-interface for Atmosphere, Land, Earth, and Ocean netCDF" (GALEON) Interoperability Experiment supports open access to atmospheric and oceanographic modeling and simulation outputs. The GALEON IE will implement a geo-interface to netCDF data sets via the OpenGIS(R) Web Coverage Server (WCS 1.0) protocol specification. The interface will provide interoperability among netCDF, OPeNDAP, ADDE, and THREDDS client/server and catalog protocols. That is, it geospatially enables important scientific data using OGC Web Coverage Service (WCS) standard (www.opengeospatial.org/projects/initiatives/galeonie).

8.2 OGC Web Services Testbed (2001-2002) OWS 1.1

After the events of September 11, the OWS 1.1 sponsors agreed to align testbed objectives to address interoperability challenges defined by officials in New York City. The OWS 1.1 demonstration scenario developed by the sponsors challenged participating technology developers and integrators to implement interoperability capabilities that address specific critical disaster management needs involving New

York City data. OWS-1 focused on open location services (mobile applications) and multi-hazard mapping requirements identified by the user community.

8.3 Critical Infrastructure Protection Initiative (2003)

Agencies from all levels of government, on both sides of the US/Canada border, came together in the CIPI-1.1 pilot project to explore emergency management data sharing. In addition to developing a reference approach to uniting communities at different levels of government to share data, the demonstration illustrated a vision of data sharing for the future based on interoperability. City, county, provincial, state, national and federal agencies showed that the different data could be brought together from different hardware platforms—from mainframes to cell phones—using software from a wide variety of vendors.

8.4 OWS-4 Testbed emergency hospital scenario

The OGC's fourth OGC Web Services testbed activity (OWS-4) involved a hypothetical scenario in which a temporary hospital and decontamination site had to be found close to the site where people were injured and exposed to radioactivity. Building information models (BIM) were available for a number of buildings in the area, which were located through a successful solution to the difficult integration/combination of BIM and geospatial information and their respective services. The BIM encoding used Industry Foundation Classes (IFC), an IAI International standard. Visual inspection and review of the online BIMs' CAD, GIS, and other data showed that one building in particular was well suited to meet the special emergency hospital requirements. Thus the search took less than an hour, and preparation of the site and transportation of patients could begin immediately.

9 Some current focus areas

9.1 CityGML

CityGML, an adopted OGC standard, provides a common information model for the representation of 3D urban objects, and it is based on the OGC Geography Mark Up Language (GML) Encoding Standard. It defines the classes and relations for the most relevant topographic objects in cities and regional models with respect to their geometrical, topological, semantic, and appearance properties. Included are generalization hierarchies between thematic classes, aggregations, relations between objects, and spatial properties (OpenGIS City Geography Markup Language (CityGML) Encoding Standard, 2008). CityGML is beginning to play an important role in Building Information Models (BIM) and other applications that require integration of various kinds of geospatial and building design data in 3D and 4D city models (http://www.citygml.org).

9.2 "Chaining" Web services for decision support

"Service chaining" involves Web services that invoke other Web services. The invoked Web service might call another Web service, just as one subroutine can call another, except that, *if the interfaces are open*, the Web services do not all have to be provided by the same software vendor as in the old paradigm. A reliable, repeatable and transparent process for "chaining" services to produce a desired result is a critical need within the SOA context. The OGC is not creating these chaining services, rather it has reached out to other standards organizations such as OASIS and the Workflow Management Coalition to demonstrate the ability of their open Standards such as BPEL and WfXML to support OGC based service chaining.

9.3 Federated geodata maintenance and update

The 2007 GeoConnections Canada Geospatial Data Infrastructure Interoperability Pilot (CGDI IP) demonstrated how OGC standards can improve the management and dissemination of geospatial data from the local to national scale. The project's collaborative activities with federal, provincial, and private-sector partners demonstrated improved mechanisms for distributing and updating framework data "closest to source" through a distributed open standards architecture. The concluding demonstration showed how products implementing the standards provide access to place names, roads, and municipal boundary data from a distributed network of 14 federal, provincial, and territorial servers. It also executed direct updates of data in provincial servers and showed how the network could assist in emergency response (OGC, 2008).

9.4 The Mass market

The term "Mass-market geo" represents the set of Web-based geospatial enabled applications that put geospatial technologies in the hands of anyone, anywhere, anytime—as long as they have Internet connectivity. Further, these are applications that can be used by users with no knowledge of GIS, remote sensing, geodata, or metadata. Examples of mass-market geo applications include the various Earth browsers, GeoRSS (www.GeoRSS.org), and location-enabled mobile applications. An important element in Mass-market geo is volunteered geographic information (VGI) (Goodchild et al http://www.ncgia.ucsb.edu/projects/vgi/products.html). VGI has important implications in Emergency Management and Early Warning. On a number of occasions, for example, people have uploaded georeferenced images to websites such as Flickr to document in real time the progress of fires and floods. The success of these applications depends on a variety of Internet and Web standards, as well as those provided by geospatial standards development organizations and initiatives. Standards are key for technology diffusion between mass-market applications and traditional geospatial technology markets.

10 Policy positions on open standards

With the realization of the value of open standards for greater interoperability, there has been a marked increase in the establishment of policy positions that further encourage standards-based interoperability. Below are a few examples showing how regional, national, and organizational policy is helping to drive greater interoperability into programs, systems, and enterprises:

10.1 INSPIRE (Infrastructure for Spatial Information in Europe)

INSPIRE (http://www.ec-gis.org/inspire/), which is now formalized as Directive 2007/2/EC of the European Parliament and of the Council (March 14, 2007) is built on five principles that cannot be realized without the kind of interoperability that OGC standards enable:

- Spatial data should be collected once and maintained at the level where this can be done most effectively;
- It must be possible to combine seamlessly spatial data from different sources across the EU and share it between many users and applications;
- It must be possible for spatial data collected at one level of government to be shared between all the different levels of government;
- Spatial data needed for good governance should be available on conditions that are not restricting its extensive use; and
- It should be easy to discover which spatial data is available, to evaluate its fitness for purpose and to know which conditions apply for its use.

INSPIRE is the first multinational effort to evaluate return on investment on participating nations' NSDI development efforts and to begin identifying the open standards on which it will build its spatial data infrastructure.

10.2 Global Earth Observation System of Systems (GEOSS)

GEO (Group on Earth Observations) is a voluntary partnership of 124 governments and international organizations. GEO is coordinating efforts to build GEOSS with the goal of enabling continuous monitoring of the Earth and access by a wide variety of researchers to a vast shared set of information resources. The OGC has led a series of GEOSS interoperability demonstrations at IEEE and ISPRS meetings, and the OGC is leading the GEOSS Architecture Implementation Pilot that has brought together technical contributions from over 120 organizations.

10.3 NATO C3

The NATO C3 adoption and implementation of the Core GIS system uses open, industry-consensus standards, including those promulgated by ISO Technical Committee 211 and the Open Geospatial Consortium, Inc. This has been done to enable

NATO to rapidly mobilize new technologies at reduced cost and risk. Because these standards are also being adopted by the militaries of the NATO countries including the United States, the United Kingdom, Canada, Germany and others, and also by civilian agencies around the world, information sharing in the military environment will be greatly enhanced and the ability to work seamlessly with civilians around the world will be part of the benefit. This widespread adoption of similar standards ensures that there will be a robust marketplace for software and services which NATO can take advantage of for decades.

10.4 US Federal Enterprise Architecture

The US Office of Management and Budget (OMB) is tasked with coordinating US Federal agencies. The Federal Enterprise Architecture (FEA), the product of an OMB initiative, is a management tool that provides a common methodology for information technology acquisition in the US federal government. A complementary goal is to ease sharing of information and computing resources across federal agencies, in order to reduce costs and improve citizen services.

To ensure that the FEA would optimally meet the cross-cutting geospatial service needs of all the agencies, the Federal Geographic Data Committee and the Federal CIO Architecture and Infrastructure Committee (AIC) worked together with others in 2006 to create the Geospatial Profile for the FEA (US Federal CIO Council Architecture and Infrastructure Committee, 2006). That profile is also known as the Geospatial Enterprise Architecture (GEA). The GEA Version 1.1 is now in active use, providing guidance to agency architects and CIOs to help them identify and promote consistent geospatial patterns in their organizational designs. OGC standards feature prominently in this document.

11 Spatial law and policy

Leaders and managers face a wide range of legal issues associated with growth in consumer and business applications for spatial technology. These issues include intellectual property rights, liability, privacy, and national security. In many cases, the existing legal and policy framework is inadequate to provide governments, businesses, and consumers clear guidance on these issues, and often they are addressed by people who do not fully understand the technology. Given the reality that laws and policy vary from nation to nation and jurisdiction to jurisdiction, the situation has potential to be extremely challenging for geospatial information and technology providers. As a result, legal and policy issues may soon impact the growth and direction of the industry.

The OGC Board of Directors recently created a Spatial Law and Policy Committee of the board to serve as an open educational forum to discuss the unique legal and policy issues associated with spatial data and technology. The Committee will work with legal representatives from OGC membership and with relevant legal groups to raise awareness of these issues within the broader legal community, and to provide feedback to the OGC process. Staff and directors anticipate that the SLPC will identify use cases that may in turn become requirements in future OGC testbeds, with the goal of ensuring that OGC standards remain agnostic to and supportive of a broad range of policy positions.

12 Other areas of current and future work

Largely due to technology convergence enabled by standards, there is a rapidly expanding set of technology domains and application domains in which geospatial information plays a role. The OGC tracks this progress to ensure that all domains benefit from a single set of compatible standards for geospatial interoperability. Areas of activity include:

- Mobile Internet, which calls for improved mobile spatial services and which addresses needs of developing nations where mobile phones are primary means of communication. This focus also anticipates a future in which there is ubiquitous wireless broadband connectivity between a wide variety of mobile and fixed devices that provide yet-unimagined capabilities as well as capabilities we now associate with TVs, computers, and cell phones.
- Semantics/Ontologies to simplify the use of geospatial information by various user communities.
- Grid computing. The OGC has an alliance and joint projects with the Open Grid Forum to work jointly on geospatially enabling grid computing.
- Digital Rights Management. The goal of the GeoDRM effort in the OGC is to develop standards that enable access to geospatial resources through a well understood and common mechanism that enables more than today's "all or nothing" protection. GeoDRM addresses the "ownership obstacle to data sharing" in spatial data infrastructure scenarios.
- Geo-enabled Business Intelligence. Leverages location in business information for improved decision making.
- Geospatial fusion, that is, the act or process of combining two or more pieces of data or information regarding one or more entities in order to improve one's capability (or provide a new capability) for detection, identification, or characterization of that entity.

All of these areas have potential to provide benefits to the EW and EW communities.

Because so many activities benefit from standards-enabled interoperability, some standards development organizations, including the OGC, are experiencing significant growth in membership and scope of activities. Many of the OGC's planned and anticipated activities will involve and benefit organizations involved in EW and EM. The OGC Web Services, Phase 7 (OWS-7) Testbed, which is being planned at the time of this writing, will develop interoperability specifications in the areas of Sensor Web Enablement, geoprocessing workflows, 3-dimensional information management, events and security architecture, aeronautical information systems, enterprise web services, mass-market geoservices and compliance testing of OGC standards. "Fusion" of data from diverse sources will be a main theme, with important implications for EW and EM. It is likely that future OGC testbeds and interoperability experiments will expand on the work of previous OGC initiatives and introduce new interoperability approaches involving the technology and application domains listed above.

13 Summary

OGC standards have been influenced heavily by interoperability requirements related to critical infrastructure protection, early warning, emergency management/response, and defense and intelligence. OGC standards promote interoperability, reduced cost, and increased flexibility for situational awareness, analysis, and decision making. Best practices and procurement policy favorable to open standards like OGC standards will further improve market adoption and user choice. There is much more yet to accomplish, and diverse, global participation in the OGC standards process is the key to ensuring that geospatial interoperability standards evolve to meet the needs of users.

References

'Development of Fixed And Mobile Debris Flow Monitoring System In Taiwan', SAO/NASA ADS Physics Abstract Service, hosted by the High Energy Astrophysics Division at the Harvard-Smithsonian Center for Astrophysics, http://adsabs.harvard.edu/abs/2008AGUFM.H42C..01Y. Also (OWS-6): http://www.google.com/url?sa=U&start=6&q=http://www.ogf.org/OGF25/materials/1526/OWS-6.ppt&ei=bOzHScGnGoHasAO14tH3Bg&usg=AFQjCNGdllql2XkwqnPsSbovOjs5c8pgdg

BuildingSmart International (2009) Model – Industry Foundation Classes, http://www.buildingsmart.com/bim

GML 3.1.1 PIDF-LO Shape Application Schema for use by the Internet Engineering Task Force (IETF) (2007), http://www.opengeospatial.org/standards/bp

Mandl D (2009) Discussion with Mark Reichardt on Earth Observation Mission 1 Satellite operations and the value of OGC standards.

Merriam Webster Online, http://www.merriam-webster.com/dictionary/communication

NOAA IOOS Data Integration Framework (2009), http://ioos.noaa.gov/dif/

OGC Catalogue Service Implementation Specification (2007), http://www.opengeospatial.org/standards/cat

OGC CGDI Pilot Demonstration video (2008) http://www.opengeospatial.org/pub/www/cgdi/index.html

OGC Geography Markup Language Encoding Standard (2007), http://www.opengeospatial.org/standards/gml

OGC Sensor Observation Service Implementation Specification (2007), http://www.opengeospatial.org/standards/sos

OGC Web Map Service Implementation Specification (2006), http:// www.opengeospatial.org/standards/wms

The OGC also has close working relationships with ISO TC 211, Web3d Consortium, Biodiversity Information Standards (TWDG), Workflow Management Coalition, Mortgage Information Standards Organization, W3C, and the Open Source Geospatial Foundation (OSGEO) (2009).
Open Geospatial Consortium Website (2009) Registered Products page. http://www.opengeospatial.org/resource/products
Open Geospatial Consortium (2002) OGC Abstract Specification Topic 12: Services. Derived from ISO 2382-1, http://www.opengeospatial.org/standards/as
OpenGIS City Geography Markup Language(CityGML) Encoding Standard, http://www.opengeospatial.org/standards/citygml
US Federal CIO Council Architecture and Infrastructure Committee (2006) Federal Enterprise Architecture Geospatial Profile Version 1.1, http://colab.cim3.net/file/work/geocop/ProfileDocument/FEA_Geospatial_Profile_v1_1.pdf

Metainformation in Crisis Management Architecture – Theoretical Approaches, INSPIRE Solution

Tomáš Řezník

Abstract

This chapter discusses interoperability of spatial data and services with an emphasis on the research of national standards, international trends and pan-European unification. The core of this study is research on both the ISO and OGC metainformation standards and implementation of the rules for INSPIRE metadata, which include their compatibility and inferential results.

Above all, the analytical part of this study is focused on the confrontation of the INSPIRE, ISO and OGC conceptual schemas; the appropriate domains; the definitions of both the code lists and the metainformation XML encoding; and related formats. INSPIRE Geoportal has been established as the highest level of the European spatial information infrastructure, according to the Draft Guidelines – INSPIRE metadata implementing rules. One component of this Geoportal that was implemented in June 2008 is the INSPIRE Metadata Editor, which is the main tool for spatial data and services metadata creation and management. However, several errors can be found in this implementation, which also include differences between the implementing rules and the implementation itself.

The results of both studies show the same insufficiencies in the concepts and the implementation, which can be considered a barrier to the reuse of metainformation, especially in the field of crisis management.

M. Konecny et al. (eds.), *Geographic Information and Cartography for Risk and Crisis Management*, Lecture Notes in Geoinformation and Cartography,
DOI 10.1007/978-3-642-03442-8_14,

1 Introduction

Crisis management is based on information; quick and effective support and the appropriate use of information are tasks for all of the scientific fields that are included in a crisis management process. The same challenges also remain in the field of contemporary cartography and geoinformatics, where the major role lies in long-term planning and administration processes that are based on spatial (geographical) information. In the first period that spatial information systems were used (also named geographical information systems), the main challenges were the possibility for different format integration and a lack of spatial data for crisis management. Since these issues were solved in the 1990s, other challenges have arisen. Up till now, the number of spatially related databases has strongly increased in all of the branches of human activity. At the same time, it has become harder for all of the participants in crisis management to find the required spatial information due to their missing descriptions. These descriptions can be used in two major ways: to obtain the information about the data (i.e. to determine if the database or service is suitable for you) and to search through all of the spatial information providers (these descriptions contain several pieces of information about, e.g. the spatial extent of the map, the creation date or the thematic field). Thus, for example, it is possible to search for all of the maps that have been created since 2003 that cover an area of the Czech Republic and contain information about water resources for fire extinguishing. It is obvious that the effective use of spatial information is tightly coupled with the description of such information and that searching between different informational resources is based on these descriptions as well.

Descriptions of geographic data sets (also called metadata) have existed for some time now. In many cases, the geographic data set descriptions for different information communities on regional and national levels have evolved in different ways and thus are incompatible. Presently, there are two main initiatives to solve global interoperability: the International Organisation for Standardisation (ISO) and the Dublin Core Metadata Initiative (DCMI; Dublin, Ohio, The United States of America). The latter is the prominent metadata initiative in the field of Information Technology (IT), see ISO 19115 (2003) and ISO 15836 (2003); a comparison between it and the INSPIRE standard (see below) is shown in Table 1.

Table 1. Overview of the characteristics and differences between the main international metadata standards ('•' indicates support of such metadata, '-' indicates that this metadata is not supported in a given metadata standard)

	Dublin Core (DCMI)	INSPIRE	ISO (19115+19119)
Number of items for description	15	35	427
Description of visualisation (portrayal)	–	–	– (enabled link to portrayal catalogue)

Spatial representation (vector, grid)	–	–	•
File format description	•	–	•
Use of thesauri	–	•	•
Data quality information	–	•	•
Language support	All world languages	23 official EU languages only	All world languages listed in ISO 639

On the other hand, specific aspects of spatial information raise metainformation activities to the next level. The Open Geospatial Consortium, which established the Catalogue Service Implementation Specification version 1.0 in 1999 (OGC 1999), is where the main concept of catalogue services was presented. A radical change occurred in 2007 (version 2.0.2) when the ISO standards were adopted and the purpose of the catalogue services was extended. According to this specification (OGC 2007), the catalogue services are the interfaces between the clients and the catalogue services, which is through the presentation of abstract and implementation-specific models. Catalogue services support the ability to publish and (full text, spatially and temporally) search collections of descriptive information (metadata) for data, services and related information objects. Metadata in the catalogues represents the resource characteristics that can be queried and presented for evaluation and further processing by both humans and software. Catalogue services are required to support discovery and binding to registered information resources within an information community. In other words, metadata represents query parameters and, at the same time, answers the queries. Therefore, metadata is a fundamental component not only for catalogue services but also for the whole spatial data infrastructure. Metadata links to data that can be visualised for crisis management purposes (see Kozel 2007).

The word 'metadata' stems from Greek and can be described as 'data about data'. In 1968, it was used for the first time in computer science literature (Moellering 1991). Although we usually talk about data description, metadata can be used as a description of a service as well.

This chapter is divided into three main parts that are related to the issues of INSPIRE and its incompatibility with other standardisation activities. The first part focuses on the background of both the concept and the implementation in order to bring a closer image of INSPIRE. The second part is related to the testing methodology of the ISO, INSPIRE and OGC metadata concepts. The last (third) part analyses the INSPIRE Metadata Editor (as it is the only official INSPIRE implementation tool) in conjunction with the requests of the ISO and OGC. This chapter does not deal with either extensive descriptions of the metadata and catalogue standards or implementation of these standards because both of these issues have been widely described in other scientific papers (see Nogueras-Iso et al. 2005, Kubicek et al. 2005). The resulting insufficiencies (complications that can emerge and their proposed solutions) are mentioned at the end of this chapter.

2 INSPIRE

The INSPIRE Directive (2008), which stands for the INfrastructure for SPatial InfoRmation in Europe, involves technologies, policies, standards, human resources and related activities that are necessary in order to acquire, process, distribute, use, maintain and preserve spatial data. INSPIRE is known as the Directive 2007/2/EC, which entered into force on 15th May, 2007. INSPIRE itself 'only' brings concepts; implementation is widely described in the Implementing Rules (hereinafter IR) for different fields of spatial information. For example, the metadata is regulated by the Metadata Implementing Rules, which are technical guidelines based on EN ISO 19115 and EN ISO 19119. It is obvious that these Implementing Rules (IR) are based on the above-mentioned standardisation activities, especially ISO 19115 (geographic information – metadata) and ISO 19119 (geographic information – services), see Table 1 above. At the same time, INSPIRE also uses other standardisation activities like the OGC catalogue services that are a part of the Implementing Rules on Discovery Services.

All of these documents have been created in different periods and therefore, several insufficiencies can be found between them. Some of them are the result of discrepancies between the ISO and/or the OGC documents on one hand and INSPIRE documents on the other hand; others result from diversity between the Implementing Rules and the implementation itself. All of these insufficiencies can be considered as a barrier to the reuse of metainformation, especially in the field of crisis management. Information can be lost due to these insufficiencies when needed in crisis situations.

2.1 INSPIRE Geoportal

The INSPIRE Community Geoportal (2008) is Europ's Internet access point, which allows access to a collection of geographic data and services within the framework of the European infrastructure (INSPIRE). The Geoportal does not store or maintain the data. It acts as a gateway to geographic data and services, which are distributed around Europe. It allows users to search, view or download (subject to access restrictions) geographic data or use the available services to derive information.

In June 2008, the Joint Research Centre (JRC) developed a Metadata Editor for the purposes of metadata creation and maintenance in a form that is compliant with INSPIRE's Implementing Rules for metadata. The Editor allows the users to validate (against INSPIRE IR) the created metadata and save the metadata record as an XML file on a local machine. Note that this version of the Editor does not support the possibility to manage existing ISO metadata without losing elements that are not part of the INSPIRE Implementing Rules.

The search function that is provided by the INSPIRE Geoportal makes it possible to search for geographic resources in the INSPIRE Geoportal metadata database as well as metadata in linked catalogues in a federated metadata search.

2.2 Methodology

The next section of this chapter will focus on discrepancies that have evolved from both adoption of different standards and the implementation process. The methodology was based on two main steps. First, it was necessary to explore conformity between the ISO and OGC standards on one hand and the INSPIRE standardisation documents on the other hand. The next logical step was to test the implementation of the INSPIRE Geoportal. This testing ought to discover differences between the metadata Implementing Rules and the implementation itself in a Metadata Editor. This verification was realised in three steps:

- validation of the XML files that were created with the Metadata Editor against the ISO standards XSD schemas,
- validation of the XML files that were created with the Metadata Editor against the GeoNetwork Open Source application,
- importation, an example of ISO 19139 XML metadata from the metadata Implementing Rules to the Metadata Editor.

The major errors that were detected are described in more detail in the next sections; thus, here, they are only summed up. *The validation of the XML files* (ISO 19139) that were created with the Metadata Editor against the ISO standards XSD schemas was not successful due to the differences between the INSPIRE and ISO concepts (see below). *The validation of the XML files that were created with the Metadata Editor against the GeoNetwork Open Source application* has been successful in one way: it has been able to only export the metadata record in an XML file and to import it into the GeoNetwork. *Importing the example ISO 19139 XML metadata files* from the Implementing Rules to the Metadata Editor was partially successful. Importing the XML file that described the services passed while importing the XML file that described the data set ended with the same error as in the case of the GeoNetwork XML file import ('Error 500: The server encountered an internal error () that prevented it from fulfilling this request'). From the analysis, it can be concluded that the structure of an example XML file from the INSPIRE IR is not well formed according to the principles of XML. Thus, it seems that this example could not be processed into any metadata platform before being published in IR.

3 Insufficiencies between the ISO, OGC and INSPIRE metadata concepts and their implementation

3.1 Language support

The ISO standards and the INSPIRE IR declare multilingual support. For that reason, the element <gmd:language> contains a value for the primary metadata language in the form of a code list. At the same time, the ISO 19115 (2003) standard declares that the allowed values of this code list should be ISO 639-2 compliant, i.e. a three-letter code that represents the name of a language (e.g. eng for English, cze for Czech, swe for Swedish, etc.) or any text like English. The INSPIRE uses the same ISO standard for language values but another subset of this standard (ISO 639-1) for a secondary language. According to this, the INSPIRE metadata must have values that are defined in IR as the two-letter code (e.g. en for English, cs for Czech, sv for Swedish). As you can see, the two-letter and three-letter values, according to ISO 639, do not respect each other and therefore, a converter for them has to be written in order to combine the ISO and INSPIRE metadata. Furthermore, support for the 23 official languages of the European Union has to be covered according to the INSPIRE IR. Unfortunately, the Metadata Editor cannot recognise whether the imported metadata is multilingual or not and thus, cannot visualise information written in other languages. In Fig. 1 (part called 'Resource language'), the imported XML metadata file has been written in Czech (primary language) and English (as the secondary language); however, only the primary language (Czech) is displayed.

3.2 Codes

In both the ISO standards and the INSPIRE IR, <gmd:code> elements in an XML source code are used to uniquely identify several parts of the metainformation (e.g. the identifier of the XML file, the identifier of the data set that is aggregated to the metadata, the reference system identifier, the geographic area identifier, etc.). The INSPIRE Metadata Editor is able to correctly distinguish the identifier of the XML file (element <gmd:fileIdentifier> – although it is not mentioned in IR) but it cannot extract the right information from the other identifiers. This is probably caused by an incorrectly written parser that sends back the <gmd:code> element at every occurrence, even if the semantics of this identifier are different (for semantic issues of data and metadata see Mulickova 2009). An example of such behaviour can be found in Fig. 1 (see the part of the Figure with the field called 'Unique resource identifier'). It is obvious that in this case it cannot be a unique identifier because we have three identifier values and three namespaces (in this case, this designation stands for the range of numerical values). The only way to find out which code is correctly related to this part is to explore the XML source code of the metadata.

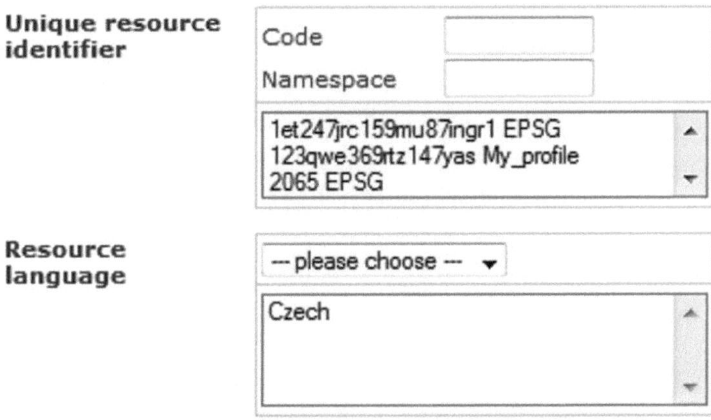

Fig. 1. The incorrect parsing of the unique resource identifier and the absence of the secondary language in the INSPIRE Metadata Editor

After the XML source code analysis, it can be concluded that the information about the coordinate system has been incorrectly parsed and, in this case, is represented by the elements <gmd:code>2065</gmd:code> and <gmd:codeSpace>EPSG</gmd:codeSpace>, which are related to another part of the XML file. These elements contain the coordinate systems information for the distribution format of the data set, which was created by the European Petroleum Survey Group (EPSG) while 2065 stands for S-JTSK (Czech national coordinate system). The second incorrectly parsed identifier is 123qwe369rtz147yas, which is in the namespace My_profile and stands for the identification of the thesaurus. The correct identifier, 1et247jrc159mu87ingr1, is in the namespace EPSG and uniquely identifies the data set. Therefore, it provides a link to the data to which the metadata belong.

3.3 Spatial extent

Also, in this case, the ISO, OGC and INSPIRE both declare that each data set must have a defined spatial extent. There are many ways to define an extent of spatial objects in geoinformatics; however, a bounding box is mandatory for all. The bounding box is the minimum bounding rectangle that encloses a set of one or more spatial objects whose sides are parallel to the axes of the coordinate system (Longley et al. 2001). A minimum bounding rectangle is defined by the two coordinates at the bottom left (minimum x, minimum y) and at the top right (maximum x, maximum y), as is shown in Fig. 2.

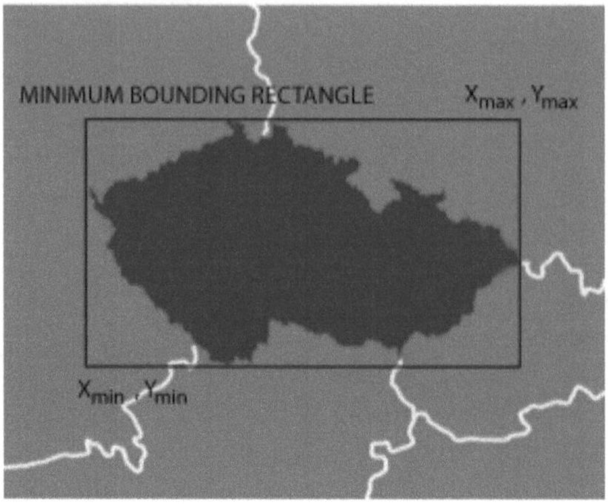

Fig. 2. Definition of a minimum bounding rectangle according to the ISP, OGC and INSPIRE IR

It is obvious from the above text that using the wrong bounding box from another part of the XML document can cause the wrong spatial indexing of the data and therefore, if searching via a catalogue service, the data will not be sent to the user as the correct answer to his request. Let us imagine the situation where two pieces of information about spatial extent are stored in one XML file in the form of a bounding box. The first is information about the extent of the whole data set (e.g. the whole state) while the second contains information about the extent of a distribution unit (e.g. district A of this state). When asked for the results of district B that is in this state, no results that match the criteria can be found; the catalogue service has only taken the wrong information about the extent of district A from the metadata editor and not the information about the extent of the whole distribution unit. The parser of the INSPIRE Metadata Editor incorrectly interpreted two semantically different occurrences of the spatial extent between the same elements (in source code <gmd:extent> and </gmd:extent>), which are found in different parts and elements of the XML metadata file. An example of such behaviour is shown in Fig. 3 (see below).

3.4 Other aspects

The main insufficiencies that can be visually shown at the current state of implementation between both the concepts and the implementation have been mentioned above. Besides them, it is possible to distinguish other discrepancies between the ISO, OGC and INSPIRE concepts. Only the most important, due to their high occurrence, will be discussed further.

The element <gmd:hierarchyLevel> contains the scope to which the metadata apply in all three of the standardisation activities (ISO, OGC, INSPIRE). In other words, the hierarchy level describes the detail of the metadata description (whether a metadata record is related to the data set, data set series, feature, attribute, software, service, etc). The ISO standards use 17 categories to distinguish between the levels of a metadata record; the OGC adds one extra level that is called 'application'. In contrast, INSPIRE has reduced the number of these categories from 17 (18, respectively) to 3: data set, data set series and service. On one hand, it is a logical step to simplify the number of levels and to build a well-arranged structure of levels. On the other hand, it does not ensure automatic parsing of the reduced number of categories. At the same time, we can say that this does not have any significant consequence on an OGC compliant catalogue service or INSPIRE.

When searching through a catalogue service, one possible query limitation can be found. The catalogue service offers full text, spatial and temporal searching. Temporal searching is enabled by query restrictions, such as 'From which time to which time shall the catalogue service search?' Temporal information is taken from the element <gmd:date>, which is in the ISO/INSPIRE structure, and put into the OGC queryable that is called 'modified'. However, an integral part of the element <gmd:date> is also a code list that helps us to better understand this temporal metainformation; it says whether on this date the data was created, published or last revised. Therefore, the element <gmd:date> can appear three times with a different meaning in one metadata file. For example, aerial photos from 1970 were scanned in 2008 and when searching via the catalogue service for aerial photos from 1960 to 1980, these aerial photos will not appear because the temporal data was taken from the wrong occurrence in the XML file (it parsed the date of the last revision instead of the creation).

Most of the other values that are written in an XML file, according to the ISO standards, are correctly parsed and visualised. At the same time, an XML file that was created in the INSPIRE Metadata Editor can be imported into any ISO compliant metadata editor that supports the namespaces (e.g. GeoMedia from INTERGRAPH, GeoNetwork Open Source or CatMDEdit Open Source). Thus far, an XML file has not been successfully imported into ArcGIS (Environmental System Research Institute).

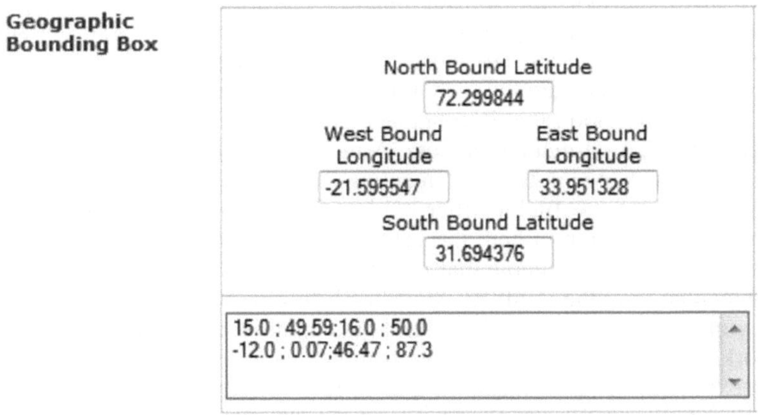

Fig. 3. Incorrect parsing of a minimum bounding rectangle in the INSPIRE Metadata Editor

4 Conclusions and future work

As has been shown in this chapter, the metainformation concepts of ISO, OGC and INSPIRE standards vary due to the different standardisation levels and the age of standardisation documents as well as the slightly different purposes of all of the above-mentioned standardisation activities. INSPIRE is based on the OGC and ISO concept; however, sometimes their structure is changed (e.g. the hierarchy level of the metadata record, the multilingualism, etc.). Metainformation that is created and maintained according to INSPIRE is well-formed and tightly coupled with the searching mechanisms. At the same time, differences between these three concepts can lead to restrictions in the technical interoperability, especially between the European and other national or international spatial data infrastructures that may affect cooperation in transcontinental crisis management situations.

The INSPIRE Metadata Editor shall become a tool that allows users to create metadata that is compliant with INSPIRE Metadata Implementing Rules. Several tests have discovered insufficiencies between the Metadata Implementing Rules and implementation of the Metadata Editor of which the three most important are the absence of multilingualism, the incorrect parsing of identifier codes and the spatial extent. Some of them are more relevant due to their effect on the catalogue service (especially spatial extent). Therefore, feedback has been given to the Joint Research Centre, which is the institution responsible for the development of this tool and its application. Furthermore, it has been found that the INSPIRE Metadata Editor is not capable of importing XML example that is a component of the Metadata Implementing Rules. Also, validation of the XML files that were created with the INSPIRE Metadata Editor against the GeoNetwork Open Source application has not been fully successful.

This situation is most relevant in the field of crisis management where quick and correct information is necessary for effective decision-making support. Therefore, in some cases, the tools for decision making that are based on the components of spatial data infrastructures (like INSPIRE) will not provide relevant answers to the queries of the users. The most important consequence can lead to an answer of the server that no data for such a user query has been found even if the data exists. Then, information that is contained in the data cannot be used in the crisis management decision-making process. A meaningful solution (according to the author) requires a change in the Implementing Rules of the INSPIRE Metadata (especially definition of the fileIdentifier – identifier of the metadata in XML, clear and unified information in both the general and detailed mapping parts and replacement of XML code example that is not valid) as well as correction of the current implementation of the INSPIRE Metadata Editor on the European Geoportal (especially multilingual support and correct parsing of imported metadata).

Future work will continue with intensive communication with the JRC as well as with the next series of XML metainformation tests. A separate validation step will be represented by conformity tests between the OGC specifications on Catalogue Services, technical guidance documents on INSPIRE Discovery Services and catalogue service implementation on the INSPIRE Geoportal.

Acknowledgements

This research has been supported by funding from Project No. MSM0021622418, which is called 'Dynamic Geovisualisation in Crises Management'.

References

Draft Guidelines – INSPIRE metadata implementing rules (2008) [online]. Ceska informacni agentura zivotniho prostredi, Praha. http://www.cenia.cz

EN ISO 15836 (2003) The Dublin Core Metadata Element Set

EN ISO 19115 (2003) Geographic Information – metadata

INSPIRE (2008) [online]. European Commision, Brussels. http://inspire.jrc.it/

INSPIRE Geoportal – the EU portal for Geographic Information (2008) [online]. European Commision, Brussels. http://www.inspire-Geoportal.eu/

ISO – International Organization for Standardization (2007) [online]. ISO, Geneva. http://www.iso.org/iso/home

Kozel J (2007) Open Contextual Cartographic Visualization. In: Proceedings of GI-Days 2007 – Young Researchers Forum. IfGI prints, Münster.

Kubicek P et al. (2005) The geoinformation infrastructure in the Czech Republic: The key role of metadata. In: Proceedings of 11th EC GI & GIS workshop, ESDI: Setting the Framework, JRC, Alghero.

Longley PA et al. (2001) Geographic Information Systems and Science. 2nd edition, John Wiley & Sons, Chichester.

Moellering H et al. (1991) Spatial database transfer standards: Current international status. Elsevier Applied Science, London.

Mulickova E (2009) Semantic annotation of geodata – a step towards interoperability and harmonization. In: Symposium GIS Ostrava 2009 Proceedings, Tanger, Ostrava.

Nogueras-Iso J et al. (2005) Geographic Information Metadata for Spatial Data Infrastructures. Springer, Berlin.

OGC (1999) Open GIS Catalogue Services Specification [online] http://www.opengeospatial.org

Open GIS Catalogue Service Specification 2.0.2 (2007) [online]. Open Geospatial Consortium. http://opengis.org/docs/07-006r1.pdf

Collaborative Post-disaster Damage Mapping via Geo Web Services

Laban Maiyo, Norman Kerle, Barend Köbben

Abstract

To mitigate the consequences of increasingly frequent disasters across the globe, better real-time collaborative post-disaster management tools are needed. The International Charter 'Space and Major Disasters', in conjunction with intermediary agencies, provides for space resources to be available to support disaster response. It is widely seen as a successful example of international humanitarian assistance following disasters. However, the Charter is also facing challenges with respect to lack of collaboration and validation, with the information flow being largely mono-directional. It is, therefore, fundamental to move away from static map data provision to a more dynamic, distributed and collaborative environment. Geo Web Services can bring together vast stores of data from heterogeneous sources, along with geospatial services that can interact in a loosely coupled environment and be used to create more suitable information for different stakeholders. The aim of this chapter is to evaluate the relevance and importance of Geo Web Services in the disaster management domain and present a suitable Geo Web Service architecture for a collaborative post-disaster damage mapping system. We focus particularly on satellite image-based post-disaster support situations, and present our ideas for a prototype based on this architecture with possibilities for User Generated Content.

M. Konecny et al. (eds.), *Geographic Information and Cartography for Risk and Crisis Management*, Lecture Notes in Geoinformation and Cartography,
DOI 10.1007/978-3-642-03442-8_15,

1 The current state of post-disaster mapping

Disaster numbers and costs have been increasing worldwide in recent years, posing an increasingly global challenge that requires timely solutions. In conjunction with better understanding of disaster risk management (DRM), including better insight into links with socioeconomic development, more global and collaborative DRM approaches have been developed. Among those are collaborative information coordination platforms, such as AlertNet, Virtual OSSOC and ReliefWeb, some already use current geocommunication means such as news feeds and alert.

An important information source for such networks is the International Charter 'Space and Major Disasters' which has been a champion in space data acquisition and delivery of image-based information to organizations involved in disaster response (Ito 2005). It aims at providing a unified system of space data acquisition and delivery to those affected by natural or man-made disasters through Authorised Value Adding Resellers (VARs) and Value Adding Agencies (VAAs) (Mahmood et al. 2008). Since its inception in 1999, there has been an increasing number of activations, aided by a recent growth in Charter membership, now including DMCii, CONAE, ISRO, JAXA, USGC and NOAA, adding their space resources to those of CSA, CNES and ESA, a major improvement in space-based disaster response and meeting disaster challenges . The bulk of the image processing has been carried out by UNOSAT, the German Space Agency's ZKI, and SERTIT.

Several other private companies and NGOs have recently become involved in post-disaster damage mapping, management, response and recovery. For example, ImageCat Inc., RapidEye, TerraSAR and MapAction focus on post disaster response, frequently linking disaster response and management efforts with the UN, the Charter and NGOs in the context of public private partnerships (PPP). These PPPs are important in bringing in a pool of resources, technology, expertise and combined efforts towards rapid disaster response. ImageCat Inc., for example, has been developing tools for more efficient image-based disaster response, most recently the Virtual Disaster Viewer (VDV) based on MS Virtual Earth, which offers an alternative method of rapid and robust damage assessment. The European Commission's Joint Research Centre (JRC) and ORCHESTRA project are also developing new disaster management tools and techniques (ORCHESTRA 2008).

2 Challenges for post-disaster mapping

Despite successes, such as an increasing number of activations, better visibility, and more reliance of decision makers and disaster response professionals on such space data, the Charter is facing, especially in meeting the changing needs of increasingly specialized players in the disaster arena. With the technology currently used, data flow is largely mono-directional: post-disaster maps are produced at the UNOSAT offices, without the opportunity to include local knowledge and additional information from other stakeholders. The resulting map products are disseminated through a website, where end users can view and download them in print-optimised PDF format. Since the charter maps are static, one-off products, there is little possibility for additional validation. It would be an important step forward to move away from static map data provision to a more dynamic, distributed and collaborative environment.

Thus an appropriate application framework has to be developed to enable multiple stakeholders in various locations to customize the post-disaster information, add value by providing feedback or access to their own information and to collaborate with other agencies involved in the disaster aftermath. This requires *geospatial collaboration* in emergency response which is technically feasible with current spatial analysis and geoprocessing tools. By making it possible to integrate different types of data and information from diverse sources, collaborative post-disaster will strengthen analytical capabilities and decision making for disaster response. When considering this new collaboration concept, however, the Charter data use remains complicated, as original imagery is not free as such and cannot be used freely after the use by the officially designated processing entity. Likewise, any information added by other stakeholders may also have access restrictions. Therefore, any distributed system architecture needs to deal with access conditions in a secure way.

Disasters can represent a challenge or an opportunity, leading to a variety of possible competing or conflicting interests since there are entities that either have a humanitarian or a commercial motivation. While originations such as MapAction may be able to focus their resources on aiding disaster response, for others, such as UNDP, disasters need to be dealt with as an additional challenge to meet development objectives. Also for UNOSAT, primarily associated with post-disaster damage mapping, disaster mapping competes for time needed for many other mapping activities. Disasters, however, can also constitute a source of prestige, be it for different disaster response websites vying to be the main platform, or different UN organizations. For example, within the UN different entities, such as OOSA, OCHA or UNOSAT, have had disagreement on who should have the right to trigger the Charter. Disaster response has become an interesting business area where humanitarian support, research and commercial interests converge.

2.1 Towards collaborative disaster mapping using Geo Web Services

A number of non-standardized frameworks for Web-based Collaborative Decision Support Services (WCDSS) amongst stakeholders already exist (Wang and Cheng 2006), but because such systems use proprietary interfaces, they are not useful for a larger user community. The solutions to collaborative environment require the use of Open Standards. Such standards have been developed and are increasingly used in Spatial Data Infrastructures (SDI), and our goal is to develop a generic architecture for such a collaborative system based on Geo Web Services. Such Service Oriented Architectures (SOA) have well-defined interfaces that interact with other loosely coupled network software applications. They fully encapsulate their own functionalities and make them accessible only via well-specified and standardized interfaces (Köbben 2008). This is achieved by encoded data in a standardized platform and application independent manner by use of encoding schemes and generic web service standards such as the eXtensible Markup Language (XML), Web Service Description Language (WSDL) and Simple Object Access Protocol (SOAP) utilized to deploy geographic web services.

There exists a range of proprietary Geo Web Services in the market. They include Google Earth/Maps, Yahoo Maps and Microsoft Virtual Earth/MultiMap. Free geobrowsers to view data through these services are available, both in 2D and 3D. Next to that, non-proprietary Open Standards have been developed in an open and participatory process, and are owned in common. Examples of Open Standards for Geo Web Services are the Open Web Services (OWS) specifications of the Open Geospatial Consortium (OGC). There are OWS specifications for most parts of the spatial data storage, analysis and delivery process: For geographic data encoding, there is the Geographic Markup Language (GML), and for spatial data delivery the Web Coverage Service (WCS) and Web Feature Service (WFS), for querying and retrieving raster and vector data, respectively. For processing of spatial data the Web Processing Service (WPS) has been defined, and Web Map Service (WMS), for data visualization in the form of maps. An emerging specification is GeoDRM, specifying Digital Rights Management of geodata.

2.1.1 Importance of Geo Web Services as a tool for collaboration

Collaborative damage mapping requires situation assessment from existing and new data sets, impact assessment with post-disaster imagery and organization of post-disaster work. Such diverse collaboration can only be supported where distributed services act as a geospatial one-stop for seamless data management. Geo Web Services as a unified system allows fast collation and analysis of distributed data set with expert knowledge, leading to a wide range of services for a long-term, comprehensive system in critical disaster response. The main focus is to design a suitable framework for a collaborative post-disaster mapping system.

Thus a Geo Web Services approach can connect the various disaster management agencies, allowing more customized delivery of data and information, and allow users to add value by providing their own information, creating new synergies in a loosely coupled environment. Despite past achievements in providing image derived information, the Charter currently lacks a framework for collaboration, synergy and feedback from major stakeholders in disaster response.

2.1.2 User Generated Content (UGC) and Neogeography tools

Apart from image analysis, emerging web services can be used to display damaged infrastructure in the field by disaster experts and volunteers by employing new interoperable Web 2.0 tools such as geotags, Flickr, GeoRSS and GeoWIKI.

Geotagging is the process of tagging images to various open layers in the form of geospatial metadata, where users can find a wide variety of location-specific information. Geotagging-enabled information services can also be potentially used to find location-based disaster-damaged infrastructure. Unlike Geotags, **Flickr** organizes images as tag clouds, referenced by place names. It offers a comprehensive web-service Application Programming Interface (API) that allows humanitarian experts to tag photographs of damaged infrastructure. **GeoRSS** is a standard for encoding location as part of an RSS feed (http://en.wikipedia.org/wiki/RSS). GeoRSS collaboration can promote interoperable web services across the disaster domain. **GeoWIKI** is a means of many people contributing to the development of a large database (crowd-sourcing), using Google Earth based GeoWIKI, designed to enable anyone to contribute or modify its content (Goodchild et al, 2007).

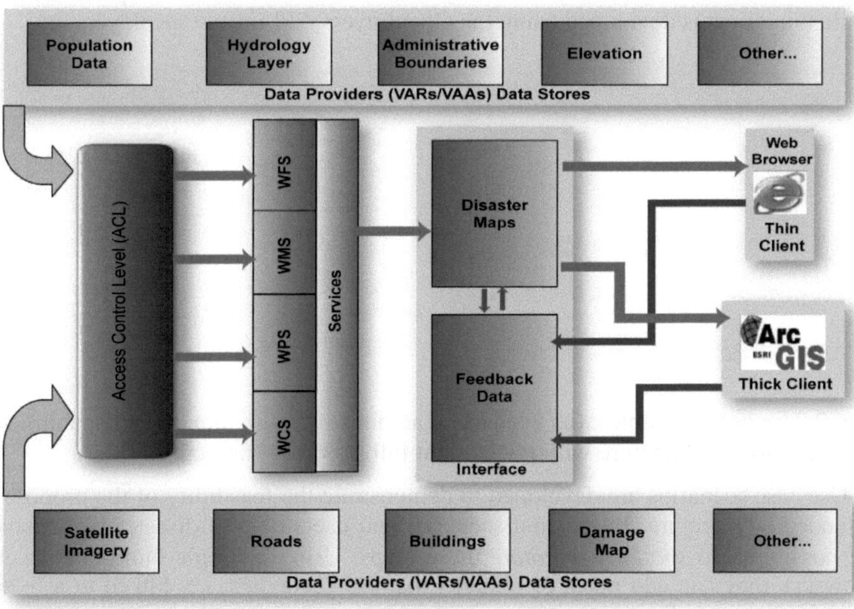

Fig. 1. The extended prototype architecture

2.2 Prototype

We develop different use case scenarios as part of a testbed for a technically feasible collaborative disaster management system. The main goal of this prototype is to demonstrate the technical concepts of a collaborative mapping system. As a proof of concept for the use of open standards for end-user access to disaster maps, we are setting up a prototype project based on appropriate service specifications. The aim is to connect to different servers hosted by VAAs/VARs and combine output of these servers in the distributed client machines via a browser or geoprocessing software as shown in Fig. 1. Data from intermediary agencies can be accessed by end-users via thin or thick client as map services through an interface, and through a regulatory Access Control Level (ACL) security mechanism.

The prototype is developed based on the concepts of distributed services. End users can employ a range of applications, from simple so-called *thin clients*, with a limited functionality (e.g. a web browser for viewing maps from a WMS) to *thick clients*, e.g. a full blown GIS system that uses the architecture to access and process the base data. The prototype is built using already available Open Source components that we use in ITC education and research projects. The Geo Web Services layer is largely based on the UMN Mapserver (http://mapserver.org), while thin client is developed using the OpenLayers API (http://openlayers.org).

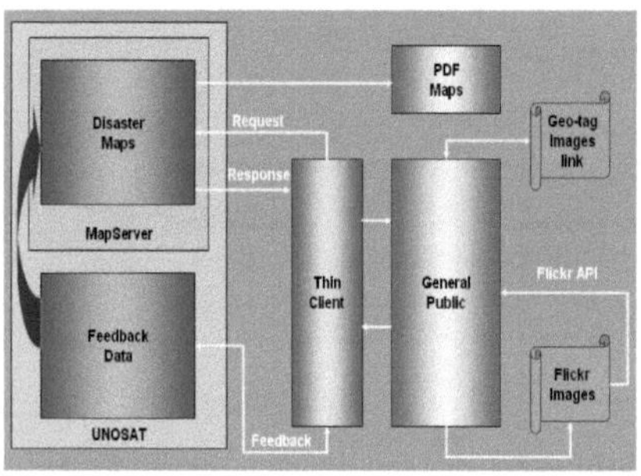

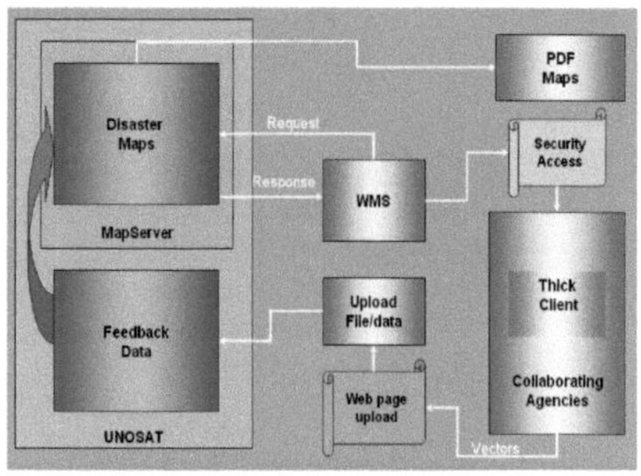

Fig. 2. Use case examples: (**a**) Scenario 1 architecture with thin client capabilities, (**b**) scenario 2 architecture with thick client full geoprocessing capabilities

Use case scenarios are developed to demonstrate the feasibility of the proposed extended architecture. In the first scenario, end users of post-disaster maps have the possibility to *spatially annotate* these maps. Using a simple thin client (web browser), they can add notes or remarks that are geotagged, i.e. linked to a fixed point in the map. These spatial annotations are made available in the web portal (see the dark arrows in Fig. 1), and therefore can be viewed by others users. They could also be used by the mapping agency to further improve their maps. Likewise, the agency can use these annotations to actively seek help, for instance by posing questions such as '*does anybody know if this building is still standing?*' or *'is this road passable?'*. The content of the spatial annotations is not limited to text, as we can employ links to existing photo-sharing services (such as Flickr (see Fig. 2a) or Panoramio) or other Geo Web Services (e.g. Google Maps). For the second scenario, we envisage a more limited user group, such as stakeholders and collaborators that are asked to collaborate actively on the production of post-disaster maps. These users require a thick client, such as QGS, uDig and ESRI's ArcGIS system, and would use that to help with data processing, in our use case delineation of damaged areas and upload it via a secure web page. These inputs are used to process the data for the final damage maps, hence a secure access and validation mechanism needs to be in place (see Fig. 2b).

3 Results

The results are the outcome of the two use case scenarios developed and the proof-of-concept output using data of the May 2006 earthquake close to Yogyakarta, Indonesia where post-event Ikonos and Quickbird images were available and several agencies produced their own maps. The designed prototype is deployed largely on OpenLayers and Geoserver running at ITC and results linked to external domains. From the results, MapServer provides a clear design by use of datastores to integrate existing Rich Internet Applications (RIAs) for damage mapping. The date and time tool is incorporated to track input data at server and client sides to accommodate location and time differences of agencies and end users.

Forms are developed using Active Server Pages (ASP) with drop-down options for attribute information input with possible URL link to photos on other sites such as Flickr or Panoramio as shown in Fig. 3 (section D). When a disaster occurs, the implementing agency sets up the system and connects the participating agencies, at the same time soliciting information from the ground. A database was created to receive the feedback data on the server side. The data are stored in a database and available as an extra layer to the end users. The performance and speed of the system is enhanced by map optimization, indexing of data, tiling of images and caching of web service. The prototype can be found at http://geoserver.itc.nl/laban/

The DEMIS online base map (www2.demis.nl) sets the projection, extent and units of the map and the disaster data act as overlay layers and end users can toggle, switch on and off layers using a checkbox list. The edit and capture tools (Fig. 3, section E) accommodate formats such as GeoJSON, GeoRSS, KML, GML and Well Known Text (WKT). At the same time, the user can define the input and output projections and associated metadata and comments. The tools to capture polygons, lines and points (Fig. 3, section A) allow feedback where end users digitize features of interests and send back the data to the database. A serialized version of the feature (section F) is available showing feature descriptions.

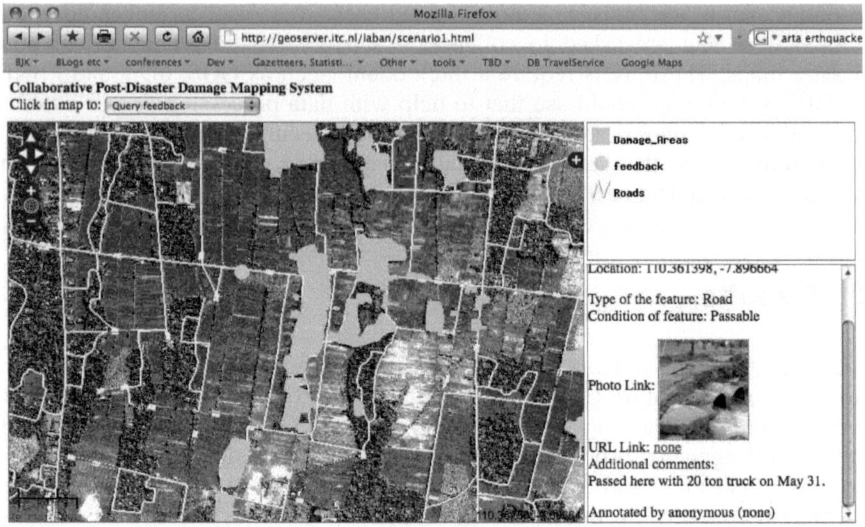

Fig. 3. Overview of the system showing damage areas, roads and imagery of Yogyakarta earthquake, May 2006, Indonesia

Styled Layer Descriptor (SLD), a standardized map styling format is used to control the visual portrayal of the data set by defining the styles and rules according to OGC standards. Section B contains the legend, while other editing tools are located in section A. More tools and features to enhance the performance and versatility of the system can be added. For example, since there is variability in geographic location, language, culture and social differences across countries and regions where disasters occur, there is need for incorporating multilingual application in the system where agencies and experts overcome language barriers.

4 Discussion and recommendations

Information systems used in the field of disaster management are often not as open and comprehensive as needed to integrate and accommodate the complex data sets and the different systems. There is currently no singly accepted architectural model for web services as a whole, although a number of groups (W3C

Architecture Working Group) are working on defining how web services will be used with their products. Interoperability as well as application-oriented integration of methods, data and systems must be improved by designing distributed software architectures (Kohler and Wachter 2006). Our proposal is more of a working dynamic system as compared to ImageCat's VDV, which is more rigidly limited to real-time post-event damage assessment. The success of the system requires well-established SDIs within disaster agencies. An SDI architecture incorporated in the service facilitates access to various information types, existing data and data coming from the field. There are generic services for SDI realization (Scholten et al. 2008) that enhance integration of information from different agencies with appropriate interfaces for different end users. SDIs are mandatory in managing dynamic information with varying agency and national data policies.

The success of this collaboration system is its scalability, reliability and quality; there should be a mechanism to limit errors and to build trust and assurance on the sourced data. The process of data provision, integration and sharing should conform to ISO, OGC and W3C standards and specifications. The utility of OGC Web Services has already been demonstrated (Kohler and Wachter 2006). Data quality and control especially in open platforms is a must, a prerogative of intermediary agencies to regulate the access, editing and integration of the data set via a common protocol. A range of tools and techniques are available to ensure data quality and integrity especially if user-generated content is of good quality. It is possible to have user-centric metadata to enable experts trace the source and quality of uploaded data set. The implementing agency should set up internal QA/QC methods. GeoDRM should be part of the collaborative quality control mechanism in post-disaster damage mapping. It is a conceptual framework, and an array of standards and software tools for guarding the rights of both producers and consumers of geospatial data and services (Lieberman 2006). Access Control Levels (ACL) are created in authentication services to manage permissions, subject to the level of access rights and privileges of heterogeneous users (Xu and Zlatanova 2007). Security can be ensured between the services and the clients by establishing HTTPS and/or Secured Socket Layers (SSL). Other best practice guides and ethical documents can be developed to ensure prudent use of disaster information. Security and compliance can be enforced to conform to identity standards such as open-access license generation and data file provenance tracking.

The proposed architecture consists of several data services that can be adopted and implemented by VAAs/VARS for real-time collaboration. Cascading and semantic chaining of disaster information by collaborating agencies can be implemented to provide a unified access to all data sources (Schmitz et al. 2006). The proposed architecture uses remote user profiles and is able to disseminate post-disaster damage maps without any major constraints. Collaboration gives emergency management organizations a pool of expertise far larger than the organization itself can provide (Siegel et al. 2008). The architecture is part of the 'mass market' initiative where many neo-concepts for UGC, crowd sourcing, VGI and ubiquitous sensor networks converge. The system itself can connect a roster of experts from any location with expertise in the disaster type, and can include a link to social and professional network sites such as Twitter (http://twitter.com/), where experts can actively participate in an emergency. The concept of citizens as sensors (Goodchild et al. 2007) allows volunteers to contribute to disaster reporting.

The implementation process should also incorporate the use of ontologies and service orchestration to enhance interoperability. Development of ontologies and ontology architectures for disaster response (Xu and Zlatanova 2007) is recommended in order to overcome semantic interoperability challenges. Ontologies are used to specify conceptualization in a domain of knowledge within different disaster risk domains (ORCHESTRA 2008), and can be mapped to enhance interoperability between convergent heterogeneous information sources in many post-disaster response scenarios. The EC's OCHESTRA, WIN and OASIS projects are developing models to overcome ontology issues in disaster management.

5 Conclusion

The Web 2.0 phenomenon has revolutionized Geo Web platforms, spanning all connected heterogeneous systems. Web 2.0 applications deliver information, consuming and mashing up data from multiple sources, including individual users, while providing their own data and services in a form that allows integration by others. The best solution to meet current post-disaster damage mapping challenges is to employ off-the-shelf geo web tools and services, in conjunction with non-proprietary tools and protocols. The process of real-time data sharing and transfer reduces the cost of travel and shipping and encourages a two-way communication channel, enhancing participatory approaches to common disaster challenges. The web service architecture allows heterogeneous stakeholders to access all available disaster information in the same geographic context. Real-time damage mapping enables distributed disaster management experts to put damage evaluation into local context, aiding in response and recovery.

Geo Web Services provide a means for analysis, augmenting both speed and precision of disaster situation evaluation. Dozens of data sources, many of them hosted by disaster management organizations, are now searchable and accessible through a portal. The data resources and data access provided by a geospatial one-stop repository will be critically important in all of these areas. This project demonstrates that Geo Web Services can supply up-to-the-minute the rapidly changing disaster thematic information. Disaster management agencies can now have additional capabilities in the areas of web-based online geoprocessing and geofusion services, an infrastructure for spatial information.

References

Goodchild M, Fu P, Rich P (2007) Citizens as sensors: The world of volunteered geography. GeoJournal 69:211–221.

Ito A (2005) Issues in the implementation of the international Charter on Space and Major Disasters. Sci. Direct 21:141–149.

Köbben B (2008) A short introduction to Geo-WebServices - Unpublished Teaching material, Enschede, The Netherlands.

Kohler P, Wachter J (2006) Towards an open information infrastructure for disaster research and management: Data management and information systems inside DFNK. Springer 2006, vol 38: pp. 141–157.

Lieberman J (2006) Upper and Lower Case on the Geosemantic Web, 5th International Semantic Web Conference, Athens, USA, 5–9 November 2006.

Mahmood C, Cubero E, Platzeck G, Bequignon J (2008) South American perspective of the International Charter 'Space and Major Disasters'. Advances in Geosciences, vol 14: pp. 13–20.

ORCHESTRA (2008) An Open Architecture for Risk Management, Report 978-3-00-024284-7, European Commission, Brussels.

Schmitz S, Glade T, Dellow GD, Greve K (2006) Thinking OpenGIS Web Services (OWS) for the dissemination of natural hazard data, XIth International Congress, Belgium.

Scholten H, Fruijter S, Dilo A, Borkulo E (2008) Spatial data infrastructure for emergency response in Netherlands. In: Nayak S, Zlatanova S (eds.) Remote Sensing and GIS Technologies for Monitoring and Prediction of Disasters. Springer, Berlin, pp. 179–195.

Siegel C, Fortin D, Gauthier Y (2008) Geocollaboration in Hazard, Risk and Response: Practical Experience with Real-Time Geocollaboration at Quebec Civil Security. In: Nayak S, Zlatanova S (eds.) Remote Sensing and GIS Technologies for Monitoring and Prediction of Disasters. Springer, Berlin, pp. 199–215.

Wang L, Cheng Q (2006) Web-Based Collaborative Decision Support Services: Concept, Challenges and Application, 8th International Conference on Intelligent Transport Systems, Vienna, Austria.

Xu W, Zlatanova S (2007) Ontologies for Disaster Management Response. Geomatics Solutions for Disaster Management. Springer, Berlin, pp. 185–200.

References

[illegible]

Decision Support for Tsunami Early Warning in Indonesia: The Role of OGC Standards

Ulrich Raape, Sven Teßmann, Andreas Wytzisk, Tilmann Steinmetz, Monika Wnuk, Manuel Hunold, Christian Strobl, Christoph Stasch, Alexander C. Walkowski, Oliver Meyer and Simon Jirka

Abstract

The December 2004 tsunami demonstrated the need for an effective tsunami early warning system for the Indian Ocean. Within the framework of UNESCO-IOC (Intergovernmental Oceanographic Commission) and its Intergovernmental Coordinating Group (ICG), various efforts on national and bilateral basis are coordinated and combined to ensure a fast and reliable tsunami warning for the whole Indian Ocean and its 27 rim countries. The work presented here is embedded in the German–Indonesian Tsunami Early Warning System (GITEWS) project. GITEWS is funded by the German Federal Ministry of Education and Research (BMBF) to develop a Tsunami Early Warning System for the Indian Ocean in close cooperation with Indonesia and is a major contribution to the Indonesian Tsunami Early Warning System (InaTEWS) which has been inaugurated by the President of the Republic of Indonesia on November 11, 2008. The system integrates terrestrial observation networks of seismology and geodesy with marine measuring sensors, satellite technologies and pre-calculated simulation scenarios. The GITEWS sensor systems integrate the respective sensor information and process it to aggregated sensor observations in real time. The processed information from all these sensor systems is transmitted to the GITEWS Decision Support System (DSS) for further processing, analysis and decision support.

This chapter describes the application of standards defined by the Open Geospatial Consortium (OGC) within the GITEWS context for integrating external sensor observation data as well as within the DSS for access to huge geodata, risk and vulnerability and sensor databases using an internal SDI. Especially the OGC Sensor Web Enablement SWE framework (Botts et al. 2006) plays a major role in

M. Konecny et al. (eds.), *Geographic Information and Cartography for Risk and Crisis Management*, Lecture Notes in Geoinformation and Cartography,
DOI 10.1007/978-3-642-03442-8_16,

sensor data management. For map display and communication with the GITEWS simulation system OGC standards are applied, too. For warning message dissemination, the Common Alerting Protocol CAP standard (OASIS 2005) is used to provide targeted regionalized messages to numerous recipients.

1 Introduction

In recent years numerous tsunami events in the Indian Ocean, and in particular along the Sunda Arc, have shown how vulnerable human society and the environment are to this sudden-onset type of disaster. Especially the December 2004 tsunami demonstrated the need for an effective tsunami early warning system for the Indian Ocean. Within the Framework of UNESCO-IOC (Intergovernmental Oceanographic Commission) and its Intergovernmental Coordinating Group (ICG), various efforts on national and bilateral basis are coordinated and combined to ensure a fast and reliable tsunami warning for the whole Indian Ocean and its 27 rim countries.

Spatial Geodata Infrastructures (e.g. GDI-DE, INSPIRE in the German and European context, respectively) nowadays are used more and more as a building block for modern Decision Support Systems whereas former systems often stemmed from locally held spatial data, metadata and accompanying other data. The latter have been implemented as standalone applications, essentially neglecting interoperability with outside systems for the provision of data and application logic. The experience gained in, e.g. NaDiNe (Natural Disaster Networking Platform) prepared for the extended use of data accumulated in different thematic contexts related to natural disaster management. On the other hand, while spatial data infrastructures are invaluable in terms of enabling the distributed collection, quality control and maintenance of data, their collection alone does not yet enable an operator to make his judgements quicker or more reproducible. Other warning systems put particular emphasis on the provision of a system for automated versatile warning dissemination, as for example in the multi-hazard Integrated Public Alert and Warning System of the US Federal Emergency Management Agency (FEMA). It is therefore considered essential to use an approach based on a-priori knowledge and the near real-time fusion and assessment of incoming information to give true decision support for the task at hand, using a standards based infrastructure for data access and message dissemination..

The system introduced here uses data which, among others, come from a multitude of sensors connected through an OGC-compliant sensor web enabled network. It attempts an integration of complex information in the sense of information aggregation and fusion. In the GITEWS DSS (decision support system) this supports an operator in quickly finding the answer to the question whether to send a warning for a given tsunami (or earthquake) event or not and what information to send to the addressee of a warning. The system integrates terrestrial observation networks of seismology and geodesy with marine measuring sensors, satellite technologies and pre-calculated simulation scenarios. It uses therefore OGC Web

Services (OWS) for an innovative fusion of information from a large network of sensors as well as the display of map and feature services in the DSS. It is thus based on standards for geospatial data, metadata and message dissemination as well as an open architecture conforming to international standards into which new sensors can be integrated flexibly.

The work presented in this chapter is embedded in the German-Indonesian Tsunami Early Warning System (GITEWS) project. GITEWS is funded by the German Federal Ministry of Education and Research (BMBF) to develop a Tsunami Early Warning System for the Indian Ocean in close cooperation with Indonesia, the country most prone for tsunamis in the whole Indian Ocean. GITEWS is the German contribution to the Indonesian Tsunami Early Warning System InaTEWS.

2 The challenge of tsunami early warning in Indonesia

What makes tsunami detection for Indonesia unique and challenging is on the one hand the extremely short time window between tsunami generation (in most cases caused by an earthquake along the Sunda Arc) and the arrival time at the nearest Indonesian coastline, and on the other hand the lack of sensor technologies that detect and measure tsunamis as such. While promising technologies are being worked on that might allow holistic tsunami detection in the future, the GITEWS project uses the best sensor technologies available today to detect indicators or evidence for a tsunami, combining those information with up-to-date modelling techniques and integrating them in a newly developed Decision Support System. Combining a priori knowledge, simulation runs and analysis results with real-time information from different types of sensors, the GITEWS Decision Support System (DSS) serves as a backbone to allow an assessment for the tsunami threat at the earliest time possible and support the decision maker whether to issue a tsunami warning or not.

Unlike classical decision-support problems, the process of combining sensor and additional information, generating situation awareness and assessing and proposing decision options is a slowly evolving process. Due to the fact that sensor information becomes available in a non-deterministic irregular sequence, initially with considerable uncertainties, in arbitrary order and with major information gaps, uncertainties will still be present when deadlines for warning decisions are reached.

3 The early warning and mitigation system

GITEWS' novel 'system of systems' concept is based on a modular and extensible architecture of different systems deployed in the BMKG Warning Center in Jakarta as part of the GITEWS Early Warning and Mitigation System (EWMS)/InaTEWS

Earthquake Information and Tsunami Warning System (EITWS). Figure 1 shows the EWMS concept which consists of following elements:

- A sophisticated Earthquake Monitoring System (SeisComP3 by GFZ Potsdam) collects real-time data from seismic sensors in the region and worldwide and is able to detect and locate earthquakes very quickly.

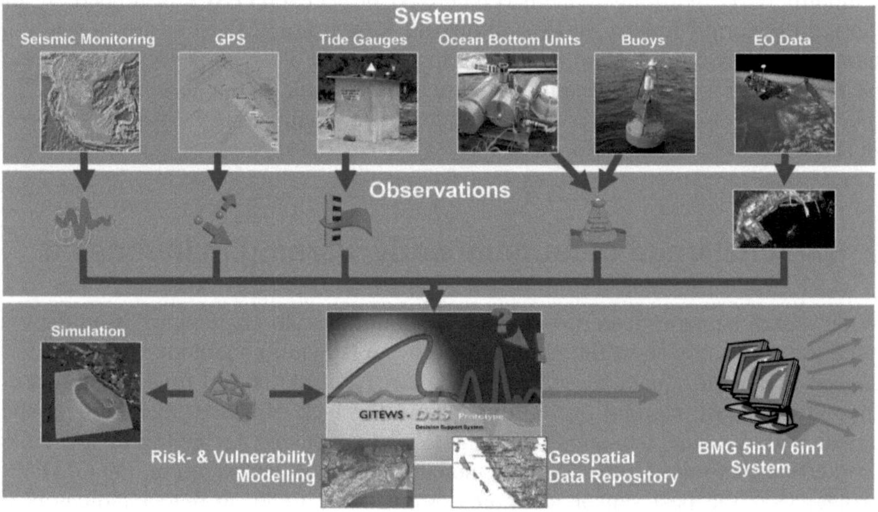

Fig. 1. The Early Warning and Mitigation System concept (*Top line* shows the incoming systems, *bottom line* their comparison/matching/fusion with data from risk and vulnerability modelling and the geospatial data repository before dissemination of a warning through the BMG 5in1/6in1 dissemination system. The intermediate layer is where observations are transmitted through the OGC SWE services to and from the DSS

- A continuous GPS System (CGPS) describes the seafloor deformation/rupture in (near) real-time based on very precise GPS measurements at smart land stations (stations equipped with GPS and other sensor technology).
- A Deep Ocean Observation System (DOOS) collects and processes sensor information transmitted from Ocean Bottom Units (OBUs, located on the seafloor underneath buoys) and Buoys equipped with tsunami-detecting instruments.
- A Tide Gauge System (TGS) collects and processes measurements of a network of tide gauges in order to detect sea level anomalies.
- An interface to future Earth Observation systems is provided.
- A central Tsunami Service Bus (TSB) collects information from the sensor systems and provides them to the DSS.
- A Simulation System (SIM) is able to perform a multi-sensor tsunami scenario selection, resulting in a list of best matching tsunami scenarios for a given set of observations.

- The Decision Support System (DSS) receives sensor observations via the TSB, requests a scenario selection from the SIM for the current set of sensor observations and communicates with the dissemination systems for message distribution and delivery.

4 The decision support system

As part of the Early Warning and Mitigation System (EWMS) the DSS provides processing, assessment, visualization, decision support, analysis, warning and management functions for the purpose of supporting disaster management related activities regarding tsunami threats for the region of Indonesia. This section will describe the resulting system requirements and solution approaches.

4.1 Status of decision support in tsunami early warning

Tsunami Early Warning Systems exist in a number of countries, e.g. the Pacific Tsunami Warning System PTWS by the PTWC (Pacific Tsunami Warning Centre near Honolulu/Hawaii) serving also the West Coast/Alaska Tsunami Warning Center (WC/ATWC) as well as the research centre in New Zealand of the National Institute of Water & Atmospheric Research (NIWA) or by the Japan Meteorological Agency (JMA).

These all differ from the system required for Indonesia, which will have to handle near-field tsunamis and enable the operators to assess the situation using a heavily aggregated situation picture with a high accuracy and certainty. The system used for WC/ATWC for instance can be assumed to be used for early warning for tsunamis travelling as long as several hours before they hit the coast. Therefore information aggregation can almost be done manually or at least there is a lot more time to consider the effects and consequences of a wave before taking the decision whether to warn or not. A decision maker/operator in the German-Indonesian GITEWS system needs to be able to send a warning or information message within a very short time span (5 min) after a potentially tsunamigenic event. Thus the whole system logic of when and whom to warn necessarily is different from the requirements with respect to decision support in the new tsunami early warning system.

4.2 Operational prerequisites

In principle, the spatial situation awareness analysis and early warning process does not require shoreline segmentation, except when limited computational resources require aggregation and priorization or when mapping products to recipients or administrational structures.

A so-called Warning Segment is a well-defined segment of the shoreline defined according to administrational boundaries and is used as smallest warnable unit for which tsunami threat information is aggregated and to which warning products may be addressed.

A coastline segmentation workflow has been developed by BMKG and DLR; the current definition of warning segments for the coastline of Indonesia along the Indian Ocean covers 125 warning segments for Sumatra, Java and Bali.

Warning segments can be set to specific states which are called warning levels in connection with the dissemination of warning products (e.g. warning messages). The warning levels depend on the expected or confirmed tsunami threat. Which warning level is assigned during the decision proposal generation process depends mainly on the height of wave at the coastline.

Table 1. Tsunami Warning levels

Tsunami Category	Warning Level	Wave Height (WH) Range [m]	Color
<none>	<none>	0.0 • WH < 0.1	Grey
Minor tsunami	Advisory	0.1 • WH < 0.5	Yellow
Tsunami	Warning	0.5 • WH < 3.0	Orange
Major Tsunami	Major Warning	3.0 • WH	Red

Wave heights of larger than 10 cm are considered to require a warning level of Advisory (yellow). Warning segments which reach wave heights from 0.5 m up to 3 m are assigned a Warning level (orange level). Warning segments with a wave height of 3 m or more are assigned the level Major Warning (red) (see Table 1).

4.3 Core DSS tasks

The decision process shall help the chief officer on duty (COOD) to become aware of a current situation, assess incoming information, exploit synergies of information fusion and analysis, assess impact and consequences and make informed decisions.

Unlike many other problems covered in the area of decision support, the situation evolves over a certain period of time, and the decision process itself must be time and space sensitive due to nature of the underlying physical phenomenon which may threaten widely dislocated places over a time period of several hours.

The core decision support loop consists of two major components:

- Situational Awareness
- Decide and Act

Situation awareness in turn comprises the steps perception (gather information), comprehension (judge information) and projection (effect estimation/projection).

In the perception step the DSS receives sensor input, including results from the simulation system. Following the sensor input will be processed and analyzed. In the comprehension step there is further analyzing of sensor input across sensor types. The projection step comprises the projection of the current situation into the future. An assessment of consequences takes place. These three steps result in an improvement of situation awareness. While situation awareness focuses on understanding the situation that evolves and its consequences, this knowledge needs to be transformed into decisions and actions. This is the focus of the second part of the core decision support loop:

- *decide* refers to the derivation of decision proposals from a given situation that the EWMS has become aware of.
- *act* refers to the implementation of the decisions that the COOD has made. Examples for such decisions are product dissemination or sensor (de-) activation.

The workflow is repeated each time new information is received by the DSS or a deadline has been reached. The workflow is terminated by the COOD if no tsunami threat exists anymore.

4.4 Additional sources of information

In addition to the collection of real-time sensor observations, the DSS can access a huge collection of a priori information and scenario data that helps interpreting the online input, assessing the tsunami threat and forecasting the consequences.

Using this approach, the information gap immanent to the first minutes of a potential tsunami is narrowed as much as possible.

The most important sources of information are:

- A geospatial data infrastructure which allows the standard-based access to large databases of geospatial baseline data, such as administrational boundaries, topographic and bathymetric data etc.
- Risk modelling and vulnerability assessment information which describe how high the tsunami risk at a particular location is and how vulnerable the people and infrastructure are. Information is also contained about the expected capability of people to respond to the event.
- The large number of tsunami scenarios contained in the Tsunami Scenario Repository (TSR) which is used by the SIM to perform the online multi-sensor-scenario selection process.

4.5 Graphical user interface (GUI)

The user interface and process workflows of the DSS have been designed for decision making under uncertainty and time pressure (Endsley 2003). Based on the large body of research literature on this topic and the results of an eye-tracking based study regarding a first DSS GUI version (FH Potsdam Interaction Design – Eyetracking Analysis), it is now available in an improved and optimized version. The GUI (see Fig. 2) consists of four displays (called perspectives) shown simultaneously to the decision maker (COOD) through which the operator can go in sequence or iteratively to gain support for the decision whether to send a warning or not. The DSS GUI was implemented as a set of plug-ins and extensions to the uDIG Open Source GIS client.

The Situation Perspective (Fig. 2, upper left) illustrates the overall situation including higher-level spatial and temporal views of all facts of interest (e.g. observations, simulation forecasts, sensor system states). For this purpose, a map view acts as spatiotemporal information display visualizing geospatial sensor data such as the event location, travel-time isochrones, estimated times of arrival (ETAs), thematic maps (e.g. borders, geologic realities), and sensor status information. A timeline view maps the incident data onto a temporal scale. All incoming sensor observations and simulation results that are relevant for the selected incident are displayed in detail in the Observation Perspective (Fig. 2, upper right). In addition, the user is provided with functionality to further explore single observations, e.g. to view parameters, time series, plots, etc.

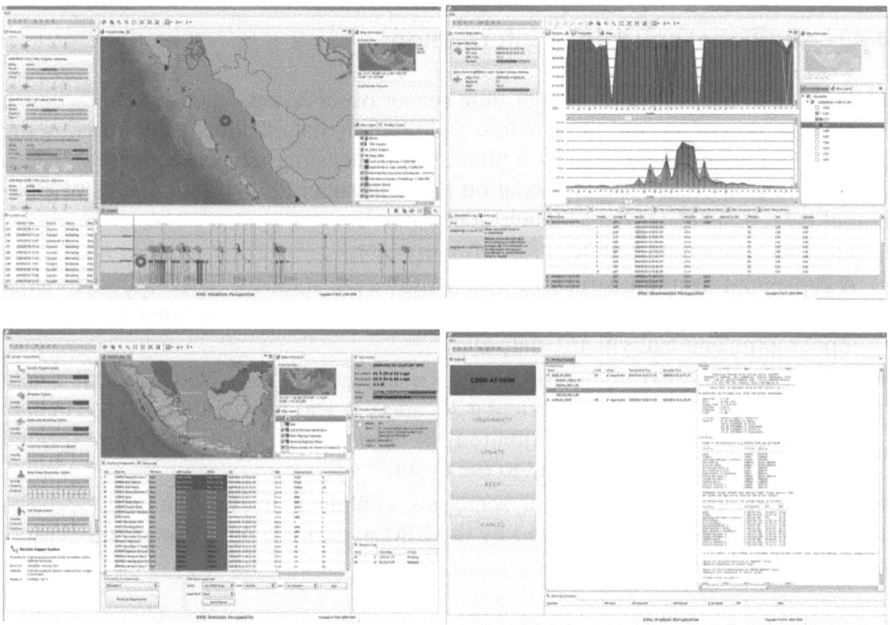

Fig. 2. DSS Graphical user interface (GUI) perspectives. Four separate screens which can be used sequentially and/or iteratively by the operator to assess an event from situation overview to dissemination of the warning message contain all the relevant observations and the assessment of the DSS in support of an ease of use in a fast decision process (From upper to left to lower right these are: situation perspective, observation perspective, decision perspective and product dissemination perspective).

The Decision Perspective (Fig. 2, lower left) contains all information that is necessary for the COOD decision-making process, including decision proposals and functionality for the configuration of warning products. This includes highly aggregated classification bars for the individual sensor systems and the SIM to support the COOD in assessing the situation. A colour code is used to represent the conclusions which may be drawn on the basis of the corresponding observations. Centred, a segment map and a list display the warning product proposals for each warning segment based on the simulation results. A similar colour code is used here to graduate the warning levels.

Additionally, indexes for risk-relevant information are shown for each of the affected warning segments: e.g. number of exposed people, number of critical facilities, and number of high loss facilities, response index and an overall risk factor.

The COOD may override the warning product proposals generated by the DSS. If the selected warning products should be sent, the button 'Ready for Dissemination' needs to be pressed. The actual execution/confirmation of actions is performed on a separate perspective where a product preview and a summary of the actions that are about to be triggered is shown (Fig. 2, lower right).

The COOD is required to confirm his choice ('Disseminate' button) in order to prevent unintended warning message dissemination; alternatively, the dissemination process can be cancelled ('Cancel' button).

4.6 Warning product generation and dissemination

Regarding the generation of warning and other products generated by the DSS, a template-based approach is applied. For all products and each of the formats and languages a product shall be provided, templates are prepared that contain fixed elements (e.g. text) and keywords. At the time of product generation, the keywords are replaced with up-to-date information of the tsunami threat (e.g. magnitude, depth and location of the earthquake, warning levels for warning segments).

The generation of products is a two-stage process:

- In stage 1 the required basic products are generated (e.g. text warning messages);
- In stage 2 these basic products are embedded into an additional message formatted according to the CAP standard (OASIS 2005).

Once the required products (warning messages) have been generated, they are transmitted to the dissemination systems which are connected to the DSS. Currently two dissemination systems are connected to the DSS: the BMKG dissemination system/5-in-1 system and the 2wcom (2wCOM) FM-RDS (Frequency Modulated – Radio Data System) based dissemination system. The DSS sends the appropriate selection of products to the individual dissemination systems and initiates thereby the dissemination process which itself is outside the scope of the DSS.

5 Standards and interoperability

A lot of effort is put into compliance to interoperability standards defined by the Open Geospatial Consortium (OGC) for geospatial and sensor data. In particular the OGC initiative 'Sensor Web Enablement' (SWE) (Botts et al. 2006) that aims at defining standards to enable the discovery, exchange, and processing of sensor observations, as well as the tasking of sensor systems as applied in the context of GITEWS.

Within the context of GITEWS SWE will be used as a basis for integrating the various external sensor systems with the EWMS and to offer sensor observation data to DSS components for further processing. This especially makes sense when additional sensors become available in the future, e.g. new tide gauge or buoy systems or even remote sensing or airborne sensor systems. By adhering to these standards GITEWS will remain open and extensible in the future. For accessing geospatial information by the DSS GUI the established OGC standards Web Mapping Service (WMS) and Web Feature Service (WFS) are also used.

5.1 OGC Sensor Web Enablement (SWE)

For realizing the DSS Sensor Data Center (SDC) presented here the SWE architecture specified by the OGC is an important foundation. Thus this section will shortly introduce the basics of the OGC SWE framework and will provide an introduction to those SWE components which were used for building the sensor data center.

The activities of the OGC, an international consortium consisting of more than 350 members, focus on the development of open standards that form the basis of geospatial data infrastructures. Within these activities the SWE initiative deals with the integration of sensors and sensor data. In order to fully integrate sensors and their data into a geospatial data infrastructure the SWE framework provides a broad range of functionalities. This includes the discovery of sensors and sensor data, the description of sensor metadata, access to sensor data (real time and historic), tasking of sensors and alerting mechanisms based on sensor measurements. For fulfilling this set of requirements a framework of standards has been developed. It comprises two aspects: the information model dealing with data formats and the service model describing service interfaces. The information model consists of the following standards:

- Sensor Model Language (SensorML) (Botts and Robin 2007): Metadata format for sensors and sensor systems
- Observations and Measurements (O&M) (Cox 2007): Data format for observations and measurements performed by sensors
- Transducer Markup Language (TML) (Havens 2007): Data format, optimized for data streaming that allows encoding sensor data and metadata.

The SWE service model provides standardized interfaces for accessing sensor data, alerting and controlling sensors:

- Sensor Observation Service (SOS) (Na and Priest 2007): Pull based access to sensor data and metadata
- Sensor Alert Service (SAS): Alerting based on sensor data and user defined criteria (for example if a measurement value exceeds a certain threshold)
- Sensor Planning Service (SPS) (Simonis 2007): Controlling sensors and their parameters
- Web Notification Service (WNS): Asynchronous communication between web services (e.g. a SAS or SPS) and/or clients.

For the DSS sensor data centre two of the above mentioned standards are of special importance: the O&M and SOS standards. Thus these two specifications will be introduced in more detail in the next two paragraphs.

The O&M specification defines a standardized format for observation and measurement results. It is based on the OGC Geography Markup Language (GML) as it specifies a GML application schema. The basic concept within O&M is the observation which is defined as an event that occurs at a certain point in time and which generates a value for an observed phenomenon. Besides time and value of measurements, O&M allows encoding further measurement properties like information about processes used for obtaining measurements or the location and quality of observations. Another important concept of O&M is the binding of measurements to features of interest (FOI). These FOIs are used for describing the objects or the measurement locations (= features) at which the measurement was performed.

Whereas O&M describes the data format for sensor data, the SOS specification standardizes an interface for accessing data gathered by sensors and sensor metadata. Thus the SOS relies on O&M as a response format for returning measurement data (for sensor metadata SensorML is mostly used). The SOS specification is divided into three profiles: core profile, transactional profile and enhanced profile. Whereas the operations of the core profile comprise the mandatory functionality the operations of the other profiles are optional. The core profile provides the essential functionality of a SOS: access to a description of the service and the available data (GetCapabilities), access to measurement data (GetObservation) and retrieval of sensor metadata (DescribeSensor). For inserting data into an SOS instance the transactional profile provides the necessary operations: registering sensors (RegisterSensor) and inserting sensor data (InsertObservation). Finally, the enhanced profile offers additional optional functionalities like retrieving FOIs or information about time-dependent data availability.

5.2 DSS Sensor Data Center (SDC)

The Sensor Data Center (SDC) is the core component of the DSS for ingestion of and provision of access to sensor data. The SDC is part of the DSS Data Management Center (DMC) which is responsible for managing all data relevant to the DSS operations (e.g. Crisis and Risk Products, Geospatial Data) and includes software components for ingestion and archive tasks. All incoming sensor observation data passes an ingestion process during which the data is validated, transformed into the DSS-internal O&M data model and forwarded to the SDC. It provides mechanisms to store and access observation data as well as metadata about the used sensor systems. By implementing the open sensor web standards and models such as the SOS, O&M and SensorML specifications, it seamlessly integrates into the existing spatial data infrastructure (SDI).

The software components Feeder and SOS used in the SDC are further developments and adaptations of 52° North open-source SWE implementations.

5.2.1 Ingestion of observation data and sensor metadata into the SDC

In situ observation data measured by the different sensor systems is sent to the SDC. After an ingestion component has verified and transformed the data it is forwarded to the Feeder component. This component acts as a Java Servlet and receives data transmitted through HTTP-POST. After validating the incoming data, the feeder determines the observation type. Depending on that type data is parsed and stored in a designated PostgreSQL database. All corresponding data, i.e. feature of interest (FOI), is being created and associated with the observation automatically. For each result element of such an observation a unique id will be generated that is returned to the plugin together with the corresponding pickup point for later service access to this specific observation.

The ingestion of SensorML data works analogously, but stores incoming SensorML files directly in a certain directory in the file system and not in a database. In both cases the SOS is being notified that new data is available by the Feeder components to update its service metadata.

5.2.2 Providing observation data and sensor metadata

Provision of observation data and sensor metadata is realized by a SOS. The service supports the implementation specification 1.0 (OGC 06-009r6) as defined by the OGC. Observation data is encoded using the O&M standard 1.0 (OGC 07-022r1). The SOS provides the operations defined in the core profile such as DescribeSensor and GetObservation. Furthermore it offers the GetFeatureOfInterest and GetObservationById operation as defined by the enhanced profile. The implementation is based on the data access object (DAO) pattern (SUN) and supports an implementation for a PostgreSQL database. Since the SOS runs as a Java Servlet, too, it is capable of receiving HTTP-GET and -POST requests. Incoming requests are analyzed and translated into SQL queries. After execution of these queries the response is encoded using the Geography Markup Language (GML) for feature requests or O&M for observation requests. SensorML documents are returned directly as stored in the file system and associated with the requested procedure. The SOS interface offers the possibility to query observation data by id or allows applying certain filters such as temporal or spatial constraints. Due to performance issues there is one SOS instance for each observation type.

5.2.3 Access to observation data

One of the main advantages of the OGC standards applied here is having a unified access layer for the retrieval of spatial data which can be reused by server-side components as well as the GUI. Access to spatial data is realized in GITEWS by using uDig [10] GIS (Geographical Information System). A plugin based on the 52° North OWS (OGC Web Service) Access Framework (OX-Framework) for uDig allows retrieving observation data from a SOS. It provides mechanisms for creating SOS requests and parsing the O&M responses of the service. The encapsulated data is transformed into the uDIG data model and as such are accessible for further actions.

Besides in situ data of the sensor systems the SDC is also able to store and provide simulation results. These contain a huge collection of a priori information with best matching tsunami scenarios for a defined set of input parameters.

5.3 Common Alerting Protocol (CAP)

The warning messages generated by the DSS are provided in the Common Alerting Protocol (CAP) format, an open standard for disaster management message exchange (Incident 2008, OASIS 2005). CAP defines a standard for alerts and notifications in the public warning domain independent of the hazard type and the technology used for transmitting warning messages. The CAP XML structure allows the inclusion of event data, text, images and geospatial references. The Federal Emergency Management Agency (FEMA) of the US Department of Homeland Security has, among others, explained their commitment to use a profile of the CAP protocol in their Integrated Public Alert and Warning System (IPAWS, FEMA 2008)

The DSS Dissemination Component offers services for creating, updating and disseminating Warning Products. For those dissemination systems connected to the DSS which are able to parse CAP the message (which includes related geospatial references) is encoded and transmitted in CAP XML.

6 Conclusions and outlook

As part of the German contribution to InaTEWS and embedded in an open and modular 'system of systems' approach capable of integrating additional sensor and tsunami scenario sources, the Decision Support System DSS presents a novel approach to support the tsunami early warning process. The decision maker is able to assess the situation and take decisions in a manner and based on information quantity and quality not possible before. He is supported through the entire process of information acquisition through information fusion to decision support and dissemination of the warning. Since the incoming information for the DSS is continuously updated, triggering an update of the decision proposal, the decision maker is supported through the entire lifetime of an incident. The extension paths of the DSS are numerous, reaching from additional sensor and data sources to international coverage and functional extensions. The use of OGC standards for implementing the DSS Data Management Center has thus proven beneficial for comfortable data access and extensibility.

Acknowledgements

GITEWS is being developed by a number of scientists and engineers of the Helmholtz Centre Potsdam GFZ German Research Centre for Geosciences (GFZ, Project Lead), the Alfred Wegener Institute for Polar and Marine Research (AWI), the German Aerospace Center (DLR), the GKSS Research Centre (GKSS), the German Marine Research Consortium (KDM), the Leibniz Institute of Marine Sciences (IFM-GEOMAR), the United Nations University Institute for Environment and Human Security (UNU), the Gesellschaft für Technische Zusammenarbeit (GTZ) and the Federal Institute for Geosciences and Natural Resources (BGR) as well as Indonesian and international partners. The project is funded by the German Federal Ministry for Education and Research (BMBF), Grant 03TSU01.

References

52north: 52north.org, http://52north.org/

Botts M, Percivall G, Reed C, Davidson J (2006) 'OGC® Sensor Web Enablement: Overview and High Level Architecture'.

Botts M, Robin A (eds.) OpenGIS Sensor Model Language (SensorML) Implementation Specification Standard 1.0. Online resource at http://www.opengeospatial.org/standards/sensorml OpenGIS document 07-000 SensorML http://portal.opengeospatial.org/files/?artifact_id=21273

Cox Simon 'OGC® Observations and Measurements Abstract Specification'. 2006. Online resource at http://www.opengeospatial.org/standards/om 'OpenGIS document Observations and Measurements 07-022r1 – Part 1- Observation schema' http://portal.opengeospatial.org/files/?artifact_id=22466/; Online resource at http://www.opengeospatial.org/standards/om 'OpenGIS document Observations and Measurements 07-002r3 – Part 2- Sampling Features (1.0) ' http://portal.opengeospatial.org/files/?artifact_id=22467

Endsley MR, Bolté B, Jones DG (2003) 'Designing for Situation Awareness'. Taylor and Francis, Boca Raton.

FEMA Integrated Public Alert and Warning System (IPAWS) 2008 http://www.fema.gov/emergency/ipaws/

FH Potsdam Interaction Design – Eyetracking Analysis: Usability Research Study of the German-Indonesian Tsunami Early Warning System (GITEWS) http://design.fh-potsdam.de/fachbereich/transfer/kooperationen.html

FM-RDS 2wCOM: Online resource at http://www.2wom.com

GDI-DE: Online resource at www.gdi-de.org

Havens S OpenGIS Transducer Markup Language (TML) Implementation Specification Standard 1.0. Online resource at http://www.opengeospatial.org/standards/tml OpenGIS document 06-010r6: http://portal.opengeospatial.org/files/?artifact_id=19371

Incident (2008) 'The CAP Cookbook'. Online resource at http://www.incident.com/cook-book/index.php/Creating_CAP_Applications.

INSPIRE: Online resource at http://inspire.jrc.ec.europa.eu/

Na A, Priest M 'OGC® Sensor Observation Service Implementation Specification' 2007 Sensor Observation Service 1.0. Online resource at http://www.opengeospatial.org/standards/sos OpenGIS document 06-009r6: http://portal.opengeospatial.org/files/?artifact_id=26667

NaDiNe Natural Disasters Networking Platform Online resource at http://nadine.helmholtz-eos.de/nadine_en.html

OASIS (2005) 'Common Alerting Protocol, v.1.1'. Oasis Standard CAP-V1.1, October 2005. Online Resource http://www.oasis-open.org/committees/download.php/15135/emergency-CAPv1.1-Corrected_DOM.pdf

Simonis I Sensor Planning Service 1.0. Online resource at http://www.opengeospatial.org/standards/sps OpenGIS document 07-014r3 : http://portal.opengeospatial.org/files/?artifact_id=23180

SUN 'Core J2EE Patterns – Data Access Object'. Online resource available at http://java.sun.com/blueprints/corej2eepatterns/Patterns/DataAccessObject.html

uDig GIS: http://udig.refractions.net/

A Service-Based Architecture for the Interoperability of Heterogeneous Sensor data: A Case Study on Early Warning

Valentina Casola, Luca D'Onofrio, Giusy Di Lorenzo and Nicola Mazzocca

Abstract

One of the main open issues in the development of applications for sensor network management is the definition of interoperability mechanisms among the several monitoring systems and heterogeneous data. Interesting researches related to integration techniques have taken place; they are primary based on the adoption of sharing data mechanisms; furthermore in the last years, the service-oriented architecture (SOA) approach has become predominant in many sensor networks projects as it enables the cooperation and interoperability of different sensor platforms at a higher abstraction level.

In this chapter we propose an architecture for the interoperability of sensor networks; it is based on web services technologies and on the definition of a common data model enriched with semantic concepts and annotations.

The proposed architecture allows the integration of heterogeneous data and the implementation of Web Services to let data (raw or aggregated) be available to authorized end-users. Finally, we will present an early warning service with the definition of a common data model for a specific class of sensors.

1 Introduction

One of the main open issues in the monitoring and management of environmental risks is the integration of heterogeneous data from different sources.

The wide diffusion of sensor systems, together with their numerous applications, has led to a huge heterogeneity that makes it difficult to interface and collect data

M. Konecny et al. (eds.), *Geographic Information and Cartography for Risk and Crisis Management*, Lecture Notes in Geoinformation and Cartography,
DOI 10.1007/978-3-642-03442-8_17,

from these systems. As for Wireless Sensor Networks, ad hoc programming languages (i.e. nesC, (Gay 2003)) have been developed to support base station programming and to express the application processing in terms of messages exchange among near nodes. Alternatively, different middlewares based on macro-programming models have been proposed in order to bridge the gap between the application and the underlying hardware layers. Middlewares for WSNs provide a system abstraction so that the application programmer can focus on the application logic without having to deal with the lower level implementation details (Hadim 2006). They are commonly used when a single application operates over a single WSN, while the application development for multiple WSNs is a rather cumbersome work.

Nowadays the interaction with multiple sensor systems is required by a lot of applications; monitoring applications of wide geographical areas have highlighted the research problem of the integrated management and correlation of data coming from various networks that cooperate for a common objective.

Specific integration frameworks for accessing different data sources are needed; they should hide the heterogeneity between different sensor systems in terms of sensor, networking or middleware technologies and provide a standard way to access them.

Interesting researches related to integration techniques for heterogeneous sensor networks have taken place, but nowadays only few architectures have been proposed. Most of them try to define a common exchange mechanism among different sensor systems in order to facilitate the integration and provide a software integration layer which allows different sensor systems to collaborate for the same purpose. These solutions are often tightly related to the referenced technologies and lack of a real implementation. In the last years, the service-oriented architecture (SOA) approach has become a cornerstone in many sensor networks projects. SOA-approach enables the cooperation and interoperability of different sensor platforms as it provides support for discovering, accessing and sharing services, data and computational and communication resources by the adoption of open standards.

The new open standards as well as the adoption of a formalized approach towards knowledge definition, formalization and sharing is one of the main features of such architectures and technologies that strongly support the definition of cooperative environments. As an example, the OpenGeospatial Consortium's Sensor Model Language (SensorML) (OpenGIS 2007) standard provides an XML schema for defining the geometric, dynamic and observational characteristics of sensors. The purpose of SensorML is to (1) provide general sensor information in support of data discovery, (2) support the processing and analysis of the sensor measurements, (3) support the geolocation of the measured data, (4) provide performance characteristics (e.g. accuracy, threshold, etc.), and (5) archive fundamental properties and assumptions regarding sensor.

In conclusion, thanks to technological improvements in computer architectures, network infrastructures and sensor technology, it is now possible to build a knowledge-based sensing architecture which can acquire huge data sets from sensors on a wide geographical scale, process terabyte of data on computing resources belonging to various administrative and political domains, store computation results and historical data in secure and accessible distributed databases. Within this context, the Campanian Region (south Italy) is holding innovative environmental sensor networks to gather real-time and near real-time information on natural (i.e. seismic, volcanic, slope, stability, pluviometric, floods and meteo-marine) and civil/industrial structural/infrastructural vulnerability. Campanian region is a densely populated area and a natural case study of the multi-hazard problem: for these reasons, the development of an early warning system as well as infrastructures devoted to risk management and to the realization of a decision support system become an interesting challenge. It may allow to process and provide in real-time (pre-event warning) or in a near real-time manner (within few minutes after the emergency) all information about the risk of damages. A risk management system may have a twofold target of monitoring high environmental risk zones and intervening as soon as possible with preventing operations such as sending alarm signals to emergency services (i.e. civil protection) and taking safety measures (i.e. control of emergency power generators).

In this chapter, a comprehensive architecture is proposed to provide pre- and post-event warning information by integrating the above-listed networks, and defining a common data model. To grant interoperability among multi-regional systems, a formal data model is presented, it specifies the data relations, terminology and meanings. It has been implemented through a set of ontologies and formally described in OWL (Web Ontology Language). Furthermore, to face interoperability issues, we have exploited Service Oriented Architectures (SOA).

As a result, we present an interoperable architecture for risk management that is completely based on services; it allows the integration of heterogeneous data and the implementation of applicative standard Web Services to let data (raw or aggregated) be available to authorized end-users and/or applications. Moreover, we describe an early warning service, based on the definition of a common data model for a specific class of sensors.

The remainder of the chapter is structured as follows: in Sect. 2 some related works are reported to assess the state of the art of many methodologies and technologies involved in facing interoperability issues among heterogeneous data. In Sect. 3 we will present the architectural model of an early warning system by illustrating its main layers and their functionalities. In Sect. 4 we illustrate the early warning system and the data model enabling the interoperability. Finally in Sect. 5 some concluding remarks are given.

1.1 Related works

As already said, one of the main open issues in the monitoring and management of environmental risks is the integration of heterogeneous data from different sources. To reach this goal, there are many obstacles carried out by the need of guaranteeing interoperability among several monitoring systems. In the literature (EU framework 2003), interoperability issues have been categorized into three different dimensions: organizational, technical and semantic interoperability.

Some solutions to specific, but not complete, aspects of interoperability are available in the literature. The OGC consortium established Sensor Web Enablement (SWE) to model sensor characteristics and services.

In particular, the suite includes different standards, among them (i) Sensor Model Language (Sensor ML), (ii) Observation \& Measurement and (iii) Sensor Observation Service, (iv) Sensor planning service and (v) Service alert service. They allow to model sensors and sensor observations, data retrieval mechanism and web services (for access of the sensor data via web); it is possible to specify information as coordinates and timestamps, but they do not allow to state the semantics of the data and the meaning of sensor observations, making difficult the interoperability, the evaluation of the phenomena and the detection of situation awareness.

A first attempt of semantics definition for sensor Web has been proposed in the SSW framework (Henson 2008) in which enhanced meaning for sensor observations is given by adding semantic annotations to existing standard sensor languages of the SWE, in order to increase interoperability and provide contextual information for situation awareness.

Several efforts have been done in the data modelling field too: different ontologies for heterogeneous sensor data representation are proposed. An ontology presented in (Ceruti 2004) models different concepts, as platforms and sensors, such characteristics tangible and intangible, and relationships and concepts such as data combinations. Another ontology for the sensor networks, presented in (Eid 2006), provides semantic representation of sensor networks data, aiming at interpreting unambiguous structured information.

From the technical interoperability point of view, the current leading approach is based on service oriented architectures (SOA). In (Kobialka 2007) the authors introduce Web Service Resource Framework (WSRF) mechanisms into the core services implementation of the NICTA Open Sensor Web Architecture (NOSA). WSRF enables to handle simultaneous observational queries to heterogeneous Sensor Networks. Moreover, the GeoICT group at York University have built an OGC SWE compliant Sensor Web infrastructure; they have developed a Sensor Web client capable of visualizing geospatial data, and a set of stateless Web Services called GeoSWIFT (Liang 2005).

Furthermore, some European projects based on service-oriented architectures (SOA) have been proposed, among the others: WIN (Wide Information Network) (Alegre 2004) (Alegre 2005), ORCHESTRA (Chu 2007) and SANY (Sensors Anywhere) (Schimak 2009) projects.

WIN aims at developing an open and flexible platform to support multi-hazard and risk domains at European level, integrating national and sub-national (regional) data flows in the frame of web service architecture. It proposes a set of generic services and standard data modeling components that facilitate the deployment of several cases. The WIN metadata model is based on existing standards (such as Dublin Core, GML and ISO19115) and provides some additional specifications for WIN.

ORCHESTRA adapts the ISO/IEC 10746 Reference Model for Open Distributed Processing to service-oriented architectures. It implements geospatial SOA by adapting the Reference Model for geospatial service networks on a process model, compliant with the ISO standard RM-ODP. The ORCHESTRA architecture uses w3C Web services platform and the Geography Mark-up Language (GML) to implement web services.

Finally, SANY is the most recent approved project, it focuses on interoperability of in situ sensors and sensor networks. SANY architecture promises to provide a quick and cost-efficient way for reuse of data and services from currently incompatible sensor- and data-sources.

All these projects underline the importance of adopting web services for the interoperability of sensor networks. In addition to available standards and platforms, we have defined a common data model that is particularly suitable for the analysis of natural hazards (with pluviometric, flood, marine and seismic sensors), furthermore it is extensible and enriched with semantic annotations.

2 An architecture model for early warning

To design an open system to manage heterogeneous sensor data sources, we have defined an architectural model that is based on Web Services. The model, named Web Services Early Warning (WS-EW) offers several services and functionalities; these can be classified into two main categories: (a) functions to elaborate data from heterogeneous sources and store; (b) functions to manage all data sources (sensors, databases, simulators, etc.).

We adopted a service-oriented architecture (SOA) paradigm that introduces the concept of service as fundamental element for developing distributed applications in a network based heterogeneous environment and a layered service-oriented Architecture is adopted (Booth 2004).

The development of web technologies and standards such as HTTP, XML, SOAP, WSDL and UDDI enables pervasive adoption and deployment of web services to reach interoperability.

In the design of complex data-centric systems, the decomposition of the whole system in different layers is the most used approach. Each of these layers offers different services and is characterized by a deep functional specialization. Each layer is able to communicate with the others via standard WSDL interfaces.

The proposed architecture is structured in layered functional services and there is also a set of transversal non-functional services as illustrated in Fig 1.

The layered approach in the WS-EW facilitates the distribution of the applicative responsibilities on more objects. It implies that the most complex functionalities are built upon the collaboration of more elements of smaller complexities.

We have located three horizontal layers: Data Collection Service Layer, Integration Service Layer, Application Service Layer, and some vertical layers whose services are transversal to the other layers, as Security, Management and Interoperability services.

Such approach concurs to build a more robust, scalable and maintainable architecture, with a better logical structure.

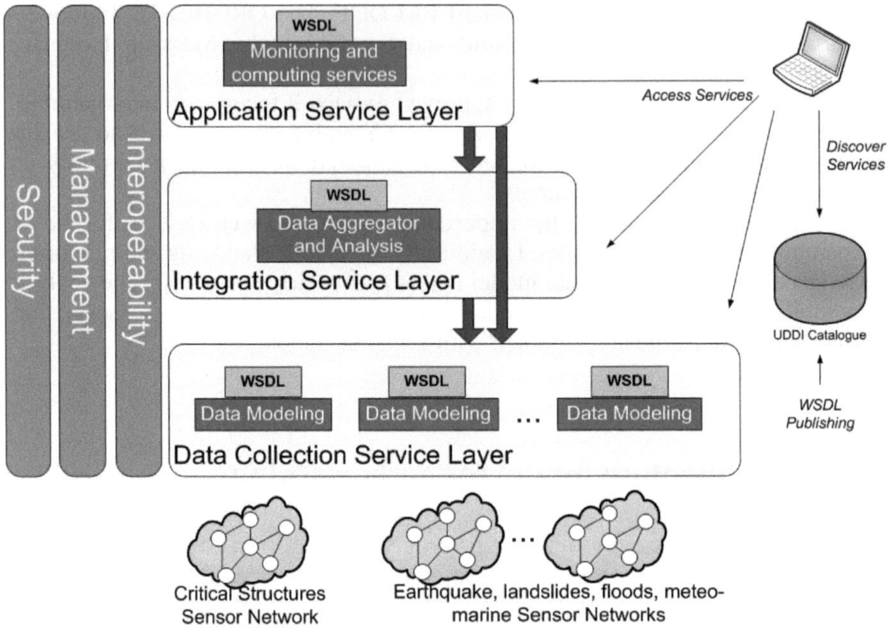

Fig. 1. An overall view of the WS-EW architecture

The layered architecture is a good solution also from a security point of view as each service can have a specific security policy to control access and preserve data privacy.

In the following we provide a brief description of layers and their specific services:

Data Collection Service Layer: in order to map the raw sensor readings onto physical reality, and to provide to the upper layer an homogeneous view of the networks, a model of that reality is required to complement the readings. The proposed approach enriches sensor querying with ontological modelling techniques, by associating an XML/RDF data model to raw sensor data sets. The main goal of this layer is to translate proprietary data format in a common data model that gives a homogeneous and standard view of the networks, sensors and observations and can be accessed by other services.

Integration Service Layer: while the Data Collection Service Layer is focused on interactions with a single resource, this layer contains services that capture interactions across collections of networks. For this reason, we refer to this layer of the architecture as the Integration layer.

The services of this layer provide aggregated data sets and numerical analysis values to the application level. It clusterizes network data sets according to the data model defined.

Application Service Layer: different services monitor and elaborate complex data structures from heterogeneous sensor network. Different kinds of applications exist in this context; for example, applications to implement warning threshold models or real-time event notification, applications to simulate events, applications to format sensor data according to specified standards and so on. Applications are built by invoking and composing services defined in other layers. Each service has its own security policy and it is published in a public registry.

In Fig. 1, each block represents an architectural macro-component offering/using available services; every arrow represents information flows (e.g. control and sensor data) associated with components (i.e. services) interaction. Further details on components' internal functional blocks as well as interfaces to access services pertain to the specific application service and to lower level services. The services in Fig. 1 are provided with a WSDL interface, they have been conceived to allow a standard interaction among modules. Furthermore, the interfaces are published in a UDDI catalogue. In this way any user can discover the services it needs and access them according to the access rights it has. Thanks to this protocol and standard adopted in our architecture, WS-EW is able to grant a high level of interoperability and reveals a high degree of scalability even considering different kinds of sensor network.

Private and sensible data has to be managed only by authorized users. The **Security services** enforce WS-Security standards to grant access (authentication and authorization) only to those authorized users. These services also deal with confidentiality and data integrity of the messages exchanged, non-repudiation of requests of messages and resilience to denial-of-service attacks.

Other management services are offered by the **Management and Interoperability services**.

3 An early warning system

In this section we illustrate how WS-ES architecture works in the practice by illustrating an early warning system related to earthquake damages prevention. In Fig. 2 we have reported a typical multi-risk application scenario where different sensor networks are located in seismic areas and on critical buildings to monitor seismic activities and building deformation information. For each network, a Data Collection Service is associated, it acquires data and formalizes them according to the common data model; the data are so available to the Early Warning application service for further elaboration.

The following paragraphs deal with the logical description of the application scenario and the data modeling methodologies adopted.

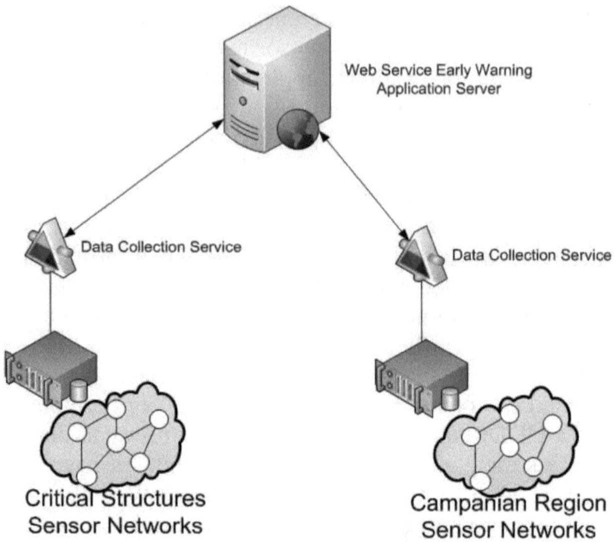

Fig. 2. Campanian Region application scenario

3.1 Application service layer for the Early Warning System

An early warning service should be able to

- Monitor sensor networks placed on critical buildings for data acquisition: this activity is a soft real-time phase, in which the attention focuses on sensor best management and access rather than on elaborations or transmitting time;
- Compute the environmental data for the localization of all buildings under risk;
- Notify alarms in case of risk.

The service is made of three main components as illustrated in Fig. 3.

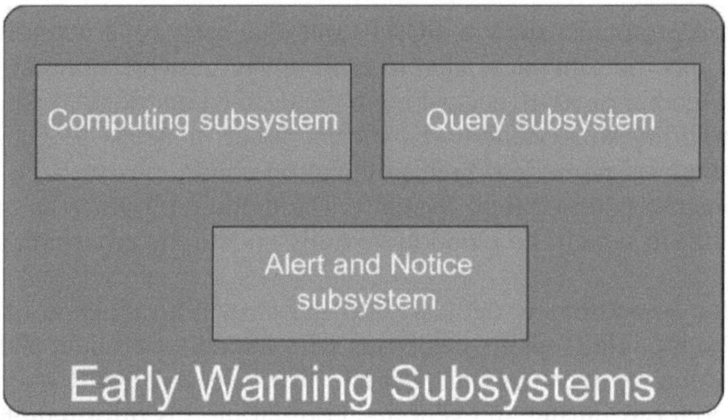

Fig. 3. WS-EW components

The *Computing Subsystem* is responsible to locate and characterize the seismic environmental hazards (F1). It is also responsible to evaluate the maximum distance (F2) where a significant damage may happen to monitored structures such as hospitals, bridges). It deals with the elaboration of sophisticated deterministic and stochastic earthquake models, computing

$$F2 = f(F1).$$

The elaboration of F2 also requires simulated data as illustrated in the architecture of Fig. 4.

A *Query Subsystem* retrieves and stores data from the building sensor networks. It also retrieves data from historical and simulator databases to evaluate the buildings structural deformation (D).

$$D = f(F1, F2, Sensors, HistoricalData, Simulations).$$

Finally, the *Alert and Notification Subsystem* elaborates the information from the previous subsystems and sends alarms to all the structures under risk for the revealed hazard and *within the distance* (F3) of propagation of the environmental effects:

$$F3 = f(F2, D).$$

According to the previous considerations, the EW-WS can be made up by composing different services provided by the proposed model: a data modelling and access service, a computing and simulating service, an alert and notification service. As illustrated in Fig. 4, all services communicate each other's trough Web Services interfaces. The mechanism of service calling and data interchanging is the same as a typical service oriented architecture: services are completely described by their WSDL interface. This interface describes, among other things, the methods to access data and the location of the service. Data interchange is made upon standard protocols, such as SOAP, and data sets are structured in XML, according to a specific schema related to each sensor network domain. Each data set is wrapped together in RDF, according to the defined ontological data model.

The Computing subsystem is the most critical: it needs to guarantee system availability both on numerical precision and on continuous working even in critical conditions (Mission-Critical System). The module provides an updating phase (non-real-time) in which the elaborative units update the environmental historical data.

The Query subsystem is dedicated to the structural analysis of one or more buildings; the module feeds its databases with the collected data and establishes a relation between an event and the generic structure involved by reporting all the structural variations during an event and correlating this information with the historical data previously archived or with simulated data sets.

The Alert and Notification subsystem selects from the structures on risk the ones that are located within the range of the hazard (like an earthquake); then it sends alarm signals to the interested buildings and lets people take the best safety reactions.

We explicitly note that the Computing subsystem is directly connected to the sensors related to the hazard (i.e. a seismic network) while the other subsystems retrieve data even when a hazard is not occurring to evaluate a possible value of the deformation and feed the databases with the historical data (this is not a real-time data). The early warning system does not work by evaluating online if a building is dangerous but pre-evaluating if a hazard of a certain magnitude can be dangerous for a specific building within the range of the just sensed hazard.

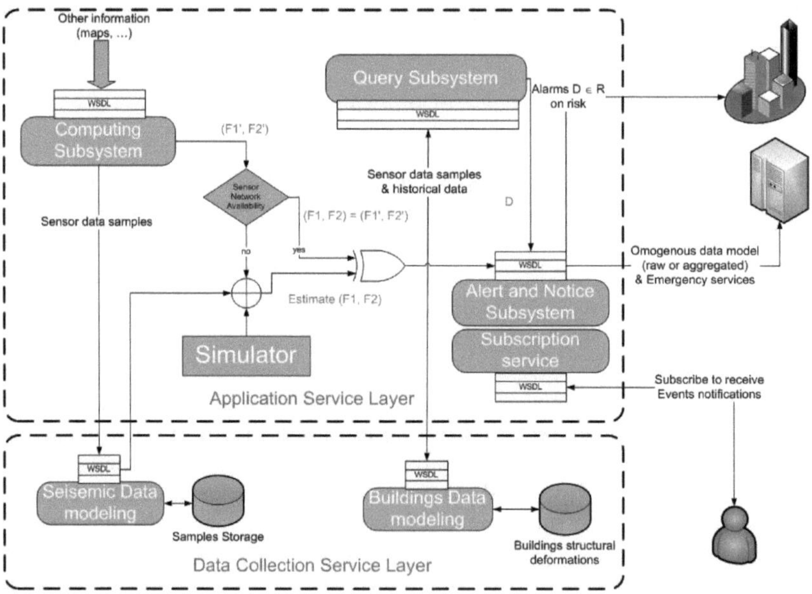

Fig. 4. WS-EW seismic application: logical model and data flows

3.2 Data Collection Service Layer: a Common Data Model

The Data Collection Service Layer aims at supplying the needed mechanisms for the interoperability among heterogeneous sensor data.

When integrating data from heterogeneous sources, the basic problem to be dealt comes from syntactic, schematic and semantics diversities of the schema. Syntactic heterogeneity refers to differences among paradigms used to format the data such as Relational DB, XML or RDF documents. Schematic integration refers to different aggregation or generalization hierarchy defined for the same real words, facts and finally the semantic integration regarding disagreement on the meaning, interpretation or intended use data (Noy 2004), (Hakimpour 2001).

Moreover in a sensor network context spatial, temporal and thematic information are essential for discovering and analyzing sensor data. Spatial metadata provides information regarding the sensor location and data in terms of either a geographical reference system or local reference. Temporal metadata provides information regarding the time instant or interval when the sensor data is captured. Finally, thematic metadata describe a real-world state from the sensor observation, such as object and events (Henson 2008).

In this section we describe the proposed Data Collection Service Layer, which allows either to resolve syntactic differences converting the proprietary data format into XML documents and then the semantic heterogeneity defining a common data model formally described using a unified ontology. Moreover, additional ontologies have been defined for modeling the sensors, sensor data and sensor observations.

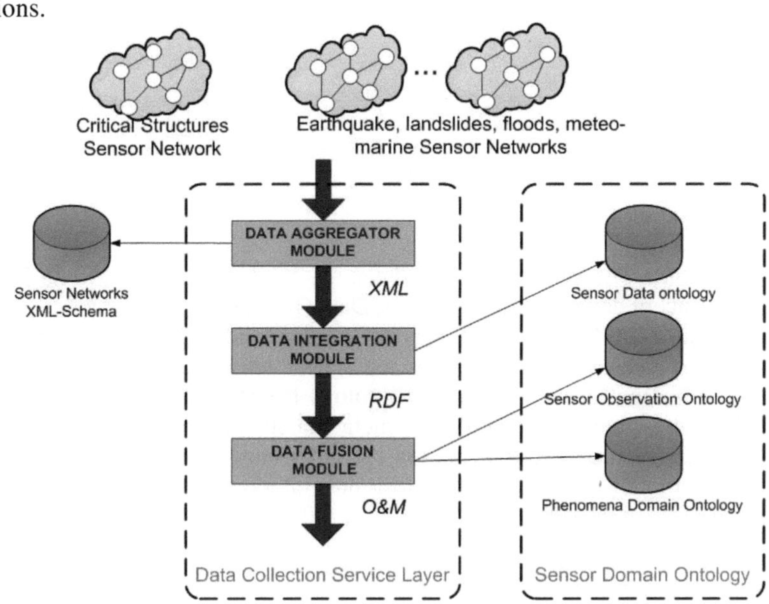

Fig. 5. Data Collection Layer

The Data Collection Service Layer provides the resources to access heterogeneous sensor network data and to translate proprietary data format into the defined common data model. To achieve this goal, three main subsystems are needed: (1) *Data Aggregation Module*, (2) *Data Integration Module* and (3) *Data Fusion Module* (see Fig. 5).

The **Data Aggregation Module** acquires raw sensor data from heterogeneous sensor network middleware and encodes them into an XML document. The mapping is done using the XML-schema defined for sensor networks and stored into the Sensor Network XML-Schema Repository. As an example in Fig. 6 the original format of the Pluviometric sensor data (right side) and the relative XML-schema (left side) are shown. The raw data are stored in a text file, which collects two days of data.

The data file, depicted into the Fig. (6a), contains the following fields: day (dd), month (mm), year (yy), hour (hh), minutes (mm) and sensor identification number. The sensor identification number column specifies the accumulated precipitation in 10 minutes.

dd	mm	yy	hh	mm	12255
3	10	8	0	20	0.0
3	10	8	0	30	0.0
3	10	8	0	40	0.0

a)

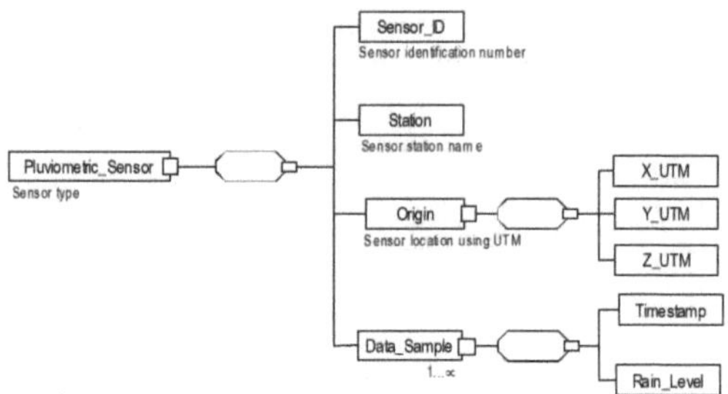

b)

Fig. 6. Pluviometric Data

The *Data Integration Module* translates the XML document into an RDF one using the concept defined into the Sensor Data Ontology.

The Sensor Data Ontology includes knowledge models for the sensors and sensor data. The Sensor Class describes the sensor attributes, capabilities and also spatial information. A snapshot of the ontology is depicted in Fig. 7. In particular, the model for a given sensor contains metadata such as Location, SensorStation, Measurement, Accuracy and Type. On the other hand, the Data class defines the formal data model needed for modeling and integrating sensor data. A Data is composed of a set of Measures, is stored in a Proprietary Format and is characterized by a set of Parameters.

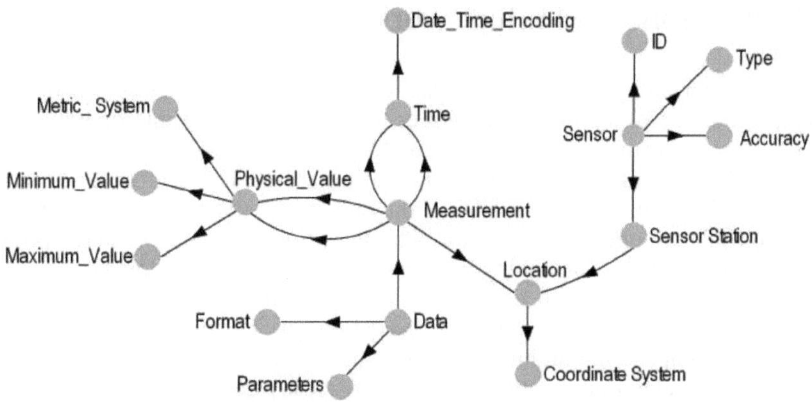

Fig. 7. A snapshot of the ontology

After this phase, the data sets have a standardized format and conflicts in data content are removed. Therefore, heterogeneous data appears to have been generated from the same source.

Finally, the *Data Fusion Module* builds a new RDF model from the data model obtained from the Data Integration Module, adding inside concepts and relations between entities gained by reasoning on the Sensor Observation Ontology and Phenomena domain Ontology. That information allows understanding the sensor information in order to eventually identify situation awareness. In particular, the Phenomena Domain Ontology contains concepts regarding the domain elements and phenomena. As an example, weather domain ontology could contain concepts describing the weather domain such as wind, air and snow and weather phenomena that should appear such as liquid precipitation and storm. Besides, the Sensor Observation Ontology describes the context of the sensor with respect to its spatial and temporal observations, and thematic information. The sensors are grouped into classes, each of one is associated with a specific domain described in the Phenomena Domain Ontology. For instance, a temperature sensor, a humidity sensor and a wind sensor may collectively monitor weather, therefore they may belong to the *weather sensor class*.

The defined ontologies are needed to semantically describe the sensor data and observations from the spatial, temporal and thematic point of view (Sheth 2008).

The use of a set of ontologies is motivated by the fact that some recent works on the use of process ontologies showed an increase in the precision of service discovery/queries when semantic representations were used over syntactic representations (Klien 2001). Finally, the RDF is are accessible through Web Services.

4 Conclusions and future work

The management and elaboration of data sensed by heterogeneous sensor networks represent one of the main open issues to face for defining interoperable services. At the state of the art, a lot of ad hoc solutions to solve the interoperability problem have been installed, but new interesting solutions based on open standards and architectures are now available; they are primarily implemented through the adoption of new open standards for data definition and data exchange protocols. In particular, SOA is becoming very popular in this research domain.

On the basis of the experiences made in the Campanian Region, in this chapter we have presented an architectural model based on service layers. Three different service layers compose the architecture; they manage data collection, data aggregation and value-added applications. The proposed layered approach allows any developer to design different kinds of applications, by invoking low-level services through their standard interfaces and by composing them.

To show the advantages of the proposed architecture, we have presented an Early Warning System based on data collection and aggregation services; furthermore the modelled services propose a data model that has been formalized according to specific domain ontologies. The data model has been built on the basis of many available sensor networks. In a future work we intend to extend the implementation of such a model, to supply other application layer services. Moreover, the description of such services through standard interfaces lets the services be accessible from public services repositories. We intend to provide our services at European level, too, within some European projects that proposed some European public registries for international plans.

References

Alegre C, Monfort C, Lazaridis P, WIN: A new Service Oriented Architecture for risk management. *www.vfdb.de/riskcon/paper/Interschutz%20Conference%20-WIN%20Paper.pdf*

Alegre C, Sassier H, Pierotti S, Lazaridis P (2005) A New Geo-Information Architecture for Risk Management. Geo-information for Disaster Management., Springer, pp. 543–550.

Booth D, et al., Web Services Architecture, W3C Working Group, February 2004. http://www.w3.org/TR/ws-arch/

Ceruti MG (September 2004) Ontology for Level-One Sensor Fusion and Knowledge Discovery. Knowledge Discovery and Ontologies Workshop at ECML/PKDD, Pisa, Italy.

Chu X, Buyya R (2007) Service oriented sensor web. Sensor Networks and Configuration, pp. 51–74.

Eid M, Liscano R, El Saddik A A novel ontology for sensor networks data. CIMSA 2006 – IEEE International Conference on Computational Intelligence for Measurement Systems and Applications La Coruna – Spain, 12–14 July 2006.

European Interoperability Framework 1.0, 2003. http://ec.europa.eu/idabc/en/document/3782/5584

Gay D, Levis P, Culler D, Brewer E, nesC 1.1 Language Reference Manual, http://nescc.sourceforge.net/papers/nesc-ref.pdf, March 2003.

Hadim S, Mohamed D, Middleware: Middleware challenges and approaches for wireless sensor networks. IEEE Distributed Systems Online, March 2006.

Hakimpour F, Geppert A (2001) Resolving semantic heterogeneity in schema integration. Proceedings of the international conference on Formal Ontology in Information Systems, October 17–19, 2001, Ogunquit, Maine, USA, pp. 297–308.

Henson C, Sheth A, Sahoo S (2008) Semantic Sensor Web. IEEE Inter Comput 12(4):78–83.

Klien M, Bernstein A (2001) Searching for services on the Semantic Web using process ontologies. 1st Semantic Web Working Symposium (SWWS-1), Stanford, CA.

Kobialka T, Buyya R, Leckie C, Kotagiri R (2007) A sensor web middleware with stateful services for heterogeneous sensor networks intelligent sensors, Sensor Networks and Information, ISSNIP 2007.

Liang S, Coritoru A, Tao C (2005) A Distributed Geo-Spatial Infrastructure for Smart Sensor Webs. J Comput Geosci 31.

Noy FN (2004) Semantic integration: A survey of ontology-based approaches, ACM SIGMOD Record, 33(4, December 2004).

OpenGIS (2007) Sensor Model Language (SensorML), Implementation Specification. http://www.opengeospatial.org/standards/sensorml.

Schimak G, Havlik D (2009) Sensors Anywhere – Sensor Web Enablement in Risk Management Applications. Ercim News, 01/2009, 76, pp. 40–41.

Sheth A, Perry M (2008) Traveling the semantic web through space, time, and theme. IEEE Inter Comput 12(2):81–86.

[illegible]

Development of a Coupled Geoinformation and Simulation System for Early Warning

Eva Ortlieb, Wolfgang Reinhardt and Franz-Xaver Trauner

Abstract

Recurring disastrous landslides cause great damage worldwide and many people were affected during the last decades. Obviously there is a strong demand for developing and improving early warning systems to save lives and properties. Although strong efforts were made in the last decade, the understanding of the hazards and the forecasting of critical events are still particularly weak points of the early warning chain. In an ongoing research project a new approach to improve these critical points in the field of early warning systems of landslides is pursued: complex finite element (FE) simulations of landslides are coupled with geoinformation systems (GIS). This chapter researches the interconnection between the GIS and the FE-Analysis system. Further, two main operational modes, the learning system and the decision support system mode, for such a coupled system are introduced and a workflow for these system is proposed and investigated in detail.

1 Introduction to the coupled system

"Early Warning Systems include a chain of concerns, namely: understanding and mapping the hazard; monitoring and forecasting impending events; processing and disseminating understandable warnings to political authorities and the population and undertaking appropriate and timely actions in response to the warnings"(NDMA 2008). Over the past years the evaluation of natural danger has been nationally and internationally identified as an important task and is still responding to a growing interest (Alexander 2006, Dilley et al. 2005, Fuchs et al. 2005).

M. Konecny et al. (eds.), *Geographic Information and Cartography for Risk and Crisis Management*, Lecture Notes in Geoinformation and Cartography,
DOI 10.1007/978-3-642-03442-8_18,

Nevertheless the understanding of the hazards and the forecasting of impending events are still particularly weak points of the early warning chain. A number of studies exist for the early warning of volcanic eruptions with sensor net approaches and for the early warning of floods (Handmer 2001, Plate 2002, Werner-Allen et al. 2005) Early warning systems for landslide hazards are hardly researched. Some approaches exist with particular sensors, e.g. ground-based SAR interferometer (Casagli et al. 2002) or other sensors (Zan et al. 2002).

In order to advance research in the field of early warning systems for landslides, the joint project "Development of suitable Information Systems for Early Warning Systems" was launched. The project aims at the development of components of an information system for the early recognition of landslides, their prototypical implementation and evaluation (Breunig et al. 2007, 2008) One subproject of the joint project addresses the coupling of complex finite element simulations with geo information systems (Ortlieb et al. 2008, Trauner et al. 2008). Numerical simulations are set up to examine the physical processes of landslides induced by various scenarios and to improve the understanding of the causes of slope instability and triggers of ground failure. This allows for the evaluation of unstable slopes and their imminent danger for human infrastructure. The coupling with the GIS allows for a user-friendly preparation of the complex simulation results in the GIS for decision support.

At present the FE simulations of landslides are a subject of research. Due to their complexity the corresponding simulation systems are predominantly used by experts and scientists. For disaster prevention and management such tools are currently not available, but would obviously be very helpful. An FE simulation of a landslide requires detailed input information. Therewith the configuration of the simulation input data is very complex and usually not sufficiently supported by the simulation system. On the other hand simulation outputs are extensive and complex and the interpretation of the simulation results is usually only weakly supported by the simulation system. For a broader use of simulation systems of landslides their handling should be more intuitive and user-friendly. GIS with their ability to store, manage and visualize geographical information provides as good basis for setting up the inputs of an FE simulation, analyzing and integrating the outputs and finally support a decision.

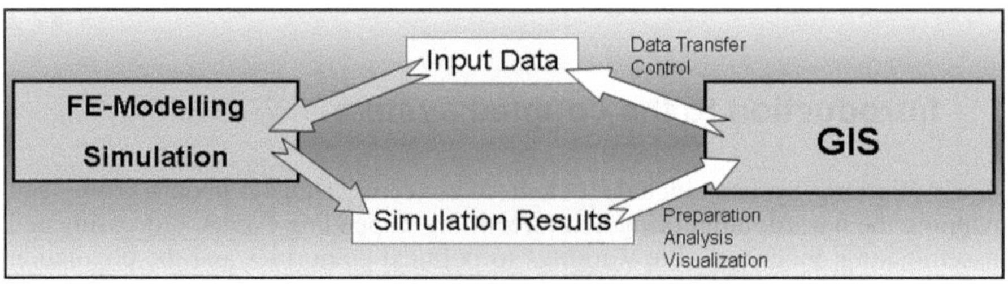

Fig. 1. Interconnection between Simulation System and GIS

The interconnection between simulation system and GIS is schematically shown in Fig. 1. The process starts with the selection of relevant parameters which are needed for the simulation. These parameters include basically geometry, the subsoil structure and several boundary conditions (Sect. 2.3). The parameter transfer is controlled by the GIS. Within the simulation system the modelling of the slope and the simulation of the landslide is executed.

After the simulation the results are transferred to the GIS for visualization, assessment and for processing into a form which is understandable for decision makers. Furthermore, stability indices and movement vectors can be calculated from the simulation results to assess the slope stability, the likely system behaviour in future and the potential risk scenario. Uncertainties in the data used in the simulation and in subsequent processes should also be modelled and visualized in the GIS. In particular for the support of the user in the decision-making process the uncertainties have to be recognizable, in order to allow for validation of the results by the user. Additionally, rule-based GIS components support the user in the decision whether to issue an early warning or not.

Because the simulation of landslides is complex and requires a number of manifold and extensive input information, the whole dataflow between the simulation system and the GIS is complex. In this chapter, a proposal for a possible workflow is made and a detailed insight is given.

2 Workflow of the coupled system

Comprehensive and exhaustive simulations are complex and computationally intensive and can be in case of an early warning decision too time consuming. Therefore two main operational modes of the coupled system with differing computational costs are identified:

1. Learning system for better understanding of landslide movements
2. Decision support system (DSS) for reaction after a hazardous event.

In the following sections a workflow of the coupled system is proposed. Some components of the workflow are only used by one operational mode, but there are also components which are used together.

2.1 Learning system workflow

In the learning system mode the user can evaluate the consequences of many scenarios to learn how certain parameters (e.g. certain amounts of rainfall) influence the stability of slopes. This allows for a better understanding and prognosis of landslides. Furthermore the learning system enables to compare observed historical events with simulated ones. This enables the calibration and refinement of simulations and supports the improvement of the understanding of the geotechnical characteristics of the slope. Further the comparison of historical events with simulated ones allows for the determination of critical events (e.g. a critical flood discharge). This allows for the announcement of warnings at an early stage when the critical event is expected or forecasted (e.g. intense rainfall).

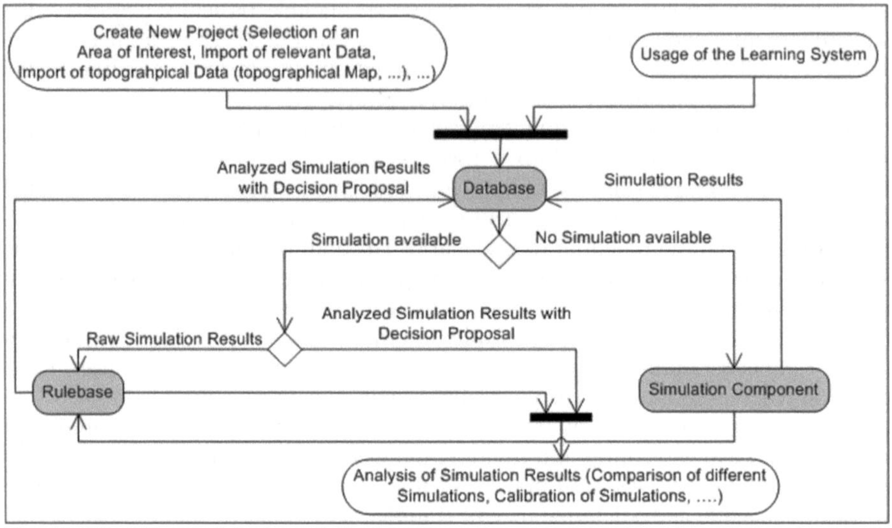

Fig. 2. Workflow of the learning system

The learning system workflow is schematically shown in Fig. 2. If the user wants to investigate a scenario, he has the possibility to check whether there have already been simulations for this scenario in the past. Therefore the user has to first query the database. In the database several results from previous simulations and previous analyses are stored. To inform future users of the available simulations they are stored with their simulation parameters and metadata. Simulation parameters include all parameters, which were used to perform the simulation (Sect. 2.3). Metadata include e.g. the person responsible for the simulation and the date of the simulation.

If there are simulations available in the database they are either transferred to the rulebase for linkage with decision rules to provide a decision proposal for the user or, if the simulation results are already linked with decision rules, directly to the learning system. If there are no simulations available a new simulation has to be calculated in the simulation component related part of the workflow (Sect. 2.3).

2.2 Decision support system workflow

In contrast to the learning system mode, the DSS mode is used if an acute danger exists. This can be the case if an ascertained event, e.g. an intense rainfall event, occurs, which may destabilize the slope and causes a potential risk. This occurrence requires a fast decision whether to issue an early warning. Therefore the focus is put on performance and fast assessment routines, to allow for the announcement of a warning timely to the critical event.

The DSS workflow is schematically shown in Fig. 3. To assess, whether the ascertained event is critical or not, the user has to adapt actual measured values of this event. Because in most cases there is no time for complex and comprehensive and therefore time-consuming numerical modelling of the slope and simulation of landslide hazard, it is of particular interest that there are already simulations existing in the database. In case there has been a simulation calculated before, the simulation results are either transferred to the rule-base for preparation or, if the simulation results are already prepared, they can be transferred for visualization in the DSS. If there hasn't been a simulation before a new simulation has to be carried out (Sect. 2.3).

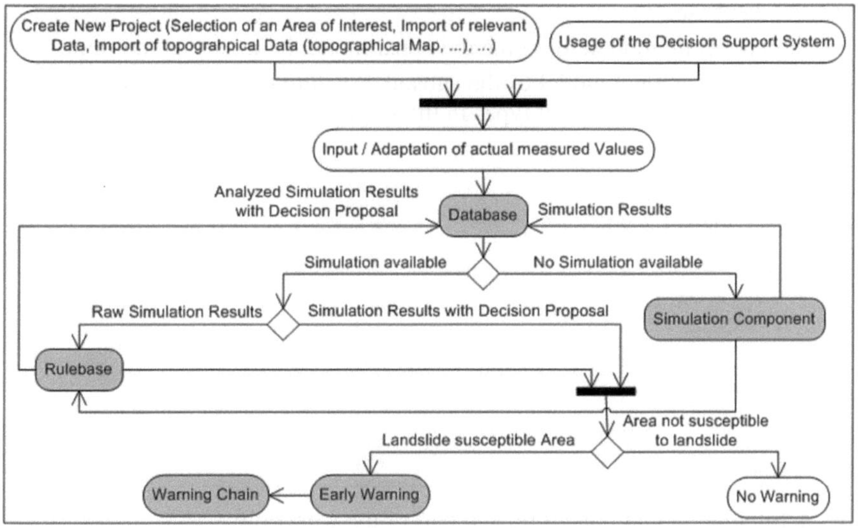

Fig. 3. Workflow of the decision support system

In the DSS the landslide susceptibility of the slope and the associated uncertainties can be visualized. In case the area of interest is landslide susceptible the user has to trigger an early warning to activate the early warning chain. Subsequently, a warning is disseminated to authorities and to the population and appropriate actions can be undertaken.

2.3 Shared components of the workflow

As shown above, for the DSS and the learning system a specific workflow has been defined. But there are also components, which are used in both cases. This includes the simulation, the database and the rulebase component related parts of the workflow. For further information about the database and the rulebase component, see Breunig et al. (2009) and Ortlieb et al. (2009a). The simulation component related part of the workflow is described in the following paragraphs.

The process in the simulation component related part of the workflow starts with the selection of the required input data (Fig. 4). For a numerical simulation various kinds of data are needed. Some of them are user-defined parameters and some can be imported from the database. The user-defined parameters include

- The area of investigation,
- The event, which influences the slope shall be examined (e.g. rainfall),
- The dimension of the event (in case of rainfall litre per square metre).

A fundamental input is an FE-mesh, which represents the geometry of the slope. It consists of a collection of nodes and edges, which defines the finite elements. To generate the FE-mesh two specifications are needed, which are queried from the database:

- A digital terrain model, to define the upper model boundary of the slope,
- A three-dimensional model of the subsoil structure (geology), which shows the distribution of different soil types in the slope.

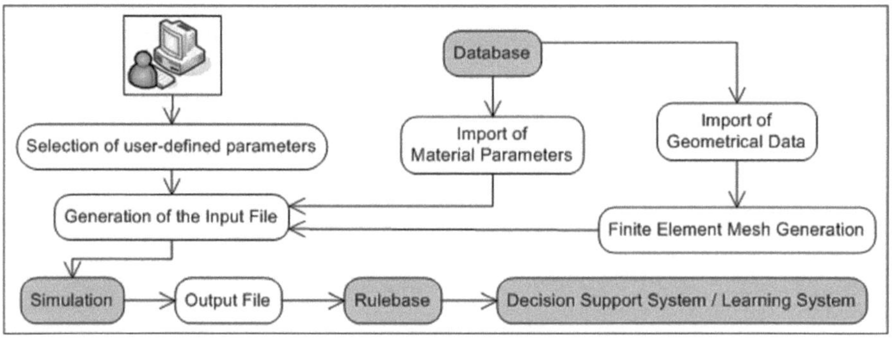

Fig. 4. Simulation component related part of the workflow

In addition to the user-defined parameters and the FE-mesh, the material of the soil layers is needed for the simulation and is therefore also written in the input file. Afterwards the input file is transferred to the simulation system. Within the simulation system the finite element modelling of the slope and the analysis of the landslide susceptibility is executed (Boley and Trauner 2009). Results of the simulation are several parameters, e.g. deformations of the nodes of the FE-mesh. Because the simulation results are very complex and extensive they have to be linked with decision rules in the rulebase to allow for a user-friendly visualization with appropriate methodologies in the learning and the decision support system, respectively (Ortlieb et al. 2009b).

3 Conclusion and outlook

In this chapter a workflow for a coupled system and two modes of the coupled system are presented. At the moment the presented workflow is tested and implemented with real mass movement scenarios. Therefore a part of the hillsides in the Isar valley in the south of Munich has been selected for exemplary simulations. In this area, the height difference of the slope reaches up to almost 40 m and potentially endangered human infrastructure is located near the edge of the slope. In the seventies there have been several landslides. After these events and because of the high risk potential several measuring devices were installed by the Bavarian Environment Agency. Today, after more than 30 years of investigations, extensive knowledge of the subsoil structure and the failure mechanism are available and can be used in the project (Baumann et al. 1988).

Future research will basically address the FE-modelling of the slope and the simulation of the landslide. Another focus is put on the preparation of the complex simulation results for the support in the DSS. Therefore they have to be linked with decision rules, which have to be defined in the presented project. Further, the simulation results have to be prepared with appropriate aggregation techniques, to allow for a user-friendly visualization in the learning and the decision support system, respectively.

Besides the user-friendly visualization of the simulation results corresponding uncertainties should also be modelled and visualized. In particular for presentation in the DSS the uncertainties have to be recognizable, in order to allow for validation of the results by the user. Both, the modelling and the visualization of the uncertainties are still a subject of research in the presented project.

Acknowledgements

The funding of the research project by the German Ministry of Education and Research (BMBF) and the German Research Foundation (DFG) under grant no. 03G0645A et al. is gratefully acknowledged. This is publication no. GEOTECH-1217 (http://www.geotechnologien.de).

References

Alexander D (2006) Globalization of Disaster: Trends, Problems and Dilemmas. J Int Aff 59:1–22.

Baumann H (1988) Bruchvorgänge in Folge der Isareintiefung südlich Münchens und die kritischen Höhen der Talhänge. Schriftenreihe des Lehrstuhls und Prüfamt für Grundbau, Bodenmechanik und Felsmechanik der Technischen Universität München, Heft 12.

Boley C, Trauner F-X (2009) GIS-gestützte numerische Analyse von Rutschhängen als Komponente von Frühwarnsystemen. Paper accepted for 24. Christian Veder Kolloquium – Stabilisierung von Rutschhängen, Böschungen und Einschnitten, TU Graz, Austria.

Breunig M, Broscheit B, Reinhardt W, Ortlieb E, Mäs S, Boley C, Trauner F-X, Wiesel J, Richter D, Abecker A, Gallus D, Kazakos W, Bartels A (2007) Development of suitable information systems for early warning systems (EGIFF). In: Stroink L (ed.) Geotechnologien Science Report No. 10, pp. 113–123.

Breunig M, Broscheit B, Reinhardt W, Ortlieb E, Mäs S, Boley C, Trauner F-X, Wiesel J, Richter D, Abecker A, Gallus D, Kazakos W, Bartels A (2008) Towards an information system for early warning of landslides. EnvironInfo 2008, Lüneburg, pp. 476–481.

Breunig B, Broscheit B, Thomsen A, Butwilowski E, Jahn M, Kuper PV (2009) Towards a 3D/4D Geo-database supporting the analysis and early warning of landslides. In: Konecny M, Zlatanova S, Bandrova T (eds.) Cartographie and Geoinformatics for Early Warnings and Emergency Management: Towards better Solutions Proceedings, pp. 100–110.

Casagli N, Farina P, Leva D, Nico G, Tarchi D (2002) Monitoring the Tessina landslide by a ground-based SAR interferometer and assessment of the system accuracy. Geoscience and Remote Sensing Symposium 2002, Toronto, pp. 2915–2917.

Dilley M, Chen RS, Deichmann U, Lerner-Lam AL, Arnold M (2005) Natural Disaster Hotspots: A Global Risk Analysis. The World Bank. Hazard Management Unit. Washington DC.

Fuchs S, Keiler M, Zischg A, Bründl M (2005) The long-term development of avalanche risk in settlements considering the temporal variability of damage potential. Nat Hazards Earth Syst Sci 5:893–901.

Handmer J (2001) Improving flood warnings in Europe: A research and policy agenda. Environ Hazards 3:19–28.

NDMA (2008) National Disaster Management Authority. Terminology of disaster risk reduction. http://www.ndma.gov.pk/BooksPublications.html

Ortlieb E, Mäs S, Reinhardt W, Trauner F-X (2008) Coupling simulations of landslides with geoinformation systems for decision support and early warning. In: Bernard L, Friis-Christensen A, Pundt H, Compte I (eds.) 11th AGILE Proceedings, Girona, Spain.

Ortlieb E, Reinhardt W, Trauner F-X (2009a) Development of a coupled geo information and simulation system for early warning systems. In: Konecny M, Zlatanova S, Bandrova T (eds.) Cartographie and Geoinformatics for Early Warning and Emergency Management: Towards Better Solutions Proceedings, Prague, Czech Republic, pp. 483–492.

Ortlieb E, Reinhardt W, Trauner F-X (2009b) Kopplung von Geoinformations- und Simulationssystemen zur Entscheidungsunterstützung. Paper accepted for 45 AgA-Tagung, Frankfurt.

Plate EJ (2002) Flood risk and flood management. J Hydrol 267:2–11.

Trauner F-X, Meyer F, Boley C, Ortlieb E (2008) Neue innovative Methoden zur Frühwarnung auf Basis von gekoppelten Informations- und Simulationssystemen am Beispiel gravitativer Massenbewegungen. Tagungsband 6. Kolloquium Bauen in Boden und Fels, Technische Akademie Esslingen, pp. 433–439.

Werner-Allen G, Johnson J, Ruiz M, Lees J, Welsh M (2005) Monitoring volcanic eruptions with a Wireless Sensor Network. Proceedings of the 2nd European Workshop on Wireless Sensor Networks (EWSN '05), IEEE Press, pp. 108–120.

Zan L, Latini G, Piscina E, Polloni G, Baldelli P (2002) Landslides Early Warning Monitoring system. Geoscience and Remote Sensing Symposium (IGARSS '02), IEEE International, pp. 188–190.

Multi-criteria Evaluation for Emergency Management in Spatial Data Infrastructures

Matthias Müller, Lars Bernard and Rico Vogel

Abstract

This chapter presents an approach for the integration of Multi-criteria Evaluation (MCE) into Spatial Data Infrastructures (SDIs) to be used in an ad hoc manner. The approach is demonstrated using two case studies: (1) a selection of action priorities in emergency management, illustrated with a specific implementation for the SoKNOS project; (2) a safe site selection problem that identifies future tasks and issues related to interoperable spatially enabled MCE Services. Special concern is given to interoperability by relating our approach to the current OpenGIS Web Processing Service specification and the ORCHESTRA and INSPIRE architecture frameworks. Finally, this chapter discusses the initial steps required for an application profile for MCE in SDI.

1 Introduction

Spatial Decision Support Systems (SDSS) can be understood as an umbrella term for systems that integrate spatio-temporal data from various sources, models, analysis, and exploration functionalities, in order to assist users in solving semi-structured problems. Frequently, solutions to decision problems are found by means of Multi-criteria Evaluation (MCE, Malczewski 1999).

Emergency management requires the integration of information from a large number of sources. SDSS can provide decision makers with refined and structured information on the spatial and temporal distribution of threats, assist in steering evacuations, or identify safe locations for emergency care.

The availability and quality of environmental data are crucial factors influencing emergency management. Spatial Data Infrastructures (SDIs) can be used in an ad- hoc manner, thus facilitating the distribution and sharing of the most up-to-date spatio-temporal data and geoprocessing functionality. Furthermore, SDIs are considered ideal means of support for efficient emergency management. The

M. Konecny et al. (eds.), *Geographic Information and Cartography for Risk and Crisis Management*, Lecture Notes in Geoinformation and Cartography,
DOI 10.1007/978-3-642-03442-8_19,

European INSPIRE Directive (Infrastructure for Spatial Information in Europe; European Parliament 2007) is currently Europe's most prominent effort to realize an operational network of national SDIs to improve sharing of geospatial data to assess the European environmental situation. With INSPIRE focusing on existing geospatial information, other European initiatives, such as GMES (GMES 2009), and national developments, such as Pegelonline (PEGELONLINE 2009), have focused on data infrastructures for sensor data obtained from remote or terrestrial sensors that provide real-time data. In this way, standardized and interoperable spatio-temporal Data Access Services are becoming available. However, the processing of geospatial data in service-oriented SDIs is an ongoing issue and provides an interesting field of research.

This chapter discusses the first attempts to specify interoperable spatially enabled MCE services that will subsequently enable distributed SDSS in SDI. Two use cases are provided as examples to study the feasibility of current specifications to support this approach and to identify related research issues.

2 Multi-criteria evaluation in Spatial Data Infrastructures

According to Pundt (2008), disaster management always includes a spatial component. Natural disasters occur at specific locations or within specific regions. For instance, a flood occurs in parts of a drainage basin and includes river floodplains with various land-use types, such as industry, business parks, settlements or agricultural areas. These disasters typically result in damage to the infrastructure, and human beings are often affected. The 1993 and 1995 floods of the Rhine River, for example, resulted in damages amounting to over 1 billion EUR (Akkermann 2004).

Emergency management helps to prevent and mitigate the impact of natural hazards or man-made disasters. Typically, a large amount of information is needed to plan the actions required for emergency response, including the spatial extent of the disaster, the amount and severity of damage caused by the event, as well as details about the affected locations and possible endangerment to neighbouring areas. SDSS can support this task of information gathering and aggregation in two ways. First, an SDSS can provide decision makers with information on potential effects of a specific decision. Second, it can provide evaluation tools to compare and rank decision alternatives according to a given goal or a given set of goals.

As there is no universal definition of an (S)DSS, Sprague (1986) used a paradigm to describe the three necessary components required to build a DSS: Dialog, Data and Modelling. Dialog components are required for user interaction, a model base is used to generate analysis results, and a database is required to contain all the data needed to solve the decision problem. A number of SDSS applications have been developed based on monolithic Geoinformation Systems (GIS) (e.g., Jankowski 1995, Jankowski et al. 2001). A recent and comprehensive survey of GIS-based SDSS applications was undertaken by Malczewski (2006). Rinner (2003) provides an overview of web-based SDSS approaches. The first efforts to link SDSS with SDI concepts based on standardized service interfaces are described in Bernard et al. (2003). In this study, a simple MCE method was encapsulated behind an OpenGIS Web Map Service Interface (WMS). The advent of the OpenGIS Web Processing Service (WPS) interface specification allowed for more sophisticated approaches to share geoprocessing functionalities in SDI (OGC 2007b). Kiehle et al. (2007) and Friis-Christensen et al. (2007) offered in-depth analyses of the potential of the WPS for distributed geoprocessing. Bernard and Ostländer (2008) presented a general framework for the integration of SDSS and SDI using standardized service interfaces and demonstrated its application for the assessment of climate change effects.

2.1 Multi-criteria evaluation

MCE is one of the most common approaches used to support spatial decision making (Jankowski 1995, Malczewski 2006). In brief, MCE helps to find solutions to decision problems characterized by multiple alternatives. These alternatives can be evaluated by the performance of decision criteria or by attributes that correspond to some predefined objective(s) and are associated with each of the decision alternatives. Several MCE methods exist to support the choosing, ranking or sorting of alternatives. An overview of these methods can be found in Figueira et al. (2005). A common goal in emergency management is the prediction of impact, relating to questions such as 'Where are most people affected by a disaster?' or 'Which areas should be evacuated to ensure the best protection of the population?'. The desired outcome of such a decision analysis usually falls into one of three categories: one recommended alternative, a set of alternatives with good performance or a simple ranking of alternatives from best to worst (Jankowski 1995).

The MCE process can be expressed as a common process model containing the elements 'decision alternatives', 'decision attributes', 'decision matrix', 'aggregation', 'sensitivity analysis' and 'final recommendation'. In this context, Starr and Zeleny (1977) distinguish between 'outcome-oriented' and 'process-oriented' approaches. The latter serves to improve insights into the underlying phenomena and is normally used in an iterative and interactive manner, as the process model requires a loop between the final recommendation and the decision matrix. Depending on the outcome of the sensitivity analysis, the evaluation result can be accepted or rejected. If an unstable result leads to an aggregation with a new set of decision preferences from the decision maker, it qualifies as a real decision support tool. In contrast, if the aggregation is run only once, the MCE process is instead used as a sophisticated assessment instrument to aggregate the considered criteria.

Due to the wide acceptance of this process model and its many existing applications, efforts have begun to transform this common procedure for spatial MCE into a process algebra for specific fields of application (e.g., Chakhar and Mousseau 2007). The work presented here focuses on analyzing the extent to which distributed SDSS processes can be realized in current SDI.

2.2 An abstract architecture

Following the methods of Rinner (2003) and transferring Sprague's (1986) (S)DSS concepts into SDI, Fig. 1 sketches an abstract service architecture. In general, this abstraction follows the architectural approaches developed in ORCHESTRA (OGC 2007a) and applied to INSPIRE (INSPIRE 2008). Most abstract service types described in the architecture can be mapped to existing platform-independent service interface specifications. For example, View Services can be mapped to the OpenGIS Web Map Service, and Data Access Services can be mapped to the OpenGIS Web Feature Service (WFS, OGC 2006) and the Web Coverage Service (WCS, OGC 2008). Service types shown with dashed lines currently lack well-defined and standardized service interface specifications (Kiehle et al. 2007, Friis-Christensen et al. 2007).

The decision attributes for spatial MCE processes are provided by Data Access Services, Sensor Services and Simulation Services (a group encapsulating simulation runs, e.g., simulation of pollutant dispersal). Hence, it is possible to evaluate present and future states of the environment. The coarse grained MCE algorithms can make use of Processing Services, such as Transformation Services (e.g., coordinate transformations) and Geoprocessing Services, which offer simple Map Algebra operators (Tomlin 1990) to calculate and aggregate spatial criteria. All services are accessible through a suitable client application that handles communication with the Web Services and provides a graphical user interface. Service instances are registered and can be found, explored and subsequently used by client applications using Discovery Services. Services might be access-controlled or charge fees. In this case, related functionality is encapsulated in a specific service layer to handle all related Digital Rights Management.

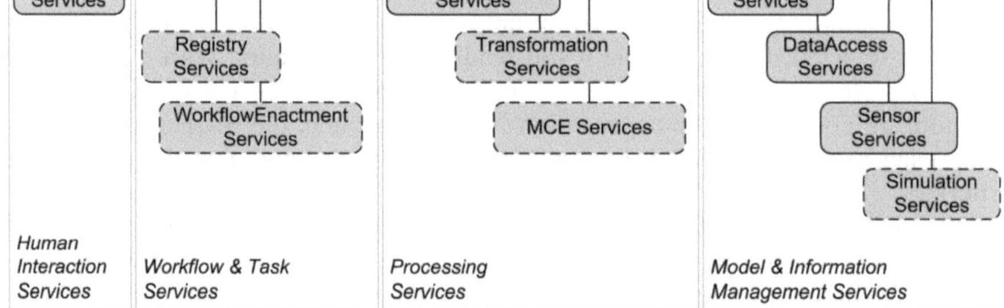

Fig. 1. Abstract architecture for a Spatial Decision Support Service Infrastructure

Furthermore, it is possible to store spatially enabled MCE processes on a Workflow Enactment Service as a chain of the necessary consecutive process executions. Such a complex process can be represented by an MCE process, connected with atomic spatial and spatio-temporal operations from the Geoprocessing Service to integrate spatial data. Schaeffer (2008) uses a transactional WPS for a similar purpose and proves his approach with a workflow for air quality assessment.

Atomic operations in MCE include classifications, normalizations or algebraic functions that contribute to the individual MCE process steps. The basic processing operations, more complex process workflows, and the schemata describing well-defined and commonly agreed-upon feature models (such as those currently being developed with the harmonized INSPIRE data specifications) are stored in repositories and made accessible via Registry Services. In this way, the Registry Services provide a common understanding about the semantics of the operations and data sources available in SDI. Lutz (2007) discusses related approaches on semantic discovery and chaining of geoprocesses.

A discussion regarding how to describe the MCE process and its elements at different levels of granularity is ongoing. Hwang and Yoon (1981) provide a taxonomy for generic MCE algorithm classes, and Malczewski (1999) translates many atomic MCE operations into their corresponding GIS functions and synthesizes geoprocessing workflows. Bernard and Ostländer (2008) apply Sprague's DSS Toolbox model to SDI services, reflecting different granularities from high-level workflows down to low-level transformation and data access operations. Chakhar and Mousseau (2007) formalize a 'decision map algebra', relying on the work of Pullar (2001) and Tomlin (1990). However, there is no interoperable architecture that helps to leverage MCE in SDI.

Current problems arise from a variety of sources. First, it is difficult to agree on a common hierarchical process model. Although there is a broad acceptance of the overall MCE process model, the decomposition of this model into atomic process steps varies by author (e.g., there is no formal creation of a decision matrix). It is also a common practice to build customized MCE processes that include the synthesis of complex decision attributes. In order to enable MCE workflows that consist of chained atomic operations or process fragments, it is necessary to agree on a strict and commonly accepted process model. Second, there is no systematic approach that we know of that leverages the full range of MCE methods in the geospatial world. Some effort is still required to create spatial data types based on both geospatial and MCE-specific data structures. The OGC abstract specifications on features and coverages (OGC 2005, Topics 5 and 6), accompanied by the data structures described in the MCE process model, can be regarded as a sound basis for this task. Third, a great amount of effort is needed to transfer processes and related operations into a Web Service infrastructure. Most Web Services function statelessly, which is a major hurdle when MCE processes will be required to provide decision support beyond a simple measurement of indicators. The necessity to branch and loop-back after the sensitivity analysis, and to provide an aggregation step with a new set of preferences, requires stateful services capable of requesting user interaction and runtime alteration of some of the input data.

3 Application example

The application of use cases within an infrastructure based on today's service interface specifications demonstrate how to realize the above given abstract architecture on a concrete technical platform. In the following section, two use cases are considered to analyze the suitability of the suggested architecture and currently available service interfaces.

The use cases are motivated by a scenario developed in the context of the SoKNOS project (Service-Oriented Architectures Supporting Networks of Public Security; SoKNOS 2009), which aims to establish a service platform as a technical basis for collaboration in emergency management. In contrast to other decision situations, the decision process in emergency management organizations results in additional system requirements, such as providing information in (near) real time, offering failsafe and secure systems, logging decision processes for later analysis, etc. These issues are also addressed in the SoKNOS-project, but are not considered in this chapter.

3.1 MCE for screening and assessment: priorities of action

North of Cologne, in the floodplain of the Rhine River, a chemical plant is damaged during a flood event and a cloud of contaminants escapes due to a subsequent explosion. In order to plan accordingly for evacuations, the decision makers are interested in the possible spatial distribution of contaminants and their health impacts. Consequently, the MCE algorithm applied in this context requires input about the pollution and the possibly affected population. Data about the spatial distribution of the contaminant can be generated by either a simulation model or by a geostatistical algorithm that relies on sensor data. To estimate the affected population, information about settlement, industry, business parks, agricultural areas, forests and surface water, as well as large-scale population distribution data, must be integrated.

In this use case, the 'priorities of action' are calculated on the basis of three criteria: 'contaminant exposure', 'population density' and 'land use'. After dividing the operational area into rectangular grid cells, each cell will be treated as an alternative for an evacuation action and will be characterized by these three criterion values.

3.2 Interactive MCE for decision making: safe site selection

In contrast to the technically simple evaluation problem presented in the former scenario, we wish to outline a more complex use case for a spatially enabled MCE workflow to point out further directions for research and show further potential of interoperability research in this field of application.

When facing major disasters, emergency response forces often decide to establish a supply camp to sustain their personnel or to medically treat the local population. It is important to select a site for this camp that is not only close to the focal point of the disaster impact but also reasonably secure from the adverse effects of the disaster. Other technical and spatial considerations, like the overall size, shape, slope or type of soil, might also be taken into account in the case of site selection.

In contrast to the straightforward assessment use case, the search for a safe site may involve several validation steps, loops and iterative adjustment of decision criteria, criterion weights and aggregation functions. Additionally, due to the increased number and complexity of the decision attributes, it might be necessary to use more complex algorithms for evaluation and incorporate an uncertainty analysis to estimate the potential error in the evaluation.

3.3 Implementing an OpenGIS web service architecture for multi-criteria evaluation

We have developed a simplified service infrastructure based on OpenGIS Web Services (OWS) to realize the 'priorities of action' use case (Fig. 2). In order to access data concerning the distribution of contaminants, Sensor Observation Services (SOS, OGC 2007c) and WPS can be established and orchestrated to realize interoperable access to the required data sets. WCS provides access to land use and population density data, which are used to indicate the presence of potentially affected populations for the MCE algorithm.

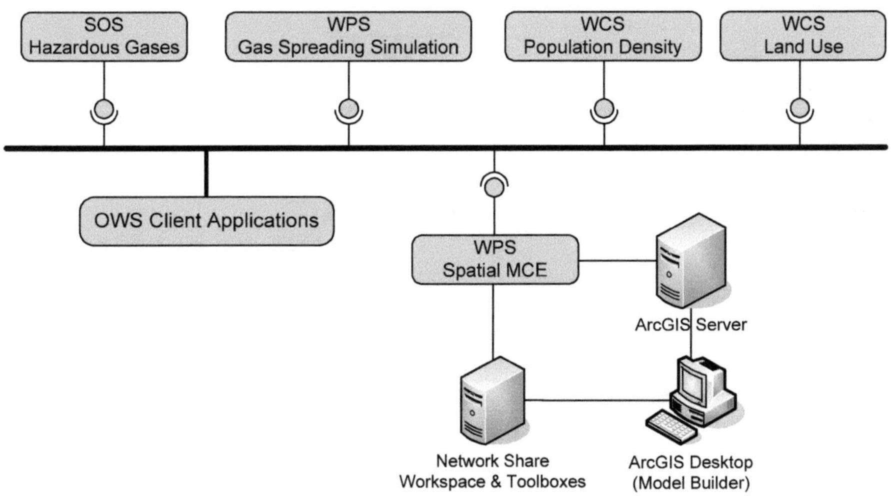

Fig. 2. Implementation architecture for spatially enabled MCE

With respect to the ORCHESTRA architecture, the spatially enabled MCE service is implemented as a WPS to ensure the potential for integration with other frameworks for spatial decision support. The 52°North WPS framework

(52°North 2009) is used for the prototypical implementation. This framework has been released under an Open Source licence and can be extended by writing custom algorithms. The MCE toolboxes containing the required MCE algorithm models are set up using the ArcGIS ModelBuilder and provided to an ArcGIS Server Instance for execution. The MCE algorithms can be started by an instance of the WPS using the processing capabilities and licences from the ArcGIS Server instance. For the 'priorities of action' use case, a prototypical MCE workflow for spatial data is accessible through a WPS interface.

The evaluation of 'priorities of action' is currently performed on the basis of a cardinal function, which indicates the performance of an evacuation at a given spatial alternative. Each alternative can be described as a cell of a geospatial raster. A high population density and a high concentration of hazardous gases and fumes yield a high performance for a spatial alternative. The performance is calculated using the weighted product method (Hwang and Yoon 1981, Malczewski 1999) in which attributes and corresponding weights for the evaluation are treated as variables and must be passed over to the MCE service (Fig. 3).

The specialized service for this use case is built of atomic operations for the generation of a decision matrix and an evaluation algorithm that calculates the ranking of alternatives. The intermediate result is a score describing the performance of an alternative and is grouped into customizable classes to enable ranking. A normalization of the input attributes is optional and is triggered by a Boolean value.

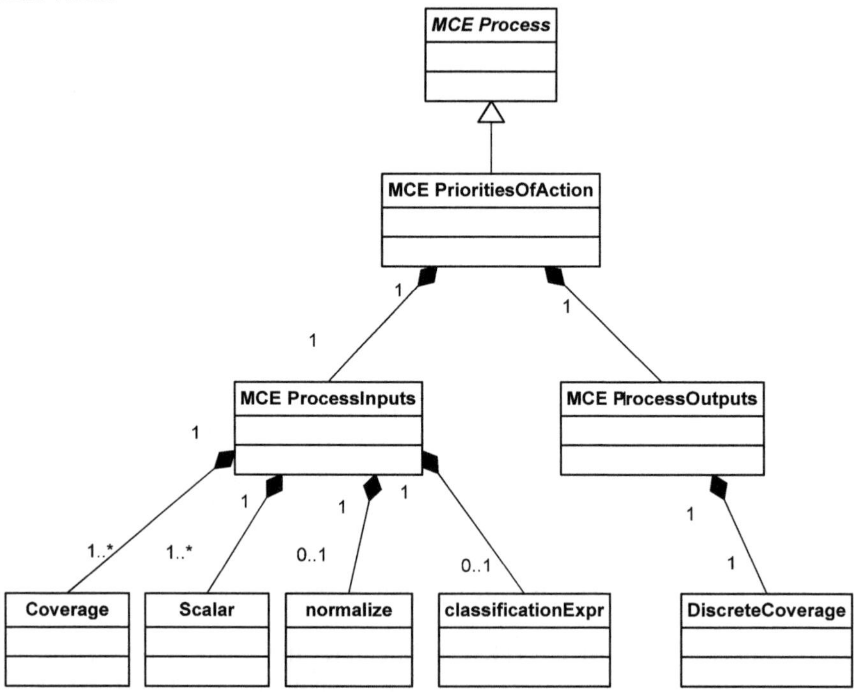

Fig. 3. Input and output of a simple spatially enabled MCE process

The service for a spatially enabled MCE must be executed while passing along a set of spatial alternatives: n spatial attributes with n associated weights. The indication of normalization and classification parameters is optional. As specified in the use case, the evaluation is based on raster data. The set of alternatives is modeled as a discrete grid-coverage, according to OGC (2005), containing one single thematic dimension. Each alternative is identified by a direct position within the coverage and its geometry is determined by the grid geometry.

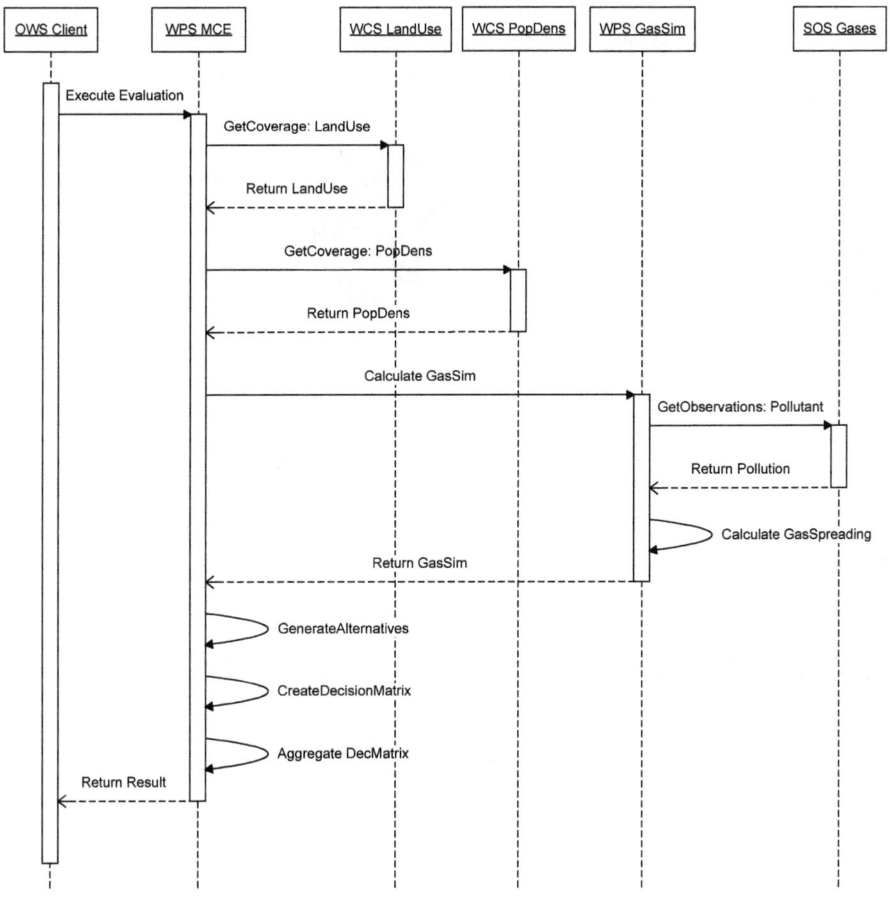

Fig. 4. Workflow of the 'priorities of action' use case

The algorithmic steps (Fig. 4) of the MCE process reflect the subsequent stages of the MCE process model. The execution command to the WPS instance forwards a reference to the necessary spatial attribute data sets and the corresponding evaluation preferences. After requesting the required data sets, the MCE algorithm begins calculating the spatial alternatives according to the cell size of the population distribution grid. Following this calculation, the spatial decision matrix is created as a set of attribute-value layers sharing the same coordinate reference system and grid geometry. If indicated by the decision maker's preferences, the attribute values are also normalized during this step. Finally, this matrix is evaluated using the desired evaluation function with the set of attribute weights and is subsequently classified if a classification expression was provided. The result is a ranking of the grid cells through a performance indicator describing the degree of threat to the population due to contaminant exposure (Fig. 5). The overall processing time, a crucial point in emergency management, amounts to about 30 seconds, depending strongly on the bandwidth of the underlying network infrastructure, the download speed of the Data Access Services and the computational performance of the GIS backend.

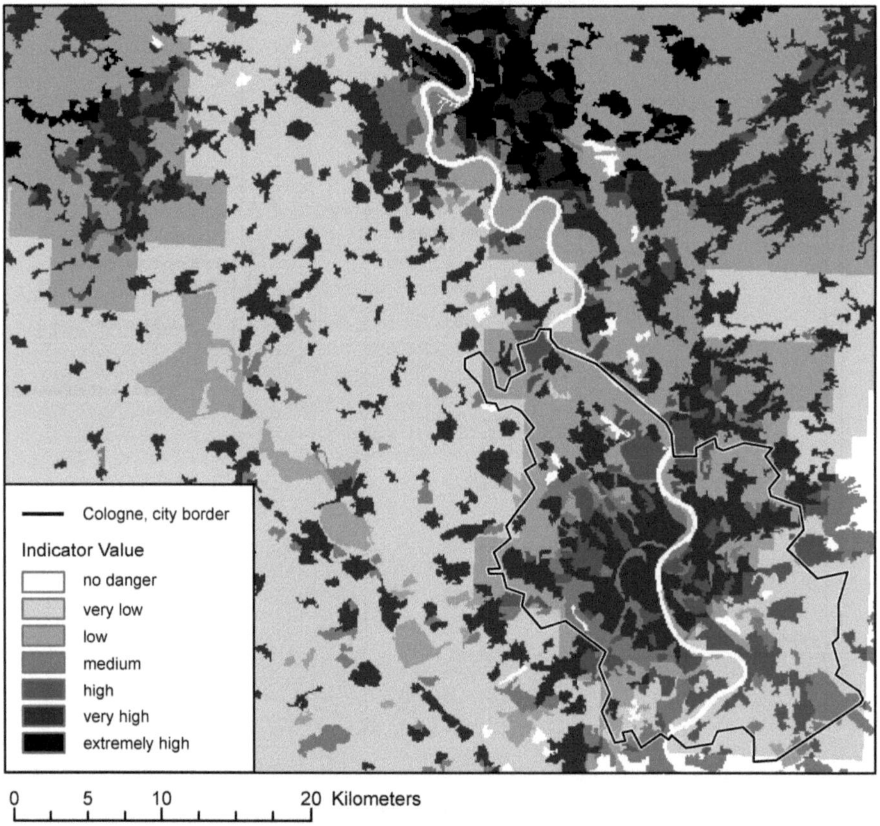

Fig. 5. Example result of the MCE process showing potential degree of threat to the population due to contaminant exposure (vicinity of Cologne, Germany)

In order to provide information about the reliability of the evaluation, a future version of this service might produce a result accompanied by a simple estimator of the result uncertainty. This step can be easily realized using a simple

Monte Carlo algorithm on top of the evaluation function that varies the input values within given intervals. A more exhaustive approach could even offer the ability to specify attribute and preference uncertainty in the execute request to the processing service. A promising candidate for this task is UncertML (OGC 2009), which provides a variety of mathematical concepts of uncertainty in a Web Service environment.

4 Conclusion and future work

As demonstrated above, it is possible to access an MCE process using spatial data inputs through a Web Service interface, thus facilitating their use in current service-based SDI. This service is currently working well in a closed environment, using coverage data. In order to develop an interoperable service for MCE of spatial data, some issues, which will be discussed in the context of the requirements of the second use case, must be addressed.

As previously discussed, the selection of an appropriate evaluation model, paired with appropriate model parameters, is crucial in obtaining a useful evaluation result. The re-implementation of MCE algorithms is a labor-intensive, expensive and error-prone task. As SDSSs are a special flavour of DSSs, it would be of value to reuse some of the functionality, especially the MCE tools, from a common DSS with spatial data, as proposed in the abstract architecture. Thus, interoperable DSSs should be further investigated to allow for efficient coupling of SDSSs and DSSs, particularly to support the generation of a spatial decision matrix and its subsequent aggregation. The preliminary and succeeding transformation steps from and to spatial data types can be accomplished using standard geo-operators for feature attribute extraction and assignment.

As emphasized by Starr and Zeleny (1977), the interactive and iterative coupling of these process elements to solve semi-structured problems cannot be adequately modelled within the SOA paradigm. The main reason for this is the stateless nature of SOA services in general, and the WPS in particular. This statelessness makes it impossible to alter parameters during runtime or to allow the service to ask back for information on how to proceed further if the sensitivity analysis fails.

The use of standards for Data Access and Processing Services is a precondition to achieve syntactic interoperability and, thus, to ensure data sharing among distributed systems. To ensure semantic interoperability, it is necessary to share a common process meta-model for MCE that represents the common problem understanding and identifies the subsequent steps of the evaluation process. For instance, the approach initiated by Chakhar and Mousseau (2007) defines dedicated input and output data types for the MCE process steps to allow for an algebra-like coupling. Typical functional components within these steps include the generation of a decision matrix from alternatives and attributes, the aggregation of the decision matrix, and the sensitivity analysis. A consistent application profile for spatially enabled MCE needs to support the components of this meta-model. Hence, it must contain not only the coarse-grained evaluation process but also meaningful process fragments and algorithmic elements, such as evaluation algorithms, normalization functions and procedures for the generation of the decision matrix. These different functional levels preserve enough flexibility for the integration of MCE functionality into customized processing chains.

Bernard and Ostländer (2008) discuss this issue by distinguishing between specific Spatial Decision Support Services and Processing Services in their SDI approach to SDSS, thus reflecting the concept of technology levels developed by Sprague (1986). Using this approach, a specific SDSS is constructed on the basis of a toolbox containing the functionality to access and process spatial data. The conflation of single tools to a decision-supporting workflow is performed by a generator component, which combines process models with the processing tools. The generation of such workflows using WPS technology and BPEL (Business Process Execution Language) has been discussed by Friis-Christensen et al. (2007), Kiehle et al. (2007), and Schaeffer (2008). They demonstrate the benefits of service orchestration for evaluation problems in emergency management and show the necessity for the ability to conduct translucent chaining in order to assemble complex workflows from simple processes and access these processes either completely or partially. A prerequisite to access processes in this manner would be the ability to enable translucent process chaining using the WPS interface, a feature which is currently not supported.

For our modelling approach to a spatially enabled MCE service, we can foster these insights and encourage further development of the specifications of Processing Services in general, and a special application schema for spatially enabled MCE in particular.

Acknowledgements

Parts of the research presented here have been conducted in the frame of the SoKNOS (project number 01ISO7009) project, funded by the German Federal Ministry of Education and Research to foster the development of service-oriented architectures for risk management. Additional thanks go to the two anonymous reviewers who helped in improving this submission.

References

52 North (2009) Web Processing Service. Home page, URL: https://52north.org/twiki/bin/view/Processing/52nWebProcessingService

Akkermann, M (2004) Beurteilung des Einflusses einer angepassten Ackernutzung auf den Hochwasserabfluss. Dissertation, Universität Hannover. URL: http://deposit.ddb.de/cgi-bin/dokserv?idn=972359346 (in German)

Bernard L, Ostländer N (2008) Assessing climate change vulnerability in the arctic using geographic information services in spatial data infrastructures. Climate Change 87:263–281.

Bernard L, Rinner C, Ostländer N (2003) Impact assessment for the barents sea region – A geodata infrastructure approach. In: Gould M, Laurini R, Coulondre S (eds.) 6th AGILE Conference on Geographic Information Science 2003. Lyon, http://agile.isegi.unl.pt/Conference/Lyon2003/AGILE2003_cd/proceedings/22.pdf, pp. 653–662.
Chakhar S, Mousseau V (2007) An algebra for multicriteria spatial modeling. Comput Environ Urban Syst 31(5):572–596.
European Parliament (2007) Directive 2007/2/EC of the European Parliament and of the Council of 14 March 2007 establishing an Infrastructure for Spatial Information in the European Community (INSPIRE). Official Journal of the European Union 50 (2007), S. L 108/1–14. http://eur-lex.europa.eu/LexUriServ/site/en/oj/2007/l_108/l_10820070425en00010014.pdf
Figueira J, Greco S, Ehrgott M (eds. 2005) Multiple Criteria Decision Analysis: State of the Art Surveys. Springer Science+Business Media, New York.
Friis-Christensen A, Lutz M, Ostländer N, Bernard L (2007) Designing service architectures for distributed geoprocessing: Challenges and future directions. Trans. GIS 11(6):799–818.
GMES (2009) GMES (Global Monitoring for the Environment and Security). Home page, http://www.gmes.info/.
Hwang C, Yoon KP (1981) Multiple Attribute Decision Making: Methods and Applications. Springer, Berlin.
INSPIRE (2008) INSPIRE Network services architecture. V 3.0. Network Services Drafting Team, http://inspire.jrc.ec.europa.eu/reports/ImplementingRules/network/D3_5_INSPIRE_NS_Architecture_v3-0.pdf
Jankowski P (1995) Integrating geographical information systems and multi-criteria methods: A case study. Int J Geoinfo Syst 9(3):251–273.
Jankowski P, Andrienko N, Andrienko G (2001) Map-centred exploratory approach to multiple criteria spatial decision making. Int J Geogr Inform Sci 15(2):101–127.
Kiehle C, Greve K, Heier C (2007) Requirements for next generation spatial data infrastructures-standardized web based geoprocessing and web service orchestration. Trans. GIS 11(6):819–834.
Lutz M (2007) Ontology-based descriptions for semantic discovery and composition of geoprocessing services. Geoinformatica 11:1–36.
Malczewski (1999) GIS and Multi-Criteria Decision Analysis. John Wiley & Sons, New York, USA.
Malczewski (2006) GIS-based multicriteria decision analysis: A survey of the literature. Int J Geogr Inform Sci 20(7):703–726.
OGC (2005) The OGC abstract specification, Version 5. http://www.opengeospatial.org/standards/as
OGC (2006) OpenGIS Web Feature Service Implementation Specification. Version 1.1.0, OGC 04-094 & OGC 06-027r1.
OGC (2007a) Reference Model for the ORCHESTRA Architecture (RM-OA) V2 (Rev 2.1).
OGC (2007b) OpenGIS Web Processing Service, Version 1.0.0.
OGC (2007c) Sensor Observation Service, Version 1.0, OGC 06-009r6.

OGC (2008) Web Coverage Service (WCS) Implementation Standard, Version Version: 1.1.2, OGC 07-067r5.
OGC (2009) Uncertainty Markup Language (UncertML) Discussion Paper. Version 0.6, OGC 08-122r2.
PEGELONLINE (2009) PEGELONLINE home page. http://www.pegelonline.wsv.de/
Pullar D (2001) MapScript: A map algebra programming language incorporating neighborhood analysis. Geoinformatica 5(2):145–163.
Pundt H (2008) The semantic mismatch as limiting factor for the use of geospatial information in disaster management and emergency response. In: Zlatanova S, Li J (Eds.) Geospatial Information Technology for Emergency Response. Taylor & Francis, London, pp. 243–256.
Rinner C (2003) 'Web-based spatial decision support: Status and research directions'. J Geogr Inform Decis Anal 17(1):14–31.
Schaeffer B (2008) Towards a Transactional WPS (WPS-T).In: Proceedings of the 6th Geoinformation Days, Münster, pp. 91–116.
SoKNOS (2009) SoKNOS home page. http://www.soknos.de/
Sprague RH (1986) A framework for the development of decision support systems. In: Sprague RH, Watson HJ (Eds.) Decision Support Systems. Putting Theory into Practice. Prentice Hall, Englewood Cliffs, NJ, pp. 1–26.
Starr K, Zeleny M (1977) MCDM – State and future of the arts. In: Starr MK, Zeleny M (eds.) Studies in the Management Sciences. Multiple Criteria Decision Making. North Holland Publishing Company, Amsterdam pp. 5–30.
Tomlin CD (1990) Geographic Information Systems and Cartographic Modelling. Prentice Hall, Englewood Cliffs, NJ.

Field Tracking Tool: A Collaborative Framework from the Field to the Decision Makers

Alessandro Annunziato, Simone Gadenz, Daniele A. Galliano, Roberto Guana and Francisco Igualada

Abstract

The JRC, in conjunction with the UNHQ Cartographic section, designed and developed a collaborative mapping framework focused on supporting the information-management process and aimed at assisting the decision makers during emergencies. Compared with several tools already available on the market, the JRC development was intended to answer three clear requirements, in particular to support the whole information flow in an emergency situation, to facilitate the users' experience and to ensure quick information dissemination from the field to the mission's headquarters. All the information is stored in a common repository and shared among all the crisis players through a web portal and other geographic-aware systems such as Google Earth. The system underwent a real test scenario during the post-2008 hurricane season PDNA exercise in Haiti where it was demonstrated to be a valuable tool to support the decision-making process in all its stages.

1 Introduction

Between 1974 and 2003 more than 255 million people were affected by natural disasters globally each year, with an annual average death toll of 58,000 lives (Guha-Sapir et al. 2004). The estimated damage cost during the last decade amounts to US$ 67 billion per year on average. There are scientific predictions and evidences indicating that the number of disasters per year will increase in the near feature. Following the usual pattern, the poor countries have more victims, while the rich countries have greater damage (Guha-Sapir et al. 2004).

M. Konecny et al. (eds.), *Geographic Information and Cartography for Risk and Crisis Management*, Lecture Notes in Geoinformation and Cartography,
DOI 10.1007/978-3-642-03442-8_20,

During the last decade ICT tools have impacted dramatically on the information-management process during emergencies. It is true that an increase in the circulation of disaster information does not always correspond to new assimilation of disaster knowledge (Marcinoni 2007) and also that the waves of raw data reaching the decision maker during the emergency is like a 'cyber-replication' of the disaster (Amdahl 2001).

In a post-disaster scenario, generally in any emergency, several organisations and bodies collaborate to plan the response and to manage the crises in the most efficient way. The interorganizational coordination of the response operations and the assessment of the ongoing crisis situation are managed through a situation room, where the disaster managers take the best decision according to the information they have available, that must be reliable and up to date.

As for the type of information needed to support the decision-making process, it is important that information from all the available fields of application or thematic domains are made available to the manager. In fact, most of the time the information is maintained in different and geographically spread repositories; the access to the needed knowledge takes too long. In some real crisis scenarios, such as after 9/11, the lack of data readily available to integrate and use in the analytical systems, affected the timeliness of the rescue teams and/or the response operations (Fickes 2003). Other studies demonstrated that substantial technical and political/social issues exist in collecting, accessing and sharing the spatial data to support post-disaster management (Sanjay Jain and McLean 2003, SNDR 2002, Rajabifard and Williamson 2003).

It is important to collect and keep up to date the needed data sets before the occurrence of a disaster; it is likewise important to efficiently manage the flow of updates from the field to the situation room.

As today, considering the management, the political and the economic issues, there is no one organization that can cover the role of unique coordinator and data manager, the data collection and management for disaster response should be done jointly through a collaborative effort of all the bodies participating in the operations (Mansourian et al. 2006).

Without any doubt, the global widespread use of Internet technologies, the almost ubiquitous mobile communication networks and the market availability of spatially aware devices set the technological structure for the development of a new set of collaborative mapping frameworks (Parker and Stileman 2005). At the same time spatial tools pervaded different aspects of every day life; people became more used to maps and to reasoning spatially. The spatial dimension became the environment where several in homogeneous data sources could be analyzed consistently. This clear trend is not only technology related; also new trends in using geographic information began after the advent of Google Earth and car navigation systems. During the last years we have seen the beginning of a new branch of geography: Volunteered Geographic information (VGI) (Goodchild 2007).

Considering all the previously stated factors, a better way of managing and sharing information, needed in all phases of the disaster-management cycle (Wattegama 2007), is now possible.

Approaching the problem with the adoption of a Spatial Data Infrastructure (SDI) will solve most of these issues. In fact, an SDI, intended as an environment in which all stakeholders can cooperate and interact with technology to better achieve their objectives at different management levels (Chan et al. 2001), assists the users by granting a link between the human component and the data, through a series of standards, policies and networks.

The authors agree with the validity of an SDI approach to spatial data management during emergencies; but, according to their experiences in the field of disaster management, the design and the implementation of multiorganization SDI has to deal with and to solve a large series of technical, institutional, political and social issues that create barriers for the adoption of such a system.

Another issue that limits the interest in adopting a unique multiorganization SDI is that many international organizations have already invested in that direction quite large amounts of money, such as the European Commission for INSPIRE and the UN for UNSDI. Such organisations would be hardly interested in changing data structure, procedures and tools toward the adoption of a unique and shared GSDI.

This chapter focuses on a georeporting system to support information management during crises and emergencies. The tool is not intended to substitute an SDI, but to couple with existent SDIs to facilitate the collection and the sharing of up to date data in the aftermath of a disaster. In the authors' opinion the most important contribution this article gives to the literature relating to ICT in support of disaster management is the description of the experience in a real case scenario. In fact, the chapter describes our system focusing on the test performed in the aftermath of the 2008 hurricane season in Haiti.

In the next paragraph the system architecture and functionalities are described, while the results of the test run in Haiti are shown in the following section.

2 The field tracking tool

2.1 Description of the system

The Field Tracking Tool (FTT) is a software suite designed in collaboration between the Joint Research Center and the United Nations Cartographic Section to promote a quick assessment of the ground scenario and situation awareness at the management level. The FTT supports information management and sharing during all phases of the crisis and emergency decision process; it endorses the collaborative field data collection and provides a common environment for the publication of all the results of the spatial analysis and the cartographic rendering.

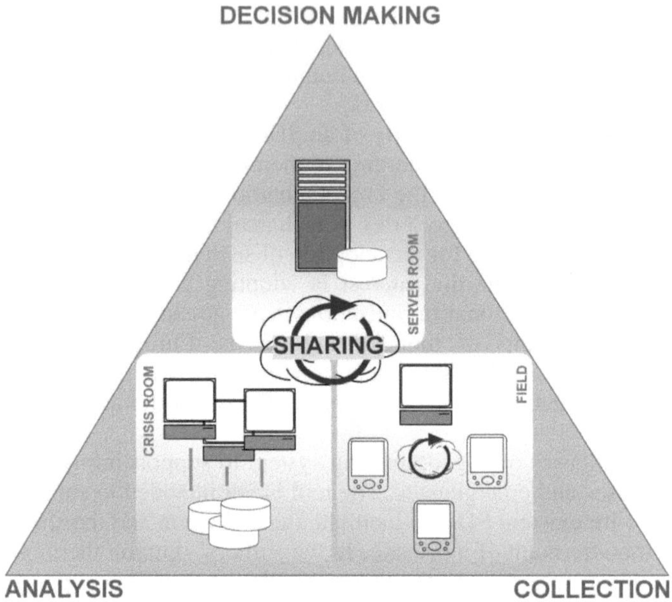

Fig. 1. The FTT in support of the decision-making process

Considering the three main user categories playing a role in the decisional process, namely the field operators, the analysts and the decision makers, the FTT allows a real-time collaborative spatial reasoning and multidirectional communication mechanism. The FTT plays a fundamental role in the spatial reasoning, because it helps with the visualization, analysis and modelling of information belonging to completely different domains of knowledge in a common collaborative environment (Fig. 1).

The field teams deployed in the crisis area have their position tracked and they are able to collect geo-referenced information and pictures. In almost real time, if any type of mobile connectivity is available in the area, the information can be uploaded to the central repository and become immediately available for all the other system users. In fact, as soon as the information is uploaded to the central shared repository, it becomes available in several different formats enabling both geographers to perform any desired spatial analysis and the decision makers to visualize the collected information using a geographic browser such as Google Earth.

The web repository is also the container for the analysis results.

2.2 System architecture and technical specifications

The FTT is made up of three components: the MobileFTT, the DeskFTT and the FTTPortal (Fig. 2).

The MobileFTT is a software application to use on PDAs, smartphones and mobile devices, the DeskFTT is a desktop application to support the preparation of the data to use in the field and to revise and/or integrate the collected information and pictures before uploading to the shared repository. The FTTPortal is a server application that provides a web interface to the information stored in the main repository and give to the authenticated users the ability of inserting, querying and downloading the available information.

The system is not a GIS software. This is a pondered design choice to give the users the ability of choosing their preferred software for the tasks they have to perform. Also, in order to eliminate all the software dependences and to ease the sharing process, the FTT uses extensively a set of standard data formats, mainly XML files like RSS or KML, and interfaces, such as the OGC's web interfaces.

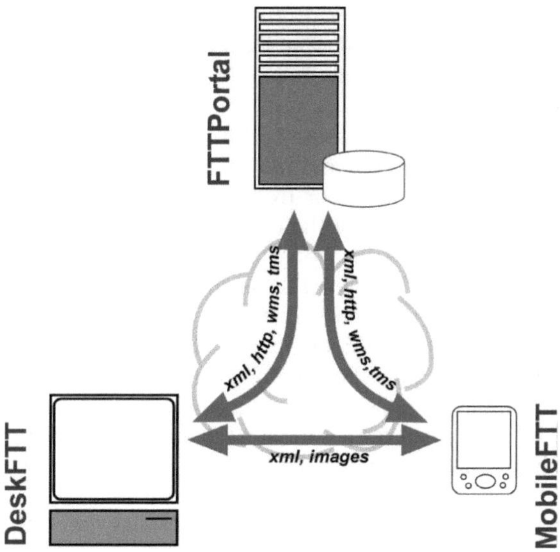

Fig. 2. System architecture

The communication between the three software components strongly relies on the Internet to exchange data and information. A network connection is needed to exploit the full potential of the system and of the collaborative functionalities. However, each single component keeps working without a connection and the easy data structure and the textual configuration files allows the exchange of information and preset applications also when offline by means of any supported memory device.

The FTT software modules are based on Microsoft .NET technology. The MobileFTT is an application written in VB.Net for Compact Framework. It is GPS aware: that is automatically detected and used if available. It can work in the background allowing the user to use other applications on the PDA, while tracking. Also, the MobileFTT exploits the built-in camera if present, to take pictures, which will be tagged with a timestamp and GPS position. Interfacing with WMS services, it can serve real-time maps for positioning and navigation purposes. It can also access huge tile sets stored on memory cards, or accessible as a TMS service over the Internet, to show fast detailed maps.

In order to speed up the acquisition process, a need usually felt when working in severe or risky environments, the application provides a series of quick-add functionalities. With only few user interactions it is possible to add geo-referenced information and pictures through a large-size pictorial button-based interface dynamically generated according to predefined domain ontology.

The DeskFTT is used for reviewing and analyzing the data both before and after the field data survey. Before deploying the field teams in the disaster area, this application serves to define the operative environment for the MobileFTT. It is possible to easily configure the available online data sources (WMS, TMS, and images) but it has also caching capabilities to store offline the data needed in a form that can be used by the PDA. The context for the field mission can also be created, adding waypoints and other useful references. Other information can be easily retrieved from the FTTPortal. The DeskFTT is then used to define the domain ontology for the future field mission and to organize quick-shot buttons to be used on the MobileFTT.

After the field campaign, the collected information can be reviewed, edited, organized and integrated in a simple and intuitive way, such as drag-and-drop support from one document to another or from lists to intuitive tree views and vice versa. This environment is meant for the user to add all information that is not easily managed by PDA interfaces like long texts, to refine data and to benefit from the work performed on many MobileFTT systems.

The moderating activity is improved by the capability to use high-resolution images and maps, both from WMS or tile sets. During this process, the information manager can integrate the data coming from the MobileFTTs, and can also feed them new information gathered elsewhere. It can also be used to design further activities to be performed during the next days. After elaboration, data can be exported easily in common formats like KML or ShapeFile, or uploaded to the FTTPortal.

The FTTPortal is DotNetNuke based, incorporating several special-purpose modules to share contents, which are geo-referenced, such as GEO-RSS, KML/KMZ and the like. It presents on interactive maps the data collected and elaborated by MobileFTT and DeskFTT, and can provide extensive reports on the data uploaded.

The FTTPortal provides also the infrastructure for the Web Service that allows the management of the information, capable of mass downloads and uploads. A simple authentication mechanism provides the means to authorize modifications of the content.

3 The FTT in a real post-disaster scenario

3.1 The 2008 hurricane season in Haiti

During August and September of the year 2008 Haiti was swept by four consecutive hurricanes (Fig. 3).

Hurricane FAY (GDACS 2008a) made landfall onto the Dominican Republic on the 15th of August and, even if it was of category Tropical Storm in the *Sapphire-Simpson* scale, it dumped heavy rains and caused approximately 50 fatalities and great destruction.

Hurricane GUSTAV (GDACS 2008b), the seventh named storm of the Atlantic 2008 hurricane season, made landfall near to Jacmel, a city in the South-West peninsula on the 26th of August. Thousands of people were affected by floods and heavy winds. The road network between the affected area and the capital, Port-au-Prince, was heavily damaged and the aid organization had to cope with logistic challenges to assess the situation in the area. Gustav caused 75 fatalities and, approximately 6,000 people were moved into temporary shelters.

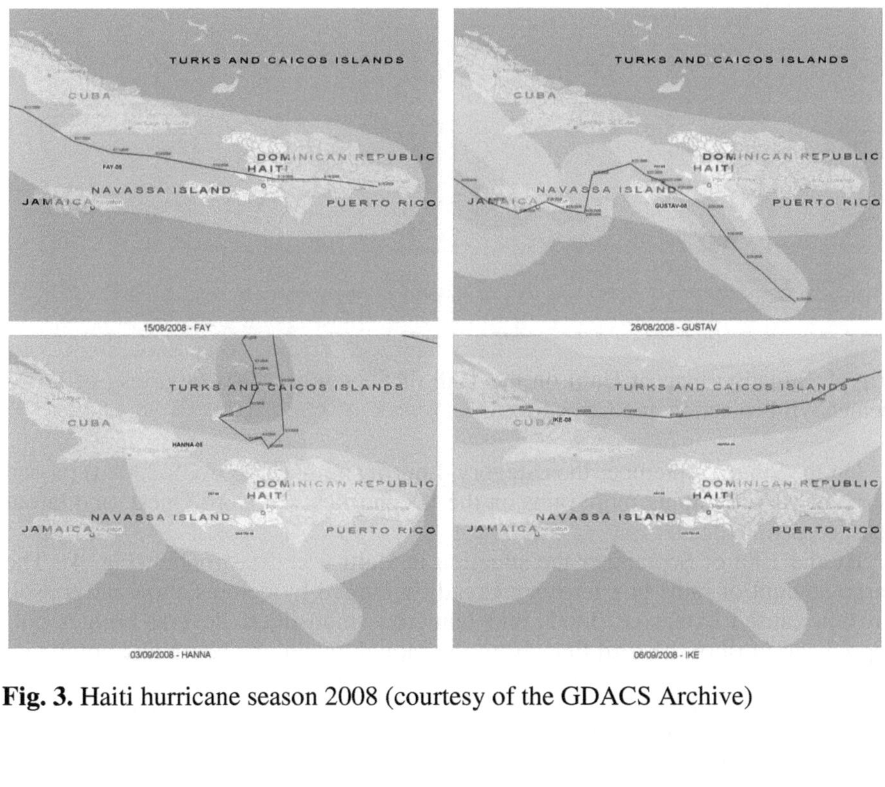

Fig. 3. Haiti hurricane season 2008 (courtesy of the GDACS Archive)

The effects of Gustav were exacerbated by the storm HANNA (GDACS 2008c) that kept moving erratically off the north coast of Haiti. Hanna's torrential rains caused several floods and mudslides in the North departments. The most affected area was the Department of Artibonite where the capital, Gonaives, was covered by two meters of water and mud. In some areas of the city the thickness of the silt layer left behind after the floodwater receded reached the roof of the houses. The UN calculated 2.5 million cubic meters of mud were deposited in the city alone and it estimated it would take the removal of about 400 truckloads of mud a day, every day for a year to clear Gonaives (Reilly 2008). Hanna killed at least 500 and left thousands of houses destroyed, flooded warehouses and hospitals.

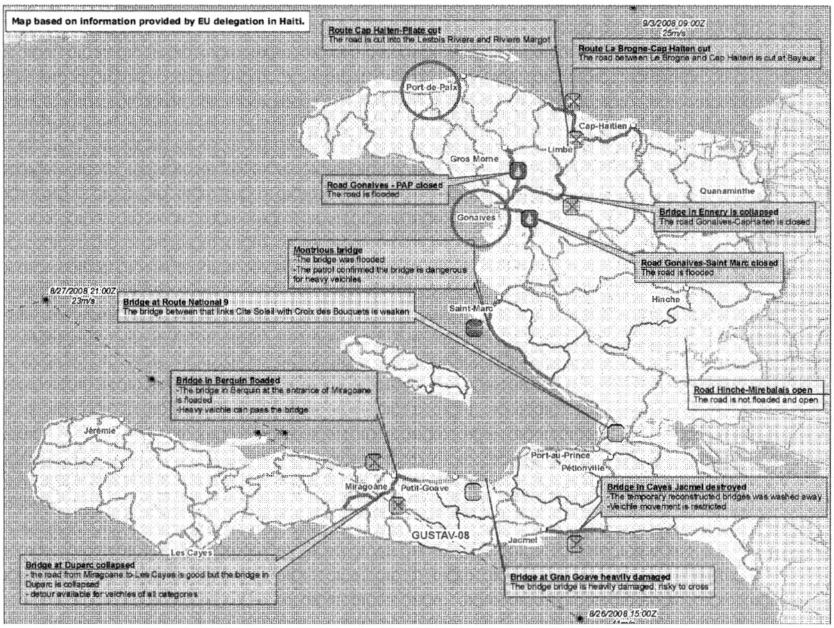

Fig. 4. Situation map of Haiti on the 10th of September 2008 (courtesy of European Civil Protection and Critech)

On the 9th of September the category 4 hurricane IKE (GDACS 2008d) passed north of Haiti dumping more rains on the 'Departments of Nord-Ouest' and left at least 65 people dead.

By the 10th of September the situation in Haiti was catastrophic (Fig. 4). The large amount of rains in a territory already scourged by deforestation and topsoil erosion caused extensive floods and numerous mudslides. Several bridges collapsed and most of the road links from the capital toward the North and South districts were interrupted. Gonaives was inaccessible by road.

For the four storms more than 800,000 people were affected, 793 killed and 310 missing. More than 22,000 houses were destroyed and approximately 85,000 were damaged (USAID 2008).

In this scenario the European Commission, the United Nations and the World Bank jointly performed a Post-Disaster Need Assessment (PDNA) exercise to estimate the extent of damage and to assess the needs in order to support the international response and funding. This was the first joint operation after an international agreement was signed in September 2008 in order to address jointly post-disaster support and recovery (Kayitakire 2008).

3.2 The operational scenario

According to a common paradigm in the discipline of Crisis Management, the FTT test workgroup was made up of four categories of experts: the field operators, the analysts, the managers/decision makers and the technical-logistic support.

Two field teams were deployed to Haiti to create a real multiuser field environment. One team focused on the quality evaluation of the damage assessment procedures based on the analysis of the satellite imagery previously performed, the other team was more focused on testing the FTT and on evaluating the real potential and the limitations of the tool in this context. The FTT was installed on three commercial off-the-shelf smartphones having a GPS receiver and a digital camera mounted. In particular the positioning system consisted of an assisted GPS for the two more recent devices, while the older one has a standard receiver. Each team were provided with two additional batteries and an additional two gigabyte memory card.

Each team's equipment included also a GPS logger and a standard digital camera to act as a failover and control system. The aim was to gather a set of reference data to validate and evaluate the quality of the information collected using the MobileFTT.

As long as the mobile connectivity is concerned, only one of the deployed teams was connected to the Internet through a GPRS EDGE connection provided by a local mobile company called Digicell. To set up a test scenario with limited connectivity, the second team was not network enabled. In this configuration the PDA devices exploit the Bluetooth connectivity for local exchange of data and only the network enabled device synchronizes with the central repository.

As a backup system the teams were given a laptop running the DeskFTT and a complete virtual map server solution providing local access to a complete set of contextual data.

The situation room was set in Ispra (VA, Italy) where one GIS expert and one manager were on duty during the ongoing survey in Haiti. Also the European Civil Protection MIC followed the evolution of the field mission from the central office in Brussels (Belgium).

The developers of all the three modules provided transversal support for all needs and the activities performed.

3.3 Test case description

As specified above, the main role of the JRC scientists within the PDNA expert group was to estimate the damage extent of the most affected urban areas by means of a methodology consisting of the analysis of satellite images and of the following validation of the results through ground truth observations. The FTT was used in this context to support the ground-truthing and it was tested to assess its real potential, and obviously its limitations, in a real post-disaster crisis environment.

The tests ran in Haiti were designed with particular attention to understand the potential of the system for collecting GPS tracks and geo-referenced punctual information and pictures, to enable real-time communications between the field teams and the situation room, to understand the grade of situation awareness achievable and to tune and refine a set of pre-defined procedures for rapid and collaborative mapping. The tests ran were also to assess and evaluate a set of system features such as:

- the performance, reliability and robustness of the MobileFTT and the PDA devices in a severe environment;
- the limits of communications among field teams and between the field team and the crisis room;
- the usability and the ergonomics of the systems;
- the learning curve for beginners.

3.4 Test outcomes

3.4.1 System software modules

The DeskFTT and the MobileFTT modules performed well during the whole set of field emergency tasks and activities.

Apart from some minor AGPS malfunctioning, it was possible to track the position of the deployed teams at any time and to collect geo-referenced punctual information and pictures. The software was demonstrated to be reliable, in the sense that even after a general system failure no data loss was registered. The data collected have the same level of accuracy as the GPS logger used as the control system. Even if no statistical analysis of the data accuracy was performed, the outcome of a visual check in urban areas showed that the error in the positioning of the acquired information was generally within the range of 5-15 meters, when compared with the geo-referenced satellite images.

The graphic user interface was demonstrated to be efficient for data collection. The quick-add pictorial buttons were the most used while collecting data when it was difficult to fill all the available fields in the form, because of the severe external environment or safety issues.

The DeskFTT was a valid support for preparing the mission and reviewing the collected information. The laptop with all the data available and a virtual map server installed was demonstrated to be a valid backup system to load data on the PDAs in any situation.

The FTTPortal performed well as a central repository of geo-referenced data and pictures and as an information-sharing platform accessed by the decision makers and the analysts. In some cases the access time to the needed information was too high when accessing through a low bandwidth connection, and it is felt that an alternate strategy should be implemented to provide efficiently the needed information through slow user connections.

3.4.2 Field devices

The commercial mobile devices used for the Haiti test session showed a lack of robustness and often the feeling was that a rugged device would be necessary. Also the cameras mounted in a this kind of device are often too slow to acquire an image when moving in a car or helicopter; and the resolution and the light sensitivity seemed poor in some contexts.

The GPS receivers mounted on the deployed devices worked well in most cases, but the GPS mounted on two of the devices experienced some problems relating to accuracy when not enough satellites were fixed. The PDA device mounting the standard GPS receiver performed well in any experienced situation.

The two gigabyte memory card was sufficient to store a full day of surveying. In order to work the whole day at least three fully charged batteries were needed and it was very helpful having the car battery charger to plug in the mobile when moving by car. Also, the desktop is a viable way of recharging the batteries when in the field.

3.4.3 Field-situation room and field–field communication

During the field mission the Digicell's EDGE connectivity allowed US to keep slow but reliable real-time communication between the mobile devices and the central repository. The upload time varied from a few seconds to several minutes according to the number and kind of punctual information collected, increasing proportionally to the number of pictures uploaded. Experienced uploading troubles, due to network failure when uploading on the move, were common but the implemented incremental uploading procedure avoided any data loss.

In general the teams deployed in the field can share information either exploiting the Internet connectivity, in this case using the server as a network bridge, or taking advantage of the built-in bluetooth antenna for a peer-to-peer data exchange. Since only one device was network enabled, the paradigm used presented a central point for uploading/downloading the information to/from the server, while the other mobile devices shared the data through bluetooth. In Haiti this was a successful configuration and it presented a set of advantages. Only one of the devices was locked during the uploading/downloading process, and the other two could be used normally for data collection in the meantime. Considering the low bandwidth available, having the data downloaded on one device it was much faster to exchange the data through the bluetooth channel than repeating the download procedure from the other devices. Moreover, this strategy reduced the cost of connectivity. As a drawback, this configuration cannot be used if the teams are deployed in very different areas: in this case each team must be network enabled and the communications between the field teams must go through the central repository.

Another architecture tested was the one where the laptop worked as a temporary field central repository and proxy, all the PDAs uploaded data to the laptop using bluetooth or a USB cable and the data synchronization was performed through the DeskFTT, leaving all the mobile devices available for field surveying.

3.4.4 Common procedures

During the test a set of common pre-defined procedures were tuned and refined to cope better with the limited, and limiting, resources available while in the field.

The procedures designed for the preparation of the data for supporting the field activities demonstrated their validity, but it took too much time to complete when in the field. In fact, considering the intense workload during the field mission, the data preparation is performed during night-time. To improve this procedure the DeskFTT should integrate components to prepare the data support more efficiently, in other words an advanced tiling system should be used.

At the same time all the hardware equipment is to be revised, the batteries charged and the data moved from the laptop disk to the memory cards. Since it is not possible to shrink the time needed to perform these activities, a parallel-processing approach should be used.

The work done for defining a rapid mapping procedure, and in particular the design of a set of map templates, the definition of symbol sets, the implementation of the analysis procedure and the automation of some data integration processing produced a high-quality rapid mapping infrastructure. Obviously, there is still substantial room for improvement but the test demonstrated this procedure to be ready to be promoted to the production level.

3.4.5 Real-time information sharing and spatial analysis

With the collected information stored in the central and shared repository, it is time to analyze the data using the common GIS programs to produce up to date situation maps or spatial analysis. As stated above, the FTT itself is not a GIS software nor is it able to perform analysis. It is an information-management tool and the users are free to use the tools they prefer, in order to perform the analysis and the elaborations of the data they need. Having then newly derived datasets, they can upload again to the portal to share the results with other authorized users.

During the test, while the mission was ongoing, the time needed for making available an updated situation map was approximately 20 minutes. A contextual example is the visit to the collapsed bridge in Montrouis, a village placed on the National Road 1, between Port au Prince and Gonaives (Fig. 4). The N1 was deviated on a secondary road where an old rail bridge survived the river flood. In a time span of about 30 minutes the data collected in the field were updated to the central repository, a new logistic map showing the new road was generated and shared with all the portal users.

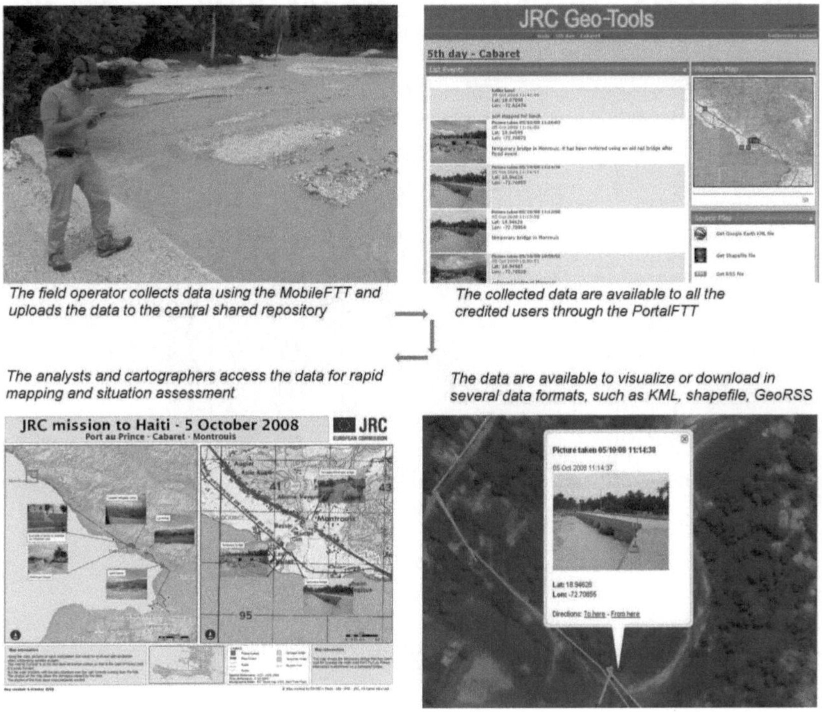

Fig. 5. Information flow during the survey of Montrouis road interruption

A similar example is the daytrip to Gonaives, the most damaged city by the events of 2008. On the 9th of October the road from Port au Prince and Gonaives was interrupted in the plain of Gonaives because of an extensive flood. The whole trip was tracked and uploaded in real time; a new map of the deviated roads was available to download in less than 20 minutes from the data collection (Figs. 5 and 6).

3.4.6 Usability, troubleshooting and learning curve

Even without any usability test done so far, the efforts to make a simple system gave the FTT a general high level of usability.

The simple data structure and configuration files make it possible also for non-IT experts to troubleshoot or customize the tool for user needs. The learning curve is really flat. A short training period is enough to make a new user productive in the field.

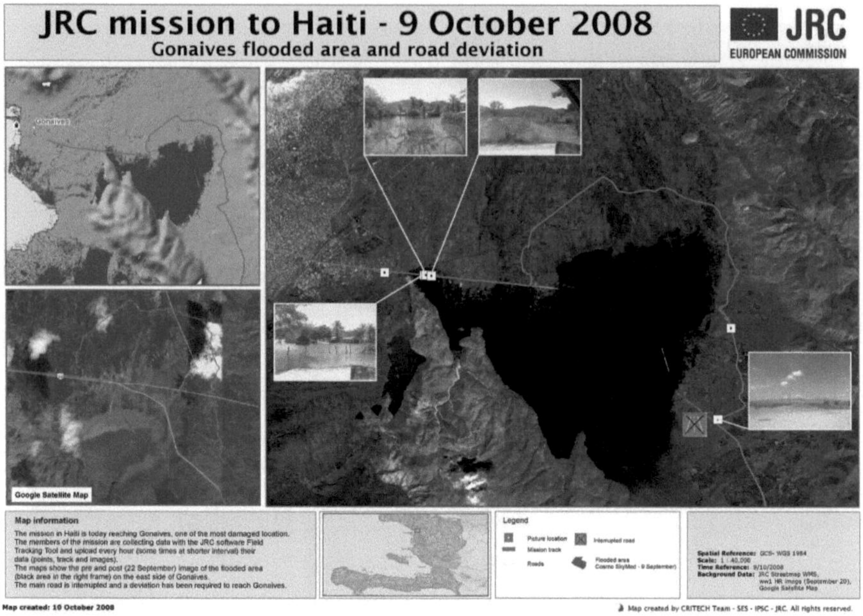

Fig. 6. Maps of the national road deviation toward Gonaives

4 Discussion

The tests performed in Haiti demonstrated that the FTT is a valuable tool to support the collaborative information management during emergencies. Obviously the system met some limitations, and some bugs were discovered but none of them produced an invalidating condition. Using the three components, the whole information flow is assisted and eased, starting from the field data collection to the decision making. Even working on low bandwidth network connections, the communication link between the field and the server was almost always possible and reliable. The network failures experienced did not cause any data loss due to the efficiency of the incremental uploading algorithm implemented. The whole system was demonstrated to have a good level of robustness and reliability. As in the case of the network failures, they did not produce any data losses.

The field test focused mainly on the MobileFTT component and on the communications among this component and the other two modules in the system. It is a logical consequence of the character of the geo-reporting tool and of the kind of field test. In fact, the server component acts mainly as a shared repository and the desktop acts as a broker between the FTT and the GIS solutions adopted in the various organizations (in our case the suite of ESRI). They both performed well assuring effective data sharing and preparation. The spatial analysis on the newly collected data are performed through the suitable GIS tools, and they were not part of the test that focused on the real-time sharing of information and analysis results between the field operators, the analysts and the decision makers.

During the test in Haiti, a group of new users were asked to use the MobileFTT while performing their field activities. As stated in the previous section of this document, even without a dedicated and well-designed usability study, the users were productive after a short training period. This outcome, together with other evidences collected, suggest that it is not yet possible to involve volunteers and citizens in the FTT activities. At least three major limitations affect the participation of those categories of users, namely the lack of field data collection experience, the availability and the cost of the smartphones/PDAs and of the Internet connectivity and the fact that training the volunteers/citizens in the next disaster affected area is very difficult, impossible in some cases, due to the social and geopolitical issues and to the frequent impossibility of forecasting where the next disaster will strike.

Concerning hardware, one of the main limitations encountered was the robustness of the commercial PDAs used. It is clear that for field use commercial mobile phones do not have the requirement of roughness and resistance needed. A more specialized and rugged device would be advisable, if not needed, for this kind of use.

The low quality of the smartphone camera is mainly due to the poor resolution, the long acquisition delay and the low light sensitivity, if compared to the cheap commercial camera available on the market, but it is not an issue in our opinion; because it is more important to have data in real time than high-resolution data that is difficult to upload when connecting through slow networks.

Several aspects can be improved. Optimizing the format of data for network transmission and also simplifying the applications serving those data would speed up the field–situation room data exchange. On the other hand, simplifying the quick acquisition functionalities would improve the speed of information collection. Designing a more integrated data preparation and context configuration would reduce the deployment time. For deployment it should be kept in mind that each device must be accurately tested before deployment in the field. To serve this, a set of guidelines for data preparation and application testing is under preparation.

The system is presently in a tuning stage to fix some bugs and to develop better and new functionalities to respond to the needs raised by the test in the Caribbean.

References

Amdahl G (2001) E Team: Emergencies and the internet. Disaster Response, GIS for Public Safety. ESRI Press, Redlands, CA, pp. 33–38.

Chan TO et al. (2001) The dynamic nature of Spatial Data Infrastructures: A method of descriptive classification. Geomatica 55(1):65–72.

Fickes M (2003) WHY WE NEED SMARTER MAPS. http://govtsecurity.com/mag/why_need_smarter/ [Accessed June 11, 2009].

GDACS (2008a) GDACS event report – Tropical Cyclone FAY-08. Global Disaster Alert and Coordination System. http://www.gdacs.org/reports.asp?eventType=TC&ID=FAY-08&system=asgard&alertlevel=Green&glide_no=&location=USA&country=United%20States&new=true [Accessed June 15, 2009].

GDACS (2008b) GDACS event report – Tropical Cyclone GUSTAV-08. Global Disaster Alert and Coordination System. http://www.gdacs.org/reports.asp?eventType=TC&ID=GUSTAV-08&system=asgard&alertlevel=Orange&glide_no=&location=HTI&country=Haiti&new=true [Accessed June 15, 2009].

GDACS (2008c) GDACS event report – Tropical Cyclone HANNA-08. Global Disaster Alert and Coordination System. http://www.gdacs.org/reports.asp?eventType=TC&ID=HANNA-08&system=asgard&alertlevel=Orange&glide_no=&location=DOM&country=Dominican%20Republic,%20Haiti,%20Turks%20%20Caicos%20Is.,%20Cuba,%20United%20States,%20The%20Bahamas&new=true [Accessed June 15, 2009].

GDACS (2008d) GDACS event report – Tropical Cyclone IKE-08. Global Disaster Alert and Coordination System. http://www.gdacs.org/reports.asp?eventType=TC&ID=IKE-08&system=asgard&alertlevel=Red&glide_no=TC-2008-000151-BHS&location=CUB&country=Cuba&new=true [Accessed June 15, 2009].

Goodchild M (2007) Citizens as sensors: The world of volunteered geography. GeoJournal 69(4): 211–221.

Guha-Sapir D et al. (2004) Thirty years of natural disasters 1974–2003, Presses University de Louvain.

Kayitakire F (2008) Rapport d`evaluation des besoins apres disastre. Cyclones Fay, Gustav, Hanna et Ike., Port-au-Prince (Haiti): Government of Haiti – The World Bank – United Nations – European Commission.

Mansourian A et al. (2006) Using SDI and web-based system to facilitate disaster management. Comput Geosci 32(3):303–315.

Marcinoni F (2007) Information technologies and the sharing of disaster knowledge: The critical role of professional culture. Disasters 31(4):459–476.

Parker C, Stileman M (2005) Disaster management: The challenges for a national geographic information provider. Geo-information for Disaster Management. Springer, New York, pp. 191–214. http://dx.doi.org/10.1007/3-540-27468-5_14 [Accessed June 15, 2009].

Rajabifard A, Williamson IP (2003) Anticipating the cultural aspects of sharing for SDI development. Proceedings, Spatial Sciences 2003 Conference. Canberra, AU.

Reilly D (2008) Reuters AlertNet – A year's worth of mud in Gonaives. http://www.alertnet.org/db/blogs/53808/2008/09/13-124438-1.htm [Accessed June 15, 2009].

Sanjay Jain, McLean C (2003) A framework for modeling and simulation for emergency response. In Simulation Conference, 2003. In: Proceedings of the 2003 Winter. Vol. 1, pp. 1068–1076.

SNDR (2002) A National Hazards Information Strategy: Reducing Disaster Losses Through Better Information. National Science and Technology Council,

Committee on the Environment and Natural Resources, Subcommittee on Natural Disaster Reduction (SNDR). Washington, DC, USA.

USAID (2008) Haiti – Storms Fact Sheet #6 (FY) 2009. http://www.reliefweb.int/rw/rwb.nsf/db900sid/TUJA-7MG2NN?OpenDocument [Accessed June 15, 2009].

Wattegama C (2007) ICT for Disaster Management. UN-APCICT, UNDP, Bangkok, Thailand.

Practical Experience with a Contextual Map Service

Jiří Kozel and Radim Štampach

Abstract

A year ago, the development of a Contextual Map Service called Sissi began. As it has been introduced in previous research articles, the purpose of this service, which is based on the OGC WMS specification, is to implement the idea of contextual cartographic visualisation into the web environment. Today, we have the first functional prototype of Sissi, which was tested in November 2008 during a field experiment called 'Transport of Dangerous Goods II'.

This chapter consists of four sections. The first section briefly introduces the terms 'context', 'elementary context' and 'contextual cartographic visualisation'. The second section deals with the main principles of the Sissi Contextual Map Service and with the practical creation of the complex context structure. The next section focuses on assigning the map's content and symbology to particular contexts, with special attention paid to map symbology that is described by the OGC Styled Layer Descriptor (SLD) specification. Finally, the chapter focuses on the course and the results of the experiment.

1 Introduction

The contextual approach to cartographic visualisation has been known since the mid-1990s; nevertheless, only a few solutions to this approach have been implemented.

Tumasch Reichenbacher (Reichenbacher 2004) introduced a solution for mobile devices that was based on a client–server architecture. The server side was responsible for the contextual cartographic visualisation; however, the client side

M. Konecny et al. (eds.), *Geographic Information and Cartography for Risk and Crisis Management*, Lecture Notes in Geoinformation and Cartography,
DOI 10.1007/978-3-642-03442-8_21, © Springer-Verlag Berlin Heidelberg 2010

was only responsible for presentation and interaction. The same idea was used in the GiMoDig project (Lehto 2003), which ensured communication between the client and server sides by extension of two specifications: the Web Map Service and the Web Feature Service. Both solutions had invariant elementary context types.

This chapter first presents practical experiences with our solution to the contextual approach, called Sissi–Contextual Map Service, which was designed two years ago (Kozel 2007). The solution follows the same philosophy as the two previously mentioned ones; however, it has several improvements. First, Sissi is not bound to a set of given elementary context types, which means that each Sissi instance is able to support different types, and therefore different contexts. Second, it is based on the Web Map Service specification with an extending request that ensures that the client can simply obtain supported elementary context types and contexts. Third, the symbology of the contextual maps is defined through the Styled Layer Descriptor specification.

2 Context theory

Context is a widely used term in many fields of science and has many meanings. In this chapter, the term is used in relation to geoinformatics and the adaptation of map applications. In this field, the term *context* was introduced by Bill Schilit (Schilit et al. 1995).

Although many definitions of context exist, none of them fully correspond to our conception. This is why we introduce another definition: 'Context is any information (or set of non-contradicting pieces of information) that is relevant to user–application interactions and that can be used by the application to affect its running or appearance.' This definition is based on the Dey and Abowd definition (Dey and Abowd 1999), which is probably the most-used, non-enumerating definition of context. For example, in relation to map applications, any information that describes the user's role, location, activity, environment, display device, etc. can be considered as a context.

From the cartographic point of view, context is closely related to the *purpose of a map*. Nevertheless, context is a much more general term than the purpose of a map in many respects. For example, while the purpose of a map influences the map's appearance, the context may affect both the map's application appearance and run.

2.1 Contextual cartographic visualisation

Contextual cartographic visualisation (or adaptive cartographic visualisation) is a process of adapting a map's application to a particular user's context by highlighting the context-relevant spatial information. The process of map adaptation includes operations that change the map's properties, e.g. content, symbology, cartographic method, coordinate reference system, extent and scale (Friedmannová et al. 2006). In other words, this process is used by context-aware map applications to change the appearance. The context is described by a user or it can be obtained, for example, from sensors (see Reznik 2007).

A user who works with a context-aware map application should be able to describe his context rather than defining the map's content and symbology (as is common in traditional approaches to cartographic visualisation). This contextual approach saves the user time because the user does not need to search through a number of map layers and choose the relevant ones. Furthermore, he does not need to know anything about access to these layers and about the definition of map symbology suitable to his context. In other words, a context-aware map application can be used by users without any cartographic knowledge because the map is (or should be) correctly composed by the application.

2.2 Special types of context

Let us define three contexts A, B and C:

- A: 'User is a fireman.'
- B: 'User intervenes in a flood.'
- C: 'User is a fireman intervening in a flood.'

By referring to the definition of context, it is clear that the set $A \cup B$ is also considered a context and is equivalent to context C. For a better orientation, it was decided to define three special types of context according to their structure.

Composite context is a context that is equivalent to the union of at least two different contexts. Context C is equivalent to the set $A \cup B$; therefore, it is a composite context.

Subcontext is a context that is a non-empty proper subset of another context. Both contexts A and B are subcontexts.

Elementary context is a context that cannot be divided into more different contexts, or its division is senseless. Both contexts A and B are also elementary contexts. Senselessness of division is a theme that is similar to atomicity of domains in data modelling.

2.3 Context types according to object of description

For simpler work with contexts, it is also necessary to classify contexts into types according to their object of description. Let us define two contexts: 'user is a fireman' and 'user is a policeman'. It is evident that both contexts describe the same object–the user's role–and therefore, they are of the same context type 'user's role'. In the rest of this chapter, the term *context type* means a set of all contexts that have the same object of description. It is possible to speak also about elementary context types, composite context types, etc.

3 Sissi – Contextual Map Service

The contextual map service called Sissi is a web-based server application that is intended to provide contextual (or context-aware) maps for map clients. In other words, Sissi is a mediator that realises the process of contextual cartographic visualisation. The communication between Sissi and map clients is ensured by an extension of the Open Geospatial Consortium (OGC) Web Map Service (WMS) specification (OGC 2002b). The extension was designed to provide contextual abilities; nevertheless, any ordinary WMS client is able to request any contextual map.

One of the main advantages of Sissi is that it is not restricted to a given set of contexts or context types, but enables its provider to define any elementary context type, any elementary context, and consequently any composite context. The only restriction that is caused by the current state of Sissi's development is that each elementary context type must be an enumeration of elementary contexts. However, it is possible to express both qualitative and quantitative data types as an enumeration, e.g. by classifying it into several classes. Each class represents one elementary context and is described by a string (e.g. 'Fireman' or '1–100 lux'). Furthermore, support for non-enumeration context types is expected to be added in the future.

Sissi obtains the content of contextual maps from one or more remote WMS servers. Furthermore, it is possible to describe appropriate symbology in the attached SLD style. The schema of the Sissi connection at both the client and server side is shown in Fig. 1.

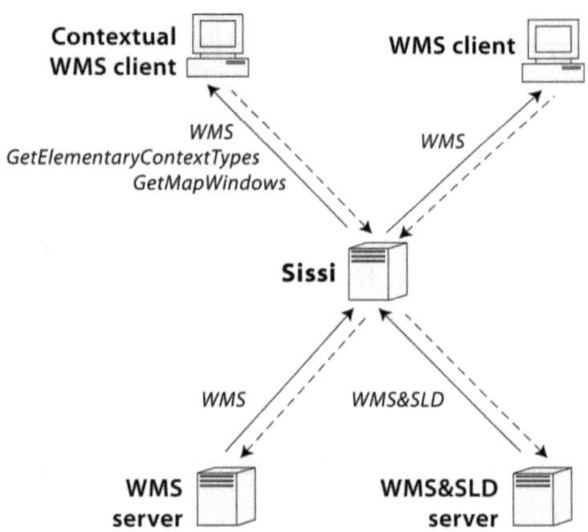

Fig. 1. Communication between Sissi, WMS clients and source WMS servers

Processing of a GetMap request (see Fig. 2) is a good example for understanding the basic functionality of Sissi. Every GetMap request that is sent to Sissi contains a special parameter that describes the user's context (see Sect. 3.2, GetCapabilities request) and two other special parameters, 'mapwindow' and 'wmssession', that specify the requested map window and the WMS session, respectively (see Sect. 4.1). First, Sissi decodes the user's context from the 'context' parameter and also the requested map window and the requested WMS session. Then, Sissi selects an appropriate context from all of the supported contexts and, together with this context, also selects the map content and the symbology definition. Finally, Sissi transforms the incoming request into one or more GetMap requests and sends them to a remote WMS server(s) that requests context-relevant content with context-relevant symbology. After obtaining image responses, Sissi merges these images together, if needed, and sends the result to the client.

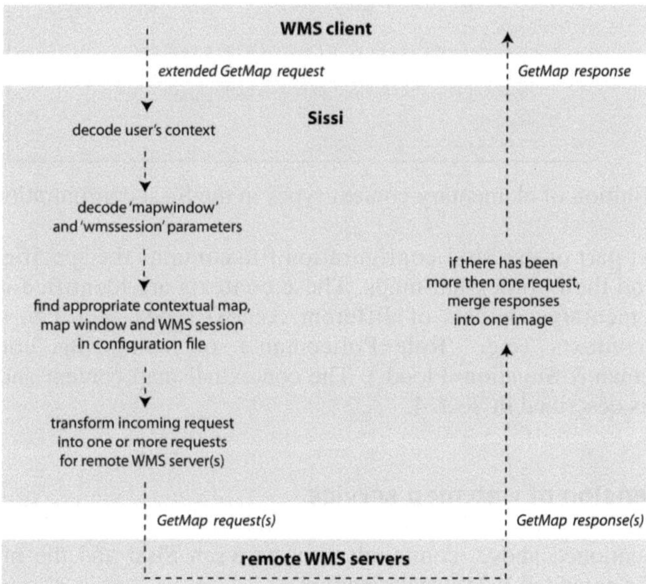

Fig. 2. Processing GetMap request with Sissi

Configuration of Sissi consists of three major steps:

1. Identify elementary context types and elementary contexts
2. Specify supported contexts
3. Define map content and symbology for supported contexts

The first and the second step are described in Sect. 3.1 while the third is described in Sect. 4.

3.1 Context structures

The first objective of the Sissi configuration is to decide which pieces of information are relevant for the provider's Sissi instance and then, to classify these pieces into elementary context types. This whole process is very similar to data modelling. For example, in crisis management, you can identify the elementary context type 'Role' as an enumeration of three contexts 'Fireman', 'Policeman' and 'Ambulance' and the elementary context type 'Situation' as an enumeration of two contexts 'Fire' and 'Flood' (of course this example is very simple). Then, the specified elementary context types are described in an XML file that is called the Sissi configuration file (see Fig. 3).

```
<ElementaryContextTypes>
  <Type name="role" title="Role">
    <Context name="fireman" title = "Fireman" />
    <Context name="policeman" title = "Policeman" />
    <Context name="ambulance" title = "Ambulance" />
  </Type>
  <Type name="situation" title="Situation">
    <Context name="fire" title="Fire" />
    <Context name="flood" title="Flood" />
  </Type>
</ElementaryContextTypes>
```

Fig. 3. Definition of elementary context types in the Sissi configuration file

The next part of the Sissi configuration file contains the specified supported contexts and their contextual maps. These contexts are identified as nonempty sets of elementary contexts of different context types. You can identify elementary contexts (e.g. 'Role=Policeman') or composite contexts (e.g. 'Role=Fireman & Situation=Flood'). The contextual map content and symbology definition is described in Sect. 4.

3.2 Extension of web map service

As was mentioned above, communication between Sissi and the map clients is based on the extended WMS specification.

3.2.1 GetElementaryContextTypes request

The first extension of the standard WMS specification is a GetElementaryContextTypes request, which shows the elementary contexts that are supported by the Sissi instance. Response to this request is a simple XML structure, which is very similar to the structure in Fig. 3.

After this request, the user should be able to choose one context, which is the most suitable for him. The more elementary contexts that are used to describe the client's context the better because a more suitable contextual map will be provided to the user.

3.2.2 GetCapabilities request

The GetCapabilities request is an integral part of the WMS specification; however, in the case of Sissi, it has one more required URL parameter: 'context'. The parameter that holds the user's context and its value is encoded as a comma-separated elementary context string (e.g. 'Fireman,Flood'). The user context is also held in the GetMap request by using the same parameter, which is a part of the GetMap URL that is obtained from the GetCapabilites response. Two other special parameters also exist, which are described below in Sect. 4.1: 'mapwindow' and 'wmssession'.

Another change in GetCapabilities is just a conceptual one. An incoming response to the request does not contain all of the Sissi map layers but consists only of map layers that are appropriate to the given context, with one appropriate symbology for each layer. In other words, the response is one complete contextual map and therefore, the user should simply display all available layers.

3.2.3 GetMapWindows request

The third extension of the standard WMS is a GetMapWindows request. A response to this request brings information about the WMS sessions of the map window and the map overview that are suitable for a given context (see Sect. 4.1). The request is not simply usable for standard WMS clients but is very valuable for extended contextual clients that are developed along with Sissi (see Sect. 5).

4 Map content and symbology definition

The most complex part of the Sissi configuration file is the definition of the contextual map content and the symbology. The contextual map content and the symbology are described separately for each supported context, as was suggested above.

4.1 Map windows and WMS sessions

Each contextual map inside of Sissi consists of one main map window and one map overview. Both the main map window and the map overview are subtypes of a map window and their content is described in the same way within the Sissi configuration file. The content of each map window is composed of graphic layers that are called WMS sessions. A WMS session is defined as a group of map layers that are sent by any WMS service as one image (i.e. as a response to a GetMap request).

An important fact is that the refresh time can be defined for each WMS session. After this time has elapsed, the map layers of the WMS session should be refreshed periodically. Hence, it is logical to place dynamic objects (e.g. cars or floods) in one WMS session with frequent refreshing while placing static objects in the second WMS session without refreshing.

Neither the WMS sessions nor multiple map windows are taken into account in the WMS specification. Therefore, in addition to the 'context' parameter, both the GetCapabilities request and the GetMap request contain the next two special parameters: 'mapwindow' and 'wmssession'. Thus, Sissi can determine which map window and which WMS session are requested. A client obtains information about the possible values for the parameters from the GetMapWindows response. If the 'mapwindow' parameter is not set, the main map window is automatically selected. If the 'wmssession' parameter is not set, all of the WMS sessions of the map window are merged together.

4.2 Definition of map content

Each WMS session element consists of map layers that are structured into thematic groups. Each layer in the Sissi configuration file corresponds with one remote layer of any source WMS server (i.e. if a client requests some layer from Sissi, Sissi requests the corresponding remote layer from the source WMS server). For every single context, Sissi is able to redefine the remote layer's name, title, scale range and other properties because the remote layer's properties do not have to be suitable for each context. Furthermore, they are generally not context-aware.

When Sissi creates the GetCapabilities response for a client, it takes layer groups and layers of all of the related WMS sessions from the Sissi configuration file and transforms this hierarchy into a standard GetCapabilites Layer element tree hierarchy. The result is always a three-level hierarchy: the zero level contains only the mandatory root layer, the first level contains the thematic layer groups and the second level contains the map layers. In other words, the structure of the map's layers in the GetCapabilities response is the same as the structure of layers and the groups in the Sissi configuration file. Every WMS client generates a structured map legend from the GetCapabilities layer structure. It means that the cartographic rules about the legend's structure should be respected even during the creation of the Sissi configuration file.

There is a slightly different situation with the order of map layers in the GetMap response image. The order is the same as an order of map layers in the Sissi configuration file by default; however, it can be changed by using SLD styles (see below). The SLD style is added to the WMS request that is sent to the remote WMS server and therefore, it controls an order of map layers that is described in the SLD specification.

4.3 Definition of map layer symbology

There are two ways to define map layer symbology inside of Sissi. First, it is possible to choose one of the styles available on the remote WMS server that provides the map layer. Second, there can be a reference to an SLD file.

Using the WMS styles (see element Style in the WMS specification) is not a very powerful way because a cartographer is limited to select one of the WMS styles that is supported on the remote WMS server. Unfortunately, WMS servers usually support only one WMS style for each layer.

SLD is an acronym for the OGC specification Styled Layer Descriptor (OGC 2002a), which describes the symbology of the WMS layers. SLD symbology is stored in the XML files that are called SLD styles. A cartographer is able to describe the most suitable symbology and is not limited to a set of supported styles. Within Sissi, it is possible to define one SLD style for each contextual map; hence, different symbology can be defined for different contextual maps. SLD styles are definitely the more powerful way; however, this way requires SLD support on the remote WMS server. This way also allows separating map content and map symbology definitions.

5 Sissi testing

The contextual map service, Sissi, was comprehensively tested during the 'Transport of Dangerous Goods II' field experiment that was held in the autumn of 2008. The experiment followed the first experiment with the same name that was held two years before (Mulíčková et al. 2007). Both experiments were organised by members of the Laboratory on Geoinformatics and Cartography, Masaryk University, Brno, Czech Republic. While the first experiment used a very simple and insufficient solution of the contextual cartographic visualisation, the second one used Sissi–Contextual Map Service. One of the main purposes of the experiment was to verify Sissi's functionality and usability.

During the second experiment, monitoring a vehicle transporting dangerous chemical substances and intervention management after its accident was simulated. The location of the vehicle and the intervening fire brigade vehicles were obtained from the GPS locators and were transferred through the wireless internet to a database. Furthermore, there were some sensors at the location of the accident, and the data from these sensors (e.g. about meteorological situation) was also transferred to the database. Data from the database was provided as map layers through a WMS server, which was one of Sissi's source WMS servers. The other source WMS servers provided topographic layers and ortophoto images.

Sissi was used for both monitoring all cars and organising the intervention. It was successfully used on different map clients at the location of the accident and in a coordination operational centre.

The only necessary software to work with Sissi was any WMS client (e.g. open source GIS software OpenJump, www.openjump.org). However, a special contextual client was developed by Jan Palas and Jakub Hrádek. The client is based on an open source JavaScript library that is called OpenLayers (www.openlayers.org) and can be accessed and used by most web browsers. The client allows easy selection of the user's contexts and shows both the map overview and the active map legend (see Fig. 4).

For the experiment, two elementary context types were defined: 'Situation' and 'Version'. The meaning of 'Situation' was the situation of the user and it consisted of two contexts. One context represented a standard supervision during transportation of dangerous goods ('Monitor') and the second represented an intervention management in the case of an accident ('Incident'). The elementary context type, 'Version', allowed switching between different types of symbology. There was one context with complex symbols for a standard monitor, one context with simple symbols for a display with a smaller resolution and there was also a context with an ortophoto layer.

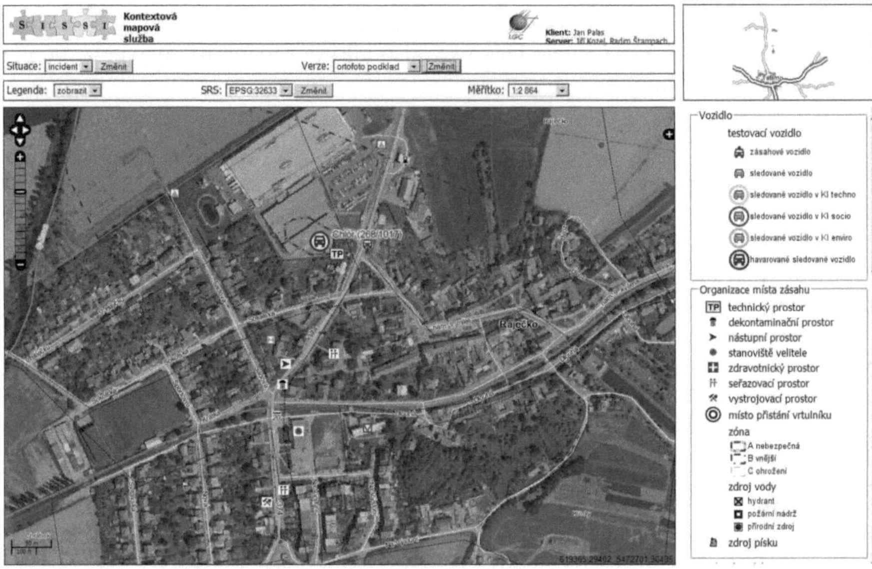

Fig. 4. An example of a contextual map that is displayed in a contextual client. The context 'Incident' is visualised with a complex version of symbols and with an ortophoto layer. The legend is shown

Five composite contexts were specified and supported as a combination of these elementary contexts. An example of one contextual map is in Fig. 4.

6 Conclusion

This chapter described the experience of development, configuration and testing of Sissi–Contextual Map Service, which is a web solution for contextual cartographic visualisation that is based on the WMS and SLD specifications. The Sissi architecture enables composition of contextual maps from the map layers of any public WMS server. By comparing with previous solutions (T. Reichenbacher or GiMoDig project), Sissi is able to support different sets of elementary context types and contexts for each instance. Furthermore, the GetElementaryContextTypes request informs the users as to which contextual maps can be requested from the specified Sissi instance.

The Sissi configuration file consists of the map's content and the symbology definition for each supported context. These definitions can be very time consuming and therefore, Sissi is especially efficient in the case of using a relatively small number of elementary context types or elementary contexts. Nevertheless, the number of supported context types and contexts is not limited by any means.

During the 'Transport of Dangerous Goods II' field experiment, Sissi's functionality and usability were tested and successfully verified. However, there have been demands for better interactivity between the client and the user. These demands mean adding the support of the WMS GetFeatureInfo request and later, also adding the general support of the OGC Web Feature Service specification. There is also a possibility to use the OGC Web Map Context specification as one format of the GetMapWindows response for better communication with ordinary WMS clients.

Acknowledgements

This research has been supported by funding from Project No. MSM0021622418 (The Ministry of Education, Youth and Sports of the Czech Republic), which is called 'Dynamic geo-visualization in crisis management'.

References

Dey AK, Abowd GD (1999) Towards a better understanding of context and context-awareness. GVU Technical Report GIT-GVU-99-22. Georgia Institute of Technology, Atlanta (GA), USA.

Friedmannová L, Konečný M, Staněk K (2006) An adaptive cartographic visualization for support of the crisis management. In: Auto-Carto Conference proceedings, Vancouver, Canada, 6p.

Kozel J (2007) Open contextual cartographic visualization. In: Proceedings of GI-Days 2007 – Young Researchers Forum. IfGI prints, Münster, Germany, pp. 223–227.

Lehto L (2003) Architecture Specification. Technical report. Finnish Geodetic Institute, Masala, Finland.

Mulíčková E, Kozel J, Kubíček P (2007) Využití kontextové vizualizace při sledování přepravy nebezpečných nákladů (in Czech). In: Proceedings of Interopsoft 2007, MSD Brno, Czech Republic, pp. 47–53.

OGC (2002a) Styled Layer Descriptor Implementation Specification 1.0.0. Open Geospatial Consortium. http://www.opengeospatial.org/standards/sld

OGC (2002b) Web Map Service Implementation Specification 1.1.1. Open Geospatial Consortium. http://www.opengeospatial.org/standards/wms

Reichenbacher T (2004) Mobile Cartography – Adaptive Visualisation of Geographic Information on Mobile Devices. Ph.D. thesis. Technical University München, Verlag Dr. Hut, München, Germany.
Reznik T (2007) Sensor Web Enablement (in Czech). In Proceedings of the 17. Cartographic Conference. Slovak Cartographic Society, Bratislava, Slovakia, 6p.
Schilit B, Adams N, Want R (1995) Context-aware computing applications. In: Workshop on Mobile Computing Systems and Applications, 1994, Proceedings. IEEE Computer Society, Santa Cruz (CA), USA. pp. 85–90.

Part III
Advanced Cartographic Visualisation

Cartography and Geoinformation for Early Warning and Emergency Management

David Fraser

Abstract

The nature of environmental disasters makes measuring, monitoring and mapping, using sophisticated geospatial analysis tools, extremely valuable in understanding the geography of disasters. Early warning systems, real-time and predictive models, along with community awareness programs, benefit from the information generated using computer systems that can integrate data from multiple sources. The common base is the geographical location and extent of the impact of the disaster. Cartographers are skilled in the analysis of spatial data and the presentation of these data as information products to support the management of disasters.

This chapter uses case studies to develop discussion into environmental disasters by considering the key information product deliverables; community vulnerability; the geospatial information creator and the geospatial information users.

1 Introduction

Early warning systems and emergency management associated with events such as forest fires, landslides, mining, hurricanes, tsunamis and volcanic eruptions require the consideration of cartographic models and associated geospatial models that draw together key characteristics of the environment so that the temporal framework for the event and the probability and impact of occurrence can be ascertained. During, or immediately after, an event it is necessary

M. Konecny et al. (eds.), *Geographic Information and Cartography for Risk and Crisis Management*, Lecture Notes in Geoinformation and Cartography,
DOI 10.1007/978-3-642-03442-8_22, © Springer-Verlag Berlin Heidelberg 2010

for the cartographer to provide sophisticated near real-time solutions for the provision of information to the emergency services dealing with the crisis.

Geospatial technologies such as Global Positioning Systems (GPS), Geographical Information System (GIS) and remote sensing all have a role to play in the development of early warning systems and emergency management geospatial resources. Cartographic products can be highly technical and require the skills of highly trained and educated individuals when production and interpretation is required. With field-based mapping such interpretation may be needed over a very short timeframe to allow the deployment of emergency services to key locations.

This chapter presents case studies associated with forest fire (Australia), landslip (Australia), mining (China), hurricane (USA), tsunami (Thailand) and volcanic eruption (Montserrat) and focuses on the creators and users of the cartographic information products that are developed for such situations.

2 Cartographers and technologies

The cartographer leading the mapping for early warning and emergency management must be highly trained, educated and skilled. His, or her, past performance will be a key factor in the selection for such a mapping task as it is directly observable and must be exemplary as competency is inferred from performance. The cartographer's observable technical performance, knowledge, problem-solving skills and work habits are all important considerations for this specialised area of mapping (Gonczi et al. 1993).

For emergency services the demand for maps online cannot be overemphasised. According to Peterson (2003) maps are now second only to weather information in the number of World Wide Web search requests. He estimates the number of maps distributed through the Internet on a daily basis to be over 200 million. Recent innovations in web site technology have revolutionised map data displays. This has been further fuelled by free satellite imagery being made available to the media and the general public by satellite image companies such as Digital Globe and Space Imaging, and GIS vendors such as ESRI. Recent global GIS awareness has been heightened by environmental impacts such as the South East Asia Tsunami and Hurricane Katrina. The media, both television and newspaper, copied GIS imagery from the Internet and rebroadcast it to hundreds of millions of worldwide viewers (Peterson, 2003). More recently in February 2009, during the horrific forest fires in Victoria, Geoscience Australia had unprecedented demand for its online national bushfire monitoring system, Sentinel, which is designed to provide timely information about fire hotspots to emergency service managers.

The demand for maps online is growing exponentially, as the Web has become the medium of preference in accessing geospatial products; accessing maps via the Web has become perhaps the first stop for the general public when they seek geospatial information. The cartographic community involved with emergency management must continually develop their knowledge and skills so that they can provide the latest information effectively and efficiently using this, and other sophisticated modes of information transfer.

As a means of exploring the different mapping requirements for early warning and emergency management a number of case studies will now be presented.

3 Case studies

3.1 Forest fire – Australia

Each year in Victoria, Australia, there is a fire season which may include days of higher than normal temperatures and strong hot northerly winds. On such days the forest, and its inhabitants, are most vulnerable to fire storms. On Saturday, February 7, 2009, Victoria experienced the worst forest fire ever recorded. Whole towns were destroyed, thousands of homes were lost, thousands of hectares of forest and farmland were destroyed and over 170 people lost their lives. The fire fighters and the public were provided with up-to-the-minute news on the location and intensity of the fires. Emergency services information was broadcast on the national radio station. Maps and satellite images of the hotspots were available over the Internet for all to view. Unlike in previous fires, not one fire fighter lost his life on this day. This, in part, was attributed to the improved management of the human resources and the more sophisticated mapping resources that were available.

3.1.1 The key deliverables

Experience from previous forest fires is used in the construction of fire behaviour models which are used to examine strategic fire behaviour and to investigate how the fires move across different landscapes, different vegetation, and under variable weather conditions.

During a fire event the forest fire spread can be simulated based on a fire behaviour model and an animation of the past and predicted movement of the fire can be generated.

Fire behaviour models, on-ground information and satellite and airborne imagery can be used to generate maps showing fire locations. This information can be generated in hardcopy form for the fire fighters on the ground.

3.1.2 Community vulnerability

Residents in fire prone areas are required to have in place a fire plan. They have two options: stay and defend their property or leave early. The Australian Broadcasting Commission dedicates the radio broadcast in a fire region to the transmission of emergency information. Residents are advised when they need to enact their fire plan. If possible they are also advised by police or emergency services officers on the ground. The broadcaster and the emergency services are informed of the location, direction of spread and severity of the fires from the mapping services at the headquarters of the fire service organisation.

3.1.3 Geospatial information creator

Development of the fire behaviour models and their simulation requires many years of research and input from past fire events. The models created before the forest fires in the Australian Capital Territory in 2002 were found to be inadequate for that event. Similarly this recent fire event in Victoria has gone beyond the limits of existing models in terms of its intensity and its spatial and temporal characteristics. Researchers are now required to rework the models to accommodate such extreme fire storms.

The detection and monitoring of forest fires is increasing in sophistication with the use of Global Positioning Systems in vehicles providing exact coordinates of spot fires and airborne imagery providing thermal scans of fire fronts. Dynamically created maps, using a web-based mapping system, can take advantage of many information sources. Handheld technology is improving and so too is our ability to report on the various aspects of fire activity and management.

3.1.4 Geospatial information users

Fire models allow key decision makers to give guidance based on the characteristics of past forest fires. Since the decision makers are dealing with life or death situations they need to be certain that the advice they are sending out is unambiguous and suitable for those at the fire front. Where cartographic products are the source of the information it is necessary for these products to be placed in the hands of those who are trained and skilled in the interpretation of the symbolic representations. Those in charge of fire fighting crews must be supplied with products that can be readily interpreted under trying conditions so that split second decisions can be made. So even though the information products may be based on years of research and development activity, they must be easy to interpret in the field.

3.1.5 Summary

These recent fires have opened up the debate on how people in the fire-affected area can be communicated with more effectively and in a timely manner. Emergency services need ready access to Internet mapping and in emergencies some critical websites are blocked from use by the general public so that the speed of access is not reduced. There is a call for government data to be open access and made freely available during times of such emergencies. Personal communication systems, such as the more sophisticated cell phones have the potential to provide early warning messages and maps showing escape routes. Forest fires can now be monitored more efficiently than in the past due to improvements in the production and distribution of geoinformation products (Geoscience Australia 2009). What is still an issue, despite the increased spatial and temporal awareness of pending fire fronts, is the civil right of citizens to stay and defend property no matter what the cost.

3.2 Landslip – Australia

This project involved the development and assessment of a mechanism for effective visualisation of community vulnerability to a perceived threat. There is an important distinction to be made between what constitutes a hazard and what constitutes a threat (Knzel 2006). In this case we are considering the concept of "*threat*" which may or may not be associated with a hazard in a geographical area, where people still feel threatened by a general awareness of risk.

The aim of this project was to assess the capability of local government to produce an effective visualisation of community vulnerability from currently available data, modelling techniques and technology, suitable for community education and hazard mitigation.

3.2.1 The key deliverables

1. A framework for Local Government to integrate currently available spatial data, models and techniques for visualisation related to community vulnerability.
2. A limited assessment of the adequacy of currently available data, models and techniques for visualisation of community vulnerability, together with an indication of the potential benefits of wider knowledge and information sharing as might be achieved through the Australian Disaster Information Network (AUSDIN) proposal.
3. A case study in the Shire of Yarra Ranges, Victoria, Australia, as an example of the degree to which an authority, with significant need for such an information product, is able to do so with generally available resources and organisational arrangements.
4. Increased awareness of community vulnerability to landslip by the Yarra Ranges community (Fraser 2001).

3.2.2 Community vulnerability

In a report produced for the Department of Industry, Science and Resources (Granger and Hayne 2001), the authors adopted five themes when assessing community vulnerability to hazard impact. The themes were: the setting (physical environment), shelter (buildings), sustenance (infrastructures), security (facilities such as hospitals) and society (community groups).

The data selected for this project were used to create visualisations of community vulnerability that had been adapted to fit the themes outlined above. Features (buildings, roads) identified to be at risk were able to be determined and an indication of the degree of community risk was identified by highlighting the areas that had a number of these features in close proximity to the risk areas.

3.2.3 Geospatial information creator

This project utilised Geographic Information System (GIS) and visualisation technology to integrate existing landslip probability assessments with social, infrastructure and economic measures to provide a visualisation of community vulnerability to landslip. The resultant information tool was then evaluated for practicality and usefulness in community education and hazard mitigation.

The information on community risk, available to officers of local government and regional authorities, is typically limited to an assessment of the risk of the hazard eventuating. However, community vulnerability to these hazards is a function of many aspects of the location. These include: proximity to the source of the hazard, proximity to the propagation path of the hazard, demographics (population density, age, income, ethnicity, education, etc.), value of land and improvements, and post-incident availability of essential services and support (a destroyed bridge may isolate an otherwise unaffected community from essential services). Information on these further variables is often available to Council, but not in integrated forms immediately useful for community education and hazard mitigation.

3.2.4 Geospatial information users

The different forms of visualisation developed as part of this project were evaluated at a local community meeting. Present at this meeting were representatives from the emergency services (Country Fire Authority, Metropolitan Ambulance Service, Police, State Emergency Services), the Shire of Yarra Ranges, the RMIT University, the local school principal and the local residents. Each visualisation met with a positive response but the simplest animation using an aerial photograph as a base was preferred by the majority of the participants. The lesson learnt is that it is often wise to build on the information products that the community already feels comfortable using, in this case the aerial photographs.

3.2.5 Summary

Geographical data stored in digital form by local shires, such as the Shire of Yarra Ranges, can be used to produce a range of visualisations suitable for use in landslip vulnerability assessment.

Geographical data stored in digital form can be structured in a manner that allows it to be processed for the following purposes:

1. Hazardous area profiling
2. Identification of areas prone to instability
3. Spatial analysis of critical facilities
4. Environmental analysis of a landslip site and its surrounds
5. Environmental modelling of a landslip site and its surrounds

The application of a number of geographical visualisation techniques to a specific landslip site has shown that photographs, maps and animations can contribute to the community's understanding of landslips and help to overcome misinformation.

The increased awareness of community vulnerability should be a staged process allowing for those most affected to be informed before the general community. This would allow these individuals to take ownership of the problem and to gain access to more detailed information. The different types of visualisation can be used to assist in informing the community at various forums depending on the audience and purpose of the forum.

3.3 Mining – China

Mining is costly both financially and environmentally. A Geographic Information System (GIS) when combined with a land suitability assessment can produce outcomes that allow decisions to be made on the suitability of particular geographical locations for mining before any on-ground development has occurred.

Key physical, environmental, social, cultural and economic considerations can be analysed and combined using the approach outlined in this project. The approach to land suitability assessment uses indices to identify the limiting factors relating to proposed mining land. The land suitability assessment procedure is used to analyse the attributes of the land and the GIS allows the results of the analysis to be displayed spatially.

The primary aim of this project was to present a geo-referenced model for mine management which integrates a land suitability system with GIS.

3.3.1 The key deliverables

- A report on the identifiable correlations between physical, economic and social parameters based on three types of mining activity (open cut, pit and underground).
- Three-dimensional geographical representation tools implemented in a GIS for mapping the mine environment. The production of interactive cross sections of geological maps should be integrated in the interactive 3D GIS.
- A report on the feasibility of placing the functionality of an existing GIS, land suitability model and imaging system into the one geospatial modelling environment that would enable the user to undertake interactive 3D modelling and spatial and temporal analysis of existing digital geographical, social and economic data.

3.3.2 Consideration

The mining suitability assessment refers to all elements of the physical environment including: climate, geology, topography, soils, hydrology and vegetation along with the economic and social attributes of the area under consideration.

Key considerations when determining the suitability of the land for mining are:

Air quality	Demographics	Drilling
Geology	Ground water	Infrastructure
Land cover	Land ownership	Land use change
Mining leases	Noise impact	Rainfall
Reclamation	Seismic testing	Soil
Surface water	Temperature	Topography
Town planning	Transport	Vegetation cover
Visual impact	Waste disposal	Water quality

The assessment relies on "systems" which define where the land exhibits consistent patterns of physical, social and economic characteristics. The system can be divided into "components" which are uniform with respect to a broad range of physical, social and economic characteristics. The components are amalgamated or subdivided when considered in the mining suitability assessment. These are the mining suitability units which can be used to rate a site's suitability for mining from Class 1 – highly suitable, to Class 5 – not suitable (Fraser 2006).

3.3.3 Geospatial information creator

For a project of this type the cartographer must draw on the expertise of many professions. Capability tables must be set up for each of the key considerations. This preliminary work can be long and protracted but, once in place, the capability tables which bring together the combined knowledge from many professions can be applied to any mining site. Map products must clearly identify geographical areas suitable for mining based on the application of the capability tables to a case study area. The skill is to bring together the spatial and non-spatial information into suitable output forms which extend beyond traditional map products.

3.3.4 Geospatial information users

The knowledge database used for this mining capability assessment is the product of specialist professionals. The user would be a specialist in only one of the disciplines but must be confident that such a system has a sound foundation. To gain this confidence a users may need to be guided through the mining suitability assessment system so that they become convinced, or otherwise, of the validity of the system. Subsequent to this workshop approach an individual would need training before such a system could be used routinely for the assessment of a potential or operational mine site.

3.3.5 Summary

The results from the land suitability assessment can be used as a stand-alone product or combined with vector data from a GIS and remotely sensed imagery. Remotely sensed imagery provides a rasterised objective view of the geographical area under study. The image can be visually interpreted and used to locate areas with specific landform and land cover characteristics. A parcel of land identified on the imagery can be analysed digitally by linking it to the vector spatial data equivalent on the relevant GIS layer. By attaching the other attributes, the spatial relationship between the physical, economic and social features for a mine site can then be analysed in the GIS and presented as a cartographic information product.

3.4 Hurricanes – USA

Emergency managers find themselves in the unenviable position of trying to make rational decisions often in the face of unpredictable natural phenomenon. The stakes may be quite high with lives and property being at risk. Resources are often inadequate and unless constant drills are undertaken the support staff may be inadequately prepared. The apparent multiple failures of emergency management, in response to the hurricane in New Orleans, were quite astonishing (Handmer 2006).

Hurricane Katrina landed on the coast at New Orleans causing the levee bank protecting the site to break resulting in the whole city being flooded. Massive property loss and dislocation of people caused the city to shut down. The follow-up response by emergency services left a lot to be desired.

Often in disaster situations there is a firm belief that technology will solve many of the problems. However, reflecting on New Orleans it is difficult to see that technology alone would have made a significant difference (Handmer 2006).

3.4.1 Geospatial information creator

Monmonier discovered that many maps associated with hurricane activity in America had not been standardised and were "a symptom of the nation's loosely structured, largely voluntary approach to hurricane preparedness" (Monmonier 1997). It makes the task of informing the public extremely difficult when a standard approach is not adopted and personnel involved in emergency services change regularly.

The challenge for the creator of any hazard map is to know at what level of technical detail the information should be portrayed for each user group. If the map is too complex then people will tend not to use it as it confuses rather than enlightens. If the map is too simple then key information may be left off and this in turn may lead the user to take inappropriate actions.

Hurricanes can be erratic and it is not possible for the cartographer to map their territory with any great certainty, such is the nature of nature.

3.4.2 Geospatial information users

Citizens may be aware that they live in an area prone to hurricane activity but human nature being as it is, they live with the belief that it won't happen to them. In the case of New Orleans the people on the whole are the poorest in the nation and education levels are very low (Handmer 2006). This low level of education and high level of unemployment results in many households just trying to survive from week to week. Informing the public in relation to hurricanes must be done at a very rudimentary level and hence map products must be theme specific and send a simple message.

3.4.3 Summary

The New Orleans hurricane event was well anticipated, planned and rehearsed. The attributes and vulnerabilities of the population were, or should have been, well known. So it would be reasonable to expect that the emergency response was at least adequate if not exemplary. As is now well known this was far from the reality with all levels of government failing (Handmer 2006).

To what extent can maps of hazards be regarded as universal? Most hazard maps may appears as if they are perfectly transferable across national boundaries. The same comment may be made about the uncritical transfer of information within countries to rather different contexts. For example, the results of risk research conducted in rural towns in the US and Australia have been applied to metropolitan areas without consideration of the differences. In regards to hurricanes there are so many local economic and social factors to consider that each location must be treated independent of another.

3.5 Tsunami – Thailand

This case study takes a different slant by commenting on the role of non-government organisations (NGOs) in facilitating economic recovery in regions of Southern Thailand devastated by the Indian Ocean tsunami. Over 70% of Thailand's people are employed in the informal sector of the economy, primarily in tourism and fishing. This is where the most Tsunami affected proportion of the population is employed. It is important that governments are aware of, and provide support for, such informal, power-weak sectors of communities. The basis of economic livelihood for a large proportion of the affected Thai population fall into this category (Coate et al. 2006).

The longer-term rehabilitation and recovery phase, in relation to the government and the NGOs response to the tsunami disaster, covers four key areas that have been broadly agreed upon, these being:

- Social protection;
- Livelihood recovery;
- Environmental rehabilitation; and
- Disaster preparedness (Handmer et al. 2007)

Resilient communities need to be built on recovery efforts that are broad based and involve community groups at the grass roots level. This is an opportunity for development plans to incorporate the needs of the community so that they take control of their future more closely (Coate et al. 2006).

3.5.1 Geospatial information creator

What role can a cartographer play in this situation? Taking each key area, as outlined above, it is the responsibility of the cartographer to consider the following:

3.5.2 Social protection

Delineate and define the property boundaries for displaced persons so that false claims cannot be made on the basis of land ownership that the people have. The cartographers may not undertake the surveying task themselves but they should be able to direct the surveyor to define the land under question. Ownership, or rights over land, constitutes a basic component of human security.

3.5.3 Livelihood recovery

It would be over valuing the cartographer's role to suggest they had a part to play in this aspect of the disaster.

3.5.4 Environmental rehabilitation

The planning and mapping of the rehabilitation, based on input from other disciplines, can provide a spatial record of the required environmental changes. Here is the opportunity to value add to the environment. This is where the mental map comes to the fore. To visualise what might be is an important component of raising the morale of the individuals affected by the tsunami.

3.5.5 Disaster preparedness

Do we learn from the past? If this is so, then we can analyse what has been, and inform the community as to more appropriate strategies for the future. Maps play a vital role in recording past events and in providing the spatial context for scenarios that may aid in preparedness for future disasters.

3.5.6 Geospatial information users

In Hawaii a tsunami evacuation map is included as part of the telephone white pages and is upgraded each year. Every home and business receives one. This map at least provides citizens with guidance on what to do in the event of a tsunami. The hazards map was developed in 1946 after the islands were hit by a tsunami resulting in the loss of 159 lives (Monmonier 1997). Is a hazard map a possibility for Thailand? What means of dissemination is available in Thailand and how can the financially poor of the country be empowered to avert disaster if a tsunami strikes again (Handmer and Choong 2006)? The process associated with a tsunami can be modelled spatially and the impact can be mapped. New techniques in geovisualisation can aid in the transfer of information to the community by transcending the language barrier.

3.5.7 Summary

People are increasingly attracted to hazardous places. The trend is to live closer and closer to the sea. Two aspects of predicted global environmental change, sea level rise and an increase in the number of cyclones associated with higher sea level temperatures, will bring more storm surges and make smaller surges more dangerous. The cartographer needs to be a key member of disaster response teams. To maintain this role the cartographers needs to remain at the forefront in terms of his/her understanding of measuring, mapping and monitoring technologies.

3.6 Volcanic eruption – Montserrat

Many natural disasters pass through a crisis point into a process of recovery very rapidly. Volcanoes on the other hand may have a more prolonged impact with precursory and eruptive activity which can last from days to years and thus deserve special consideration.

Taking Montserrat as a case study, the seismic activity began to escalate, physically manifesting itself in June 1995 as ash and steam began to vent from within the volcanic crater. Activity during 1996 led to the permanent evacuation of the south of the island. In 1997 a volcanic dome collapsed triggering a large pyroclastic flow. By 1997, the volcano was experiencing repetitive explosions with pyroclastic flows occurring in a radial direction. Activity had slowly increased in severity becoming cyclic and uncertain (Haynes 2006).

The small size of the island made emergency management problematic.

There were major issues associated with the disaster response. Some issues recognised as leading to the lack of preparedness for the disaster were the following:

Limited corporate learning; the rotation of government every 3–4 years with limited support staff reduced the capacity for the retention of corporate knowledge of scientific activities and disaster preparedness; difficulty in communicating a low-probability high-consequence event.

Many of the authorities and public held an inflated belief in the predictive powers of the scientists to provide accurate and timely warnings and reduce the uncertainty associated with the volcanic activity.

3.6.1 Geospatial information creator

Volcano hazard maps must be created at two levels. Firstly a map providing an overview of volcanic activity would represent the active volcanoes and the areas being monitored. Secondly more detailed maps could show areas which may be impacted by ash or pyroclastic flows along with evacuation routes and assembly points.

Volcanologists cannot forecast the exact forces and geographic impact of an eruption and hence hazard-zone maps must accommodate this high level of uncertainty (Hayne 2006).

3.6.2 Geospatial information users

Predictive modelling and scenario mapping are tools available to the user. Such information must be combined with local knowledge relating to infrastructure, likely flow paths, escape routes, safe areas and impact zones. A cartographer can model data and produce excellent graphical representations of phenomenon. The next step is the critical one and that is relating the model outcomes to reality. Natural processes are complex and models can only use a limited number of parameters to mimic a process. The uncertainty and imprecision associated with models of volcanic activity should be discussed at any forum where predictions and scenarios are presented. Such user feedback will allow the cyclic revision of the cartographic products to make them more information rich and usable by the targeted community.

3.6.3 Summary

Information support systems associated with volcanology need to have built into them the consideration of the social demands that may be encountered in a volcanic crisis and be flexible enough to be adapted to specific situations.

During a long-running crisis it is important to continually update and renew education and outreach activities using innovative techniques.

Risk communicators need to mix and match communication methods to suit the audience and must have confidence to relinquish some responsibility for decision making to those at risk.

4 Discussion

Humans are skilled at distilling important aspects of their environment and are trained, from an early age, to recognise patterns and analyse structures (Burrough and Frank 1995). Consequently today's emergency management teams build on this base level training and look towards computer databases to provide them with those data suitable for the production of spatial information related to natural disasters. The difficulty here is that decision makers may rely too heavily on the spatial database system to provide them with an appropriate approach to take. A poorly formulated problem description and methodology will result in poor outcomes. A computerised decision support system may provide easy access to a large volume of data but for most applications cannot be considered as a substitute for sound problem formulation undertaken by the modeller. After the problem has been carefully formulated conceptually the creator of the database must decide what features are important and what useful characteristics can and will be retained. Knowledge from those with expertise related to a particular natural disaster is required as input and an understanding of how to present the data in computer compatible form is required. According to Goldstein and Goldstein (1978) facts selected are usually based on some preconception or past theory which guides us as to what is important and what is not. A cartographers must consider first the information which is appropriate to aid emergency managers and victims during and after a disaster and consider their own professional needs last.

5 Concluding comments

We map natural hazards in our belief that the underlying science provides us with a suitable level of geographical certainty associated with the location and perhaps the temporal nature of the hazard. Sometimes this proves effective and yet at other times the mapping fails dismally. Random events, such as lightning strikes, cannot be mapped because of the geographical uncertainty associated with the event (Monmonier 1997). Fault lines associated with earthquakes can be mapped but the temporal nature and the intensity of an event cannot be mapped to the same degree of certainty.

Considerable changes in approach to hazard mapping have resulted in changes to what is expected and required from those working in the cartographic sciences. Rapid changes in technology, the growth of internationalism in trading, an increasing awareness of, and need for, digital data in various formats and a move towards more formalised data transfer standards, have been dominant factors in this forced change. This has meant that the cartographer's role has changed from that of basic hazard map production, to that of being a producer of customised geo-information products, these geo-information products being appropriate map depictions of various types of hazard on a variety of spatial and temporal scales.

This change requires that cartographic education must also be adaptable and responsive to changes in technology and employment opportunities, recognising the blurring of traditional professional boundaries and encouraging the student to become a self-motivated learner, capable of changing and growing. To achieve these goals requires a shift in the teaching of the cartographic sciences, in the way that the subject matter is both approached and presented.

The huge increase in the use of geo-information for all types of decision making means that more than ever, the cartographer must be able to supply the user with the information that they need, in the format that they require it. Implicitly then, the cartographer's role does not end with map production but with making sure that the users are getting the most appropriate information. This includes providing the necessary information on data accuracy, not only on the data set as a whole, but also about local anomalies in data quality and mismatches between data sets. Knzel's experience with mapping for rapid response, while responding to the Pakistan earthquake in 2005, was that the quality of map content varied with different maps; conflicting administrative boundaries occurred, place names differed on different maps and maps had different geographic reference systems (Knzel 2006). There is obviously a need for more education in techniques of data quality analysis and visualisation.

Maps not only help people to decide what to analyse, they also support people in formulating decisions on issues with a spatial impact and help to communicate these decisions. In the past, cartographers have placed too little emphasis on these aspects of decision making. The common function of maps produced for education, orientation and navigation, management, monitoring and scientific analysis is the act of decision making based on spatial information. Maps are documents, still images or animations that support this decision-making process and should therefore be produced in a format that enables and sustains such processes. This calls for specific designs that allow for the possibility of maps being produced quickly and quasi-automatically by the decision maker. This, in turn, calls for education in building and updating databases and accessing and using databases in a way that quality information may be derived from them.

Hazard mapping relies heavily on evidence of prior performance which leads to refinement of the cartographic products. Direct observation of the phenomenon on the ground and after the event can assist cartographers to refine the maps and models they create. New approaches test the technical skills of the cartographers but also allow their creative side to find a place to shine. Simulation of past events can be used to generate discussion on the portrayal of particular phenomenon and this questioning should be encouraged and acted upon so that cartographers remain at the forefront of mapping for early warning and emergency management.

References

Burrough PA, Frank AU (1995) Concepts and paradigms in spatial information: are current geographic information systems truly generic? Int J Geogr Inform Syst 9(2):101–116, Taylor & Francis.

Coate B, Handmer J, Choong W (2006) Taking care of people and communities–rebuilding livelihoods through NGOs and the informal economy in Southern Thailand. Disast Prev Manag 15(1):135–145, Emerald Group Publishing Limited.

Fraser D (2001) Assessment of the capability of local government for the visualisation of community vulnerability, Project number: EMA 06/2001, Shire of Yarra Ranges, Lilydale, Victoria.

Fraser D (2006) The application of GIS to the determination of land suitable for mining. In: Proceedings of the International Workshop on Geospatial Technologies and Their Environmental Applications, Xuzhou, Jiangsu, China, pp. 13–18.

Geoscience Australia (2009) Sentinel, http://sentinel.ga.gov.au/acres/sentinel/index.shtml (Accessed 23 February 2009).

Goldstein M, Goldstein IF (1978) How We Know – An Exploration of the Scientific Process. Plenum Press, New York.

Gonczi A, Hager P, Anthanasou J (1993) The development of competency-based assessment strategies for the professions. Research Paper No. 8, June 1993, Department of Employment, Education and Training, Australian Government Publishing Service, Canberra.

Granger K, Hayne M (eds.) (2001) Natural Hazards and the Risks They Pose to South-East Queensland, Department of Industry, Science and Resources, Commonwealth of Australia.

Handmer J (2006) American exceptionalism or universal lesson? The implications of Hurricane Katrina for Australia. AJEM 21(1): 29–40.
Handmer J, Choong W (2006) Disaster resilience through local economic activity in Phuket. AJEM 21(4):8–15.
Handmer J, Coate B, Choong W (2007) The Indian Ocean Tsunami: local resilience in Phuket. Chapter 8, Managing Coastal Vulnerability, Elsevier Ltd., Amsterdam, pp. 121–139.
Haynes K (2006) Volcanic island in crisis: investigating environmental uncertainty and the complexity it brings. AJEM 21(4):21–28.
Knzel W (2006) Mapping for rapid response and early recovery: lessons learned while responding to the Pakistan earthquake. In: Proceedings of InterCarto12, Berlin, pp. 233–241.
Monmonier M (1997) Cartographies of Danger – Mapping Hazards in America, The University of Chicago Press, Chicago, p. 84.
Peterson M (2003) Maps and the Internet, Elsevier, Oxford.

Process Support and Adaptive Geovisualisation in Emergency Management

Petr Kubíček, Tomáš Ludík, Eva Mulíčková, Jaroslav Ráček, and Gustav Šafr

Abstract

A multidisciplinary approach is key element in current emergency management research. This chapter illustrates issues of defining, formalising, and visualising geodata content within emergency management and combines them with process modelling and procedures analysis. We compare existing US and EU attitudes in emergency management organisational and geoinformation needs and analyse the situation within the Czech Republic. Because of the overall complexity of emergency management, the solution is proposed for a specific scenario dealing with traffic accidents and the transportation of dangerous substances. We answer the following questions: Who are the main actors within this scenario and what roles do they play? What geoinformation must be delivered in order to successfully complete the emergency response? Which geovisual techniques are appropriate? We propose adaptive cartographic representations based on an analysis of user roles and types of emergency situations.

1 Introduction

The field of emergency management (EM) usually divides any situation into four components (phases) that roughly correspond to the before (termed *preparedness*), during (termed *response*), and after phases (*recovery*) of any particular event. The last but not least *mitigation* phase has to do with minimising the effects of possible disasters by mandating policies and regulations that will lessen the effects of a hypothetical disaster.

M. Konecny et al. (eds.), *Geographic Information and Cartography for Risk and Crisis Management*, Lecture Notes in Geoinformation and Cartography,
DOI 10.1007/978-3-642-03442-8_23, © Springer-Verlag Berlin Heidelberg 2010

Many approaches have been proposed to develop a geoinformation-driven system for disasters; these concentrate primarily on the response phase. This phase deals with the immediate aftermath of an emergency, including the mobilisation of relief agencies, the delivery of aid and the provision of medical care. In the Czech Republic, there exists a common will to develop effective information support at all levels of EM. Geoinformation and process support at the tactical level and direct visualisation and updating of geoinformation in the field could simplify the decision-making process of an intervention commander and raise the quality of work of the Integrated Rescue System.

One of the important questions raised by the Committee on Planning for Catastrophe and National Research Council (NRC 2007) is how we can effectively provide assistance to those who have been affected, through the development of a common operating picture and common situational awareness shared by all emergency responders. Geospatial requirements and gaps identified in the response phase by the Committee on Planning for Catastrophe and National Research Council (NRC 2007) have been partly addressed within the scenario 'Transportation of dangerous goods' proposed in 2006 by Talhofer et al. (2007) and further developed in 2008 in a focus on verifying the dynamic geovisualisation procedures for communication reports to and from the field.

The aim of this chapter is to propose a complex geoinformation attitude for supporting decision making in the response phase of EM. Our proposal combines adaptive mapping that reflects the geospatial needs of EM users (what will be visualised and how depends on the particular activity in the specific emergency situation and the geoinformation skills of the user) and process modelling that controls the run of tasks within the activity for which the user is responsible.

The chapter is structured as follows: Section 2 discusses wider research frameworks for emergency management support, analyses activities within the integrated rescue system, and introduces the visualisation principles of adaptive mapping. Section 3 describes the selected case study and geoinformation needs during the response phase as a process model. Section 4 deals with the direct geovisualisation support of the intervention commander within the field activities, the methodology and standards used to deliver the solution. Section 5 concludes on our preliminary results and proposes further research directions.

2 Principles of emergency management

The organisation of emergency management as well as the system of civil protection varies between the EU and US. However, there are still important commonalities. The nature of emergencies means that all levels of government (federal, state, regional, local) and all sectors of society are responsible for dealing with them (NRC 2007, Diehl et al. 2006). There generally exists a 'bottom up' approach for requests for resources support that travel upward until appropriate resources are ensured and the incident stabilised. Thus, most emergency incidents are responded to at the local level. On the other hand, each country has different legislation, procedures, and obligatory documents to be followed within the EM process. There also exists no universal terminology within EM, not only internationally but even within the agencies across all government levels.

The following paragraphs describe the principles used in civil protection in the Czech Republic that are valid for designing the elaborated case study. Civil Protection is a set of prepared measures that are realised during Extraordinary Events (EE) and Crisis Situations. These measures are executed by components of the Integrated Rescue System (IRS). The IRS is a legally specified, open system of coordination, cooperation, and modelled cooperation, procedures. In the case of an Extraordinary Event, the IRS components realise Rescue and Liquidation Works (RaLW), resp. Civil Protection. In order to conduct the RaLW, there must be Forces and Means (FaM), e.g. sources of manpower, tools, technical equipment, and powers (competence), i.e. qualification for various activities within the RaLW given by law. The basic aim of the IRS is to integrate facilities of all actors who should participate in the RaLW. There are basic and other IRS components (i.e. actors) (Table 1).

Table 1. Actors in the integrated rescue system of the Czech republic

Basic IRS Components	Other IRS Components
Fire Rescue Corps of the Czech Republic (FRC)	Army of the Czech Republic
Fire Prevention Units (FPU)	Armed security corps (except the PCR)
Police of the Czech Republic (PCR)	Other rescue corps (except the FRC)
Medical Rescue Service (MRS)	Public health authorities
	Emergency, profess. and other services
	Civil Protection facilities
	Non-profit organisations and civil associations, etc.

Basic IRS components are responsible for permanent emergency phone call response (numbers 150, 155, 158, and 112), evaluating the event and immediate intervention. *Other IRS components* are used when the basic IRS components are not sufficient for RaLW. All of the IRS components are registered and their cooperation is set by the *IRS Alert Plan*.

2.1 Coordination of activities within the emergency management system

Coordination is done on three levels, tactical, operational, and strategic (Šafr 2008), and is corresponding with the generic levels as defined by ORCHESTRA (2008).

The **strategic level** of the RaLW management is realised by standing or temporary coordinating authorities of the administration, region commissioners and Ministry of Interior – General Management of the Czech Republic Fire Rescue Corps.

On the **operational level,** permanent coordination and cooperation within and between individual IRS components takes place; this includes operational centres of the basic IRS components (FRC, PCR, and FRS) and dispatching centres, standing services, and oversight centres of distributive and emergency services.

The *operational and informational centre* (OIC) manages cooperation within the RaLW with IRS documentation (e.g. IRS alert plan, emergency plan of the region, water sources survey, Model Action Activity of the IRS Components at the Common Intervention). The responsibility of the OIC includes securing the activities of the intervention commander, coordinating higher level activities, citizens' warnings, exchange of information, etc. Current geospatial support on this level is described in Pilný (2008).

The **tactical level** includes activity coordination at the place of intervention and cooperation of IRS components. The *intervention commander* proclaims the appropriate level of alert, which predetermines the needs of the FaM for the RaLW. In simple cases, the intervention commander coordinates the FaM alone. In cases requiring time-consuming and complex cooperation, the *staff* of the intervention commander is established. Its members are the IRS component leaders and eventually experts or assistants of cooperation units. In cases that are too complex or require large-scale intervention, individual sectors are set and sector commanders are nominated. The intervention commander organises the RaLW based on consultation with IRS component leaders; he or she also follows the document 'Model Action Activities of the IRS Components at the Common Intervention'. This level is currently missing any direct geospatial support for field activities.

3 Adaptive mapping

The adaptation of geographic information (GI) can be seen as an optimisation process that enables the provision of objects of high utility that satisfy a user's current situational context (Erharuyi and Fairbairn 2005); this can be carried out at different levels – e.g. data, communications, and task-specific levels. For example, GI can be adapted to a special format, adapted for transmission over a wireless network, adapted to a specific device etc. Geographic information is produced and used by people to support better informed and faster decision making. However, this potential can only be exploited adequately if the purpose (tasks) for which the user needs the data is taken as an important intervening variable (operator) for the optimisation process.

Adaptive maps have become a vital approach for modern cartography in general and map use in particular (see e.g. Erharuyi and Fairbairn 2005; Reichenbacher 2001; Wang et al. 2001). The principles of adaptation deal with the development of so-called contexts. The context is a set of determinants identifying particular cartographic representations. If something happens around the map device, its context is changed and an appropriate visual representation is selected. The basic idea of adaptable maps follows the practice of map use. There, we can distinguish many attributes of map context and their impacts. The selection of these attributes is strongly related to the overall purpose of map representation. Table 2 describes selected attributes of context in EM.

For context identification, a detailed analysis of the solved task is crucial. A generic description of the goal and necessary information is not enough. We need to know how the acting subjects perceive the reality. The same phenomenon may have a different meaning for a specialist who observes it than for people who are influenced by it. To handle such issues, we need detailed descriptions of the world views of crisis management actors and of specialists who create supporting data. Additionally, we need to create the necessary translation between these models. Questions of semantic translation within context visualisation in EM are discussed in Mulickova and Stachon (2009), while the metadata issues are described in Reznik (2008). In full accordance with Erharuyi and Fairbairn (2005), we believe that there is a need to refocus geographic information adaptation from a strictly technological to a more problem-based process, asking questions such as the following: What are the activities for which we use it? What are the tasks that constitute an activity or phase in emergency management? What actions do we need to perform within a task? Geographic information is produced and used by people to support more information-rich and faster decision making, but this potential can only be exploited fully if it accommodates user expectations and decreases cognitive overload.

Table 2. Selected context attributes and how they influence the final map representation. Adapted from Stanek et al. (2007)

Context attribute	Influence	Expressed in
USER - age	Perception abilities	Symbol complexity, size and use of colours
USER - education	Map content according to explicitness of representation (theme knowledge decreases demand of explicitness)	Number, geometric type and meaning of features
USER - map skills	Complexity of representation (people more effectively use familiar representations)	Symbology and selection of cartographic methods
SITUATION - location	Orientation in space (how familiar is the user with the place and how he or she can identify the distribution of phenomena)	Topographic background and map face extent and orientation
SITUATION - time/season	Importance and validity of objects	Symbol enhancement and object selection
SITUATION - accident type	Map content	Feature and object selection
ACTION - map purpose (task)	Map content, importance of features	Feature selection, object selection, feature symbolisation

It is very important to understand the process of reading and understanding map representations in various situations (Stachoň and Šašinka 2009). Within the research plan 'Dynamic geovisualisation in crisis management', various cognitive tests were undertaken; the results of the testing are being used to construct suitable maps for various situations. In addition, adapting the graphic user interface (GUI) is one of the present research topics, since users have different abilities to work with geoinformation technologies.

4 Case study

The adaptive visualisation of geoinformation can be seen as a tool to support cooperation on both tactical and operational levels. This chapter focuses on the tactical level of cooperation – activity of the intervention commander and decision-making support during the *organisation of intervention* in the situation *'Accident of vehicle transporting dangerous goods'*. In a wider context of emergency management, we are focusing on certain gaps in geospatial needs that, according to NRC (2007), exist in the response phase:

- Capability to track the location and characteristics of responders.
- Fleet tracking systems that provide full resource locations in a dynamic context.
- Integrated, location-based field data acquisition system linked to a central GIS.

The principle aim of this research is a cross connection of adaptive visualisation and process modelling on the domain of emergency management. The proposed solution has to do with the control of the intervention commander activity during the response phase of the accident of a vehicle with dangerous goods. The authors are aware of other attitudes that have been proposed and are often used in this area, like semantics-driven or ontology development (e.g. Klien et al. 2005, Tanasescu et al. 2007). However, this research aims at building the support in a different manner starting with a simple use case, identifying the main obstacles and possible bottle-necks for a more extensive implementation.

4.1 Analysis of IRS activities and geoinformation needs

While in the US there exists an office responsible for geospatial issues coordination (Geospatial Management Office – see NRC, 2007 for details), there is no equivalent within the EU.

Users' requirements have been investigated on three different levels of detail – on the EU level, several international projects (e.g. OASIS 2008, ORCHESTRA 2008), and studies (e.g. Diehl et al. 2006) dealing with emergency management and geoinformation have been analysed and their results synthesised into the generic requirements level. A directed interview with a complex questionnaire has been further conducted within the public administration bodies in the Czech Republic, defining the current state of the art and main demands on geoinformation support in EM (Foltynova and Stachon 2008). Both US (NRC 2007) and EU sources (ORCHESTRA 2008; Diehl et al. 2006) agree that geoinformation needs vary across the disaster phases and hazard types. Therefore, key user requirements are usually divided according to the EM phase and into categories like requirements, current capabilities, and gaps to be solved. On the other hand, the local level geoinformation requirements for a specific use case (forest fire, flash floods) are usually described within the national procedures (Diehl et al. 2006) and can serve as a pilot implementation.

The activities of the intervention commander were analysed using the document *Model Action Activities of the IRS Components at the Common Intervention,* which models the activity of the IRS component of Rescue and Liquidation Works with regard to the character of the Extraordinary Events. It defines the responsibility and activity of the involved units in nine different emergency situations, e.g. use of radioactive agents, aircraft accident. In order to analyse the activity of the intervention commander in the *accident of vehicle with dangerous goods,* the model situation *finding item with suspicion of presence B-agents or toxins* (Fig. 1) was used and modified for the case of a *vehicle accident* by analysing other IRS documents and based on discussions with experts. A few observations on Operational and Informational Centre were undertaken to gain deep knowledge about cooperation and coordination on both tactical and operational levels.

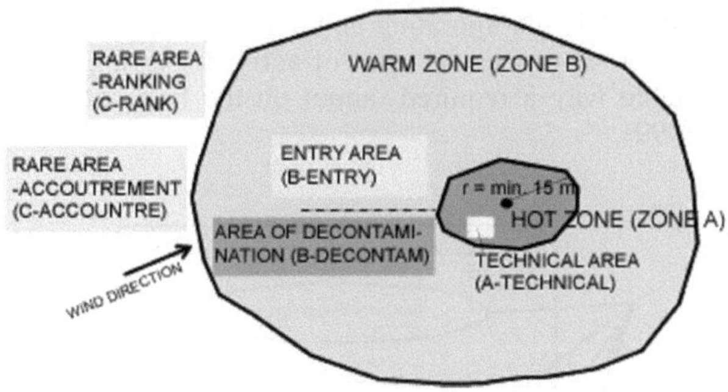

Fig. 1. Place of an Extraordinary Event and its zoning by common intervention of IRS in the situation *finding item with suspicion of presence B-agents or toxins* (adapted from FRS 2006). In brackets are the names of map features defined for the use case

4.2 Process support of the intervention commander

The event 'accident of vehicle transporting dangerous goods' can be seen in complex view represented by the UML (Unified Modelling Language) use case diagram. The main purpose of the use case diagram is to find and document the modelled system requirements (Ludík and Ráček 2008). UML is only a modelling tool for capturing use cases and does not directly relate to methodology. Diehl et al. (2006) used UML to model actors and the relations between them with respect to 25 disaster management activities (e.g. traffic control and management, disinfection of vehicles and infrastructure, and clearing and evacuation) as specified in the Netherlands. They described and formalised selected measurements that might be needed if dangerous substances are released in different environmental conditions (air, water, soil, and surface). We are following a similar procedure where each use case is based on analyses of the EM procedures in the Czech Republic. This model is also applicable for EM in other states, but other actors and use cases would probably be administered.

The border of the modelled system is defined by the Czech Fire and Rescue Act. Everything else is considered to be the surroundings of the system. By analysing the activities within the event, an actor list is created containing different roles that are assigned to persons or subjects that use the modelled system. Addressed questions are 'Who or what uses the system?' and 'Who or what communicates with the system? Identified actors are depicted in Fig. 2.

Having understood the roles of the individual actors, it is possible to start creating use cases. A use case is perceived as the specification of the sequence of activities that the system or subsystem can execute through interacting with external actors. Figure 2 illustrates five use cases in the main use case diagram as a result of grouping similar activities. Each use case can be specified by process maps (Fiala and Ministr 2007) incorporating and defining the activity sequences in the particular directives. A process is a set of activities arranged in parts that creates in a repeatable way a required output on the basis of one or more inputs (Hollingswort 1999).

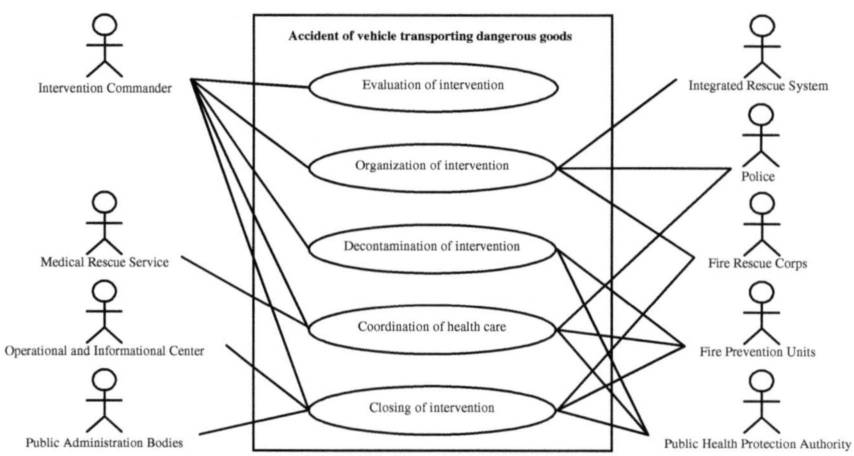

Fig. 2. The main use case diagram of an accident of a vehicle transporting dangerous goods. Actors are illustrated on the left and right sides

To illustrate a process map, the use case called *'organisation of intervention'* is processed. This directive controlled by the intervention commander consists of 10 activities illustrated in the process map in Fig. 3. In this way, the process map of organisation of intervention is created.

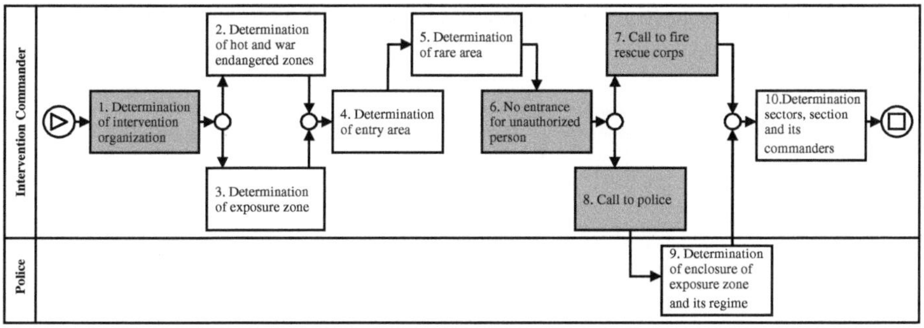

Fig. 3. The process map of *organisation of intervention.* In grey are manual activities, in white are activities automatically supported by the system

All modelled processes (process maps) are transformed to XPDL (XML Process Definition Language) format (Hollingswort 1999), where the individual process activities are assigned to the required geoinformation. The example of the resultant relationship between geoinformation and process activities is shown by the CRUD matrix in Fig. 4.

4.3 Geovisualisation support of the intervention commander

Within the activities of the intervention commander, it is possible to identify specific tasks that are more or less spatially dependent and thus require geoinformation support.

To judge *what and how* to visualise, it is necessary to decide what parameters will determine the context in which geographic information will be used. In order to simplify the case study, the following parameters were selected and populated to define the context: *USER – member of Fire Rescue Corps, ACTION – organising of intervention, SITUATION – accident of vehicle with dangerous goods, DEVICE – TabletPC.* Broadly, *ACTION* and *SITUATION* determine the knowledge necessary for decision making and thus *what* to visualise. *USER* and *DEVICE* specify *how* to visualise this data, i.e. set the visualisation criteria.

The process formalisation in the XPDL form specifies which geoinformation needs to be used and finds an appropriate way to process to. Subsequently, the processes and geodata are rebalanced using the CRUD matrix. The matrix shows which activities are executed in the process and which geoinformation is needed. In the particular matrix fields there are the following operations that can be applied to the map feature: create (C), read (R), update (U), and delete (D). The CRUD matrix is illustrated in Fig. 4, which also lists map features defined for the case study. The meanings of selected map features can be found in Fig. 1.

The map features listed in the CRUD matrix (Fig. 4) pose so-called *context specific map content* that is visualised over the background of a *topographic base.* This BASETOPO is a set of topographic features that can be reused in other contexts. BASETOPO is defined at a few scale ranges – in the case study, the use of

BASETOPO for large-to-middle scales is expected. An example of visualisation is given in Fig. 5.

	1. Determination of intervention organization	2. Determination of hot and war endangered zones	3. Determination of exposure zone	4. Determination of entry area	5. Determination of rare area	6. No entrance for unauthorized person
OBJECT-HAZARD		R	R	R	R	
OBJECT-IN_DANGER		R	R	R	R	
LOCAL		R	R	R	R	
ZONE-A		R	R	R	R	
ZONE-B		R	R	R	R	
ZONE-C			R	R	R	
A-TECHNICAL						
B-DECONTAM						
B-ENTRY				C	R	
B-COMMANDER						
B-HEALTH						
C-RANK					C	
C-ACCOUTRE					C	
ENCLOSE						
AIR_RESCUE						
EVAC_STAND						
BASETOPO		R	R	R	R	

Fig. 4. Part of the CRUD matrix of the use case *organising of intervention*. In rows are the map features and types of allowed operation (R-read, C-create), while in columns are the activities within the use case. The letters A, B, and C in the name of the map feature determine within which endangered zone (ZONE -A,-B,-C) the feature is defined

Within the framework of the research plan 'dynamic geovisualisation in crisis management', a contextual web service (SISSI) was developed to maintain adaptive mapping in a web environment. The contextual map service is based on Open Geospatial Consortium standards. A transactional Web Feature Service (WFS-T) is used for the bi-directional transfer of data and on-the-fly updating of the central database. The service is described in detail in Kozel (2007).

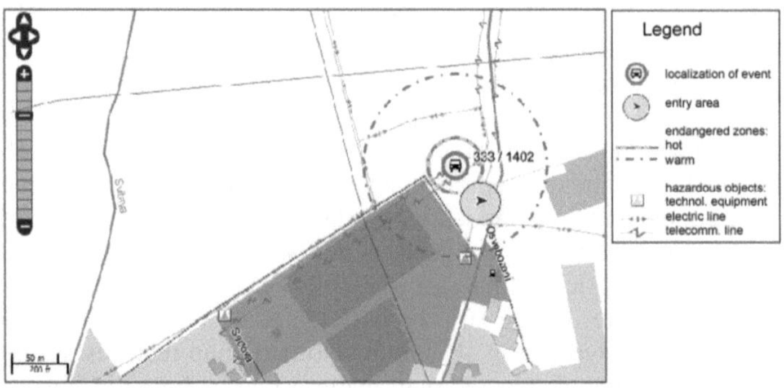

Fig. 5. Map content and its visualisation within the activity 'determination of entry area'. During this activity, the user creates the object 'entry area'. The background shows BASETOPO on the large scale

5 Conclusion and further research

The contextual web service based on previous process analysis and mapping has been used in the field experiment during which an *accident of vehicle transporting dangerous goods* was simulated and geoinformation support of the intervention commander tested. Nowadays there is no direct automatic connection between adaptive visualisation and process analysis. However, the fusion of both methods has helped to optimise the geodata visualisation rules and the amount of information transferred. Future development will consider the CRUD matrix as a direct driver for contextual service automation and further development of other emergency management use cases. The general research line here is in compliance with the directions for the future research and practice on EU emergency and crisis management as distinguished by Ekengren and Groenleer (2006). Multidisciplinarity, EU and transgovernmental level comparison, and fostering the direct cooperation of research and policymaking communities together built a wider EM research framework.

The described activities of integrated rescue system procedures are mainly localised within the Czech Republic. However, both the adaptive visualisation methodology and the contextual web service (SISSI) are generally valid, based on international OGC standards, and usable in the wider European emergency management framework. The combination of process analysis and adaptive geovisualisation has the goal of improving the visual communication to and from the field and supporting the efficient delivery of appropriate geoinformation according to the context. Despite the achieved system interoperability of the proposed solution, limitations still exist. The diversity of methodological approaches reduces the comparability of information and makes it difficult for information to be consolidated at the European level. The 'Community approach on the prevention of natural and man-made disasters' (EU 2009) published in early 2009 offers a strategic approach for disaster prevention and for the first time emphasises the importance of guidelines on hazard/risk mapping.

Acknowledgements

This research was supported by funding from Project No. MSM0021622418, which is called Dynamic Geovisualisation in Emergency Management.

References

Diehl S, Neuvel J, Zlatanova S, Scholten H (2006) Investigation of user requirements in the emergency response sector: the Dutch case. In: Proceedings of the Second Gi4DM, 25–26 September. Goa, India, CD ROM, 6 p.

Ekengren M, Groenleer M (2006) European Union crisis management: challenges for research and practice. Int J Emerg Manag 3(8): 83–90.

Erharuyi N, Fairbairn D (2005) Task-centred adaptation of geographic information to support disaster management. In: van Oosterom P , Zlatanova S, Fendel EM (eds.) Geo-information for Disaster Management. Springer Berlin Heidelberg, pp. 997–1008.

EU (2009) Community Approach on the Prevention of Natural and Man-Made Disasters. Report on Commission of the European Commission, Brussels, 23.2.2009, final version, on line at: http://ec.europa.eu/environment/civil/pdfdocs/com_2009_82en.pdf

Fiala J, Ministr J (2007) The model of process framework in Czech public government. Change Management. In: Proceedings of the 26th International Conference on Organizational Science Development. University of Maribor, Portorož.

Foltynova D, Stachon Z (2008) State of art in geodata utilization by crisis management authorities in the Czech Republic. In: Jubilee International Conference of CGS. Liberec, Czech Geographic Society, pp. 86–93. (in Czech).

FRS (Fire Rescue Service) (2006) Catalogue Set – Model Action Activities of the IRS Components at the Common Intervention STČ 05/IZS 'Finding item with suspicion of presence B-agents or toxins' VCNP 2006; MV - GŘ HZS ČR; čj. PO-2792-9/IZS-2006 (in Czech).

Hollingswort D (1999) Terminology & Glossary. Workflow Management Coalition.

Klien E, Lutz M, Kuhn W (2005) Ontology-based discovery of geographic information services – an application in disaster management. Comput Environ Urban Syst 30(1):102–123.

Kozel J (2007) Open contextual cartographic visualization. In: GI-Days 2007 – Young Researchers Forum: Proceedings of the 5th Geographic Information Days. Münster, Germany: IfGI prints, 2007. pp. 223–227.

Ludik T, Racek J (2008) Process analysis in crisis management and its usage for development applications ontology. In: Riešenie krízových situácií v špecifickom prostredí, Žilinská univerzita, Žilina, pp. 503–508 (in Slovak).

Mulickova E, Stachon Z (2009) Semantic issues of geodata integration in emergency management. Cartography and Geoinformatics for Early Warning and Emergency Management: Towards Better Solutions. Prague, January 19–22.1.2009. CD ROM, 10p.

NRC (2007) Successful response starts with a map: improving geospatial support for disaster management. Committee on Planning for Catastrophe and National Research Council, U.S. National Academies Press, Washington, USA.

OASIS (2008); Open Advanced System for crisIS management; on-line at http://www.oasis-fp6.org/

ORCHESTRA (2008); on-line http://www.eu-orchestra.org/

Pilný L (2008) Geographical data for the first responders and crisis management in the Czech republic. Seminar with EU Cooperation on Early Warning and Disaster/Crises Management, Borovets, Bulgaria, January 21–24, 2008, Proc.2, pp. 143–149.

Reichenbacher T (2001) Adaptive concepts for a mobile cartography. Suppl J Geogr Sci 11:43–53.

Reznik T (2008) Metadata in risk management. In: Geography in Czechia and Slovakia First. Brno, Masaryk University, pp. 421–426, 5p.

Safr G (2008) Structure and legislation of Integrated Rescue System, coordination and activity sequences of IRS components, Extraordinary Events and Crisis Situations (study materials); České Budějovice; on-line at www.zsf.jcu.cz/

Stachon Z, Sasinka C (2009) Perception of various cartographic representations under specific condition. In: Zlatanova S, Konecny M (eds.) Geoinformation and Cartography for Early Warning and Crisis Response. Springer-Verlag, In review.

Stanek K, Konecny M, Friedmannova L (2007) An adaptive cartographic visualization for support of the crisis management. XXIII International Cartographic Conference – Cartography for Everyone and for You, Roskartografija, Moscow.

Talhofer V, Brazdilova J, Kubicek P, Svatonova H (2007) Transport of dangerous chemical substances and its modelling. In: Proceedings of 10th Conference on Geographic Information Science, AGILE 2007. Aalborg, Denmark, pp. 257–261.

Tanasescu V, Gugliotta A, Domingue J, Villarías L, Davies R, Rowlatt M, Richardson M, Stincic S (2007) Geospatial data integration with semantic web services: the eMerges approach. In: Scharl A, Tochtermann K (eds.) The Geospatial Web, Springer-Verlag.

Wang Y, Liu Y, Chen X, Meng L (2001) Adaptive geovisualization: an approach towards the design of intelligent geovisualization systems. J Geogr Sci 11(Suppl.): 1–8.

Perceptions of Various Cartographic Representations Under Specific Conditions

Zdeněk Stachoň, Čeněk Šašinka and Václav Talhofer

Abstract

Maps are usually better abstractions of reality than other media and allow easier perception of included spatial information. This fact makes maps convenient for several specialised societal purposes, including emergency management and transportation, among others.

It is very important to understand the processes of reading and understanding map representations during a variety of situations. There are significant differences in map use by various users. Differences can be caused by variations in cartographic method or lack of time. Therefore, specific situations need specific map representations. This article addresses the problem of testing the practice of reading and understanding maps. The evaluation emphasises the externality of the testing process, meaning that results cannot be based only on subjective opinions of tested participants.

Results of the testing will be used for construction of suitable maps for various situations as well as the development of a software tool that will enable creation of standardised testing sets. Additionally, standardised research methodology will be developed to enhance exploration of cognitive map reading processes. Results are validated by statistical analysis tools using a combination of quantitative and qualitative methods.

Fundamentally, this work provides insights into the processes of perception by different groups of users, allowing increased map information transmission efficiency, especially important during crisis situations.

M. Konecny et al. (eds.), *Geographic Information and Cartography for Risk and Crisis Management*, Lecture Notes in Geoinformation and Cartography,
DOI 10.1007/978-3-642-03442-8_24,

1 Introduction

This chapter deals with the evaluation of perceptual-cognitive qualities of cartographic products, in particular considering the potential influence of stressors that occur in the field of crisis management. We describe an experiment in which test subjects performed tasks using maps displayed on computer screens. The aim was to monitor not only the influence of alternative cartographic visualisation methods on achievement but also the overall behaviour of the test subjects throughout the experiment. The goal is to use the data acquired from observation of the testing process itself and from the research interviews for development of a standardised research method that will enable map evaluation in future. Interdisciplinary cooperation is necessary in this kind of research. Psychology concentrates on the internal environment of the human mind and the factors that influence mental processes. Cartographers are specialists in maps – representations of the external environment – and are interested in how people are able to work with their products.

2 Stressors and cognitive processes

The influence of acute stress on cognitive processes has been verified in many studies. Recent work has focused on deep understanding of stress mechanisms and their influence on particular cognitive functions, such as memory, attention and decision making (see Gohm et al. 2001, Slobounov et al. 2000, Zeelenberg et al. 2006).

Vedhara et al. (2000) explored the relationship between acute changes in cortisol and memory and attention in the context of an acute naturalistic psychological stressor. They observed a significant improvement in a short-term memory task associated with a reduction in cortisol levels, whereas significant declines were observed in tasks measuring selective and divided attention and in the primacy effect[1]. No significant differences were observed in a task measuring auditory verbal working memory. In an experiment by Dror et al. (1999), participants under time pressure were more conservative at lower risk levels but were more prone to take risks at higher risk levels. These and other studies provide information on the effects expected during map work conducted under stress. However, these discoveries cannot be applied directly to perception of cartographic products since a map is a complex piece of information. It cannot be decomposed into parts, and the individual cognitive functions involved in map reading cannot be observed separately.

Hammond (2000) distinguishes two groups of stressors. Exogenous disruptions are those that arise from outside the system, for example, fire, noise, heat, and interruptions from external physical breakdowns. Endogenous disruptions are those

[1] The primacy effect is the tendency for the initial stimuli presented in a series to be remembered better or more easily and to be more influential than those presented later in the series.

that emerge from the task situation itself. Examples include time pressure or a computer program with bugs. Inadequate map representations and GIS systems used in emergency situations can be endogenous disruptions that eventually increase the stress level of the user.

Emergency situations are characterised by high dynamics, complexity and uncertainty of cues. Participants have responsibility; their wrong decisions can worsen the situation and can have fatal and irreversible consequences. These individuals almost always work under time pressure and stimulus overload.

Stressors are events that provoke stress. What will be perceived as a stressor depends on the personality traits of participants. Gohm et al. (2001) investigated the role of emotional experience and understanding in acute stress situations for fire fighters. They assessed the relationship between three individual difference variables (clarity, attention and intensity) and cognitive difficulties under acute stress.

Bourne and Yaroush (2003) introduced three methods that can be used to measure stress effects: Neuro-physiological measures (for example, EEG patterns or cortisol in the saliva of human subjects), self-reported measures (interviews and questionnaires) and performance or behavioural measures.

The most appropriate method for measuring stress during work with maps appears to be a method that provides objective data during the course of the entire experience. The SOM Biofeedback 8000 device measures EEG, temperature, pulse, skin resistance and respiration frequency simultaneously while participants work. The advantage of this technique is that the baseline values of test subjects can be measured before the test for subsequent comparison with the values obtained after exposure to various kinds of stress. As a result, we can evaluate the physiological responses associated with various stressors. More importantly, we are able to concurrently monitor and record the achievement of the map work and the physiological correlates associated with the task. Measurements of experienced stress can thus be assigned to various critical moments. The disadvantage of this method is that subjects must be connected to the device throughout the experiment, which restricts movement to some extent and thus reduces the ecological validity[2] of the experiment. The use of this device demands very special conditions (e.g., a stable microclimate), and the preparation of the participants is time-consuming.

By using this method for experimental evaluation of cartographic products designed for crisis management, we can evaluate how a stressful situation impacts performance during work with maps and how various cartographic products contribute to the stress experienced (i.e., whether the cartographic products cause endogenous disruptions).

---

[2] The settings of the experiment must approximate the real-life situation that is under investigation.

3 Spatial information transmission

Maps represent one of many possible ways of spatial information transfer. The amount of information transferred by maps can be evaluated using cartographic theories. The precursors to cartographic theories were developed in the 1960s when information and communication theories were published (Pravda 1990). One theory still being developed is the language theory, which describes maps as a special type of language. The language theory was elaborated in detail by Bertin (1967) and Pravda (1990). The process of information transfer was described by Kolá ný in 1968 (see Fig. 1).

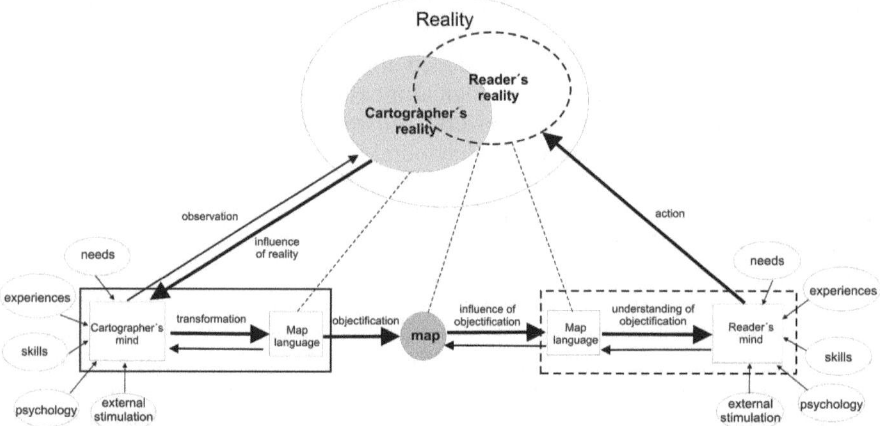

Fig. 1. Information transfer as described by Koláčný (1968)

Koláčný described the process of map creation and the transfer of information as a straight line from cartographer to user. Communication is provided by the map through map language. Stress factors can be represented by psychology (personality) and external stimulation. The theory of map communication was subsequently further developed by numerous authors (e.g., Morita (2004), Slocum (2005)). Essentially, the process of information transfer from map to reader is expected to take place on several levels, especially under stress conditions, and can be considerably influenced by cartographers through understanding of this process.

The process of information transfer from cartographer to reader was described to show where potential loss or distortion of information can be expected. A map is a simplified representation of reality that can be misinterpreted. Therefore, identification of potential problems with information transfer is very important. Comparison of complete and accurate information about the external environment with results of experimental map use can provide information about the cause and origin of misinterpretation.

4 Psychological perspectives on geovisualisation

Egon Brunswick argued (Dhami et al. 2004) that psychological processes are adapted to environmental properties. He emphasised (Figueredoa et al. 2006) that we need to understand the structure of the external environment to be able to understand and evaluate internal cognitive processes and the resulting human behaviour. Various situations initiate differences in ways of thinking. Hammond developed this concept further, forming the cognitive continuum theory. He argues (Kostroň 1997) that cognition oscillates between two extremes in accordance with the nature of a perceived problem. Pure intuitive and pure analytic judgements are the two extremes of how we can deal with the problem. Intuitive judgement is characterised by high levels of information processing and by low levels of both awareness about the process and knowledge about the situation. Analytic judgement is the opposite. This distinction means that in a given situation, one type of judgement is expected to produce better results than the other. A financial audit is a typical task relying on analytical judgement, while piloting an aircraft in an unusual emergency situation is an example of intuitive judgement.

A map has two qualities. First, it is an abstraction and a representation of an environment that offers some additional information: cues to the user for problem solving. Secondly, a map is an environment in itself with a specific structure, and the user has to understand and decode this system to find the required information. Users can be expected to reach higher achievement levels when the structure of the map corresponds to their cognitive styles[3] and/or to the structure of the whole situation (Fig. 2).

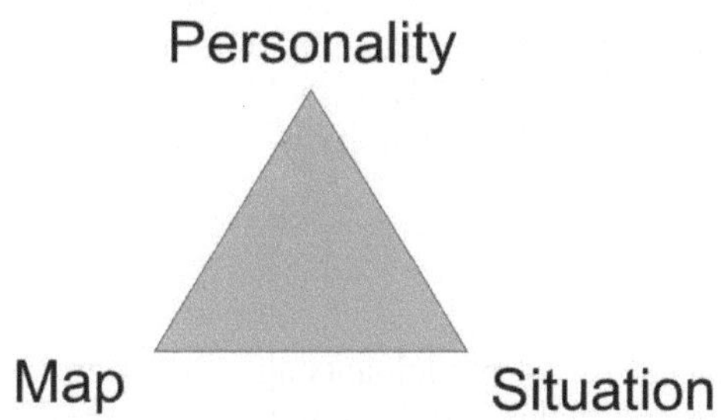

Fig. 2. Interaction of the three main variables involved in map reading

[3] Cognitive styles were described as developmentally stabilised cognitive controls that are relatively invariant across situations (Sternberg and Zhang 2001). Cognitive style describes the way individuals think and perceive information or their preferred approach to using such information to solve problems.

Brunswick also shows in his lens model[4] how humans use directly perceived cues from the environment to make assumptions about invisible phenomena. For example, an internist may assume that an invisible agent (virus, bacterial infection) causes the symptoms of an illness that he perceives directly.

When a cartographer chooses a method of spatial data representation or creates an iconic symbol, he chooses which features best represent the object based on his own experience. He also needs to know if these cues will be valid for recognition of this object by future users (c.f. Fig. 1). For example, a cartographer may believe that a cross symbolises an ecclesiastical building, but a user may find a lofty tower a better representation.

5 Methodology

Our research addressed two goals. The first was to explore possibilities and methods of testing map usability. We are currently using these results to develop special map-testing software. The second goal was to test hypotheses about the influence of three different map representations on perception.

We anticipated that different map representations would influence the process of cognition. In particular, we assumed that a dot map would stimulate intuitive thinking more than a choropleth map or a simple bar map because a dot map offers information about quantity in a form that is not readily decoded. We further hypothesised that only the choropleth representation would lead to ambiguity in the process of reasoning without a legend. The dot map and simple bars both offer cues that are closely associated with concrete physical experience: the number of dots or height of the bars represents quantity. On the other hand, a choropleth map uses colour as the cue. Following Brunswik's theory, this means the cue has a lower level of ecological validity; quantity is not naturally characterised by colour.

5.1 Test participants

The test was conducted for 74 participants aged between 10 and 41 years. Results from 59 participants (36 males and 23 females) were used in the statistical analysis. The others were tested under specific conditions, such as teamwork. The test subjects were volunteers chosen from visitors at a public event called 'European Researcher's Night 2008'. They were informed about the goal of our research.

[4] The theory assumes the perception of distal and uncertain events (phenomena) is dependent on a process of inference from immediately observable (proximal) cues.

5.2 Structure of the test

The test consisted of three variants, each using a different cartographic method. Variant A used a choropleth map, variant B used a dot map and variant C used simple bars. Each variant consisted of five simple and short tasks in which participants were asked to mark requested areas on maps displayed on a PC. We used two types of research methods. Nomothetic methods provided information about causal relationships between variables and statistical analyses showed the extent to which two variables were correlated. The advantage of using nomothetic methods is the ability to survey a large number of subjects with relatively low cost. The primary disadvantages of nomothetic methods are that we are only able to study a limited number of variables and the method is completely insensitive to any other variables. We measured the correctness and quickness of the answers. Idiographic methods allowed us to better understand the behaviour of test subjects during the tasks. We were able to record unusual (infrequent or unexpected) phenomena using this method. The disadvantage is that it is difficult to generalise data acquired in this way. During the group work on the tasks, we used unstructured observation.

In the first task of the test, test subjects were given a simple map without a legend. The participants had to determine the location of highest intensity of a given phenomenon. The second task was the same but this time with a legend. The third task required participants to mark a concrete value according to the legend. The fourth and fifth tasks also involved determining concrete values using the legend, but with additional phenomena included on the map (one additional phenomenon in task 4, two in task 5). Figure 3 shows examples of tasks 3 and 4.

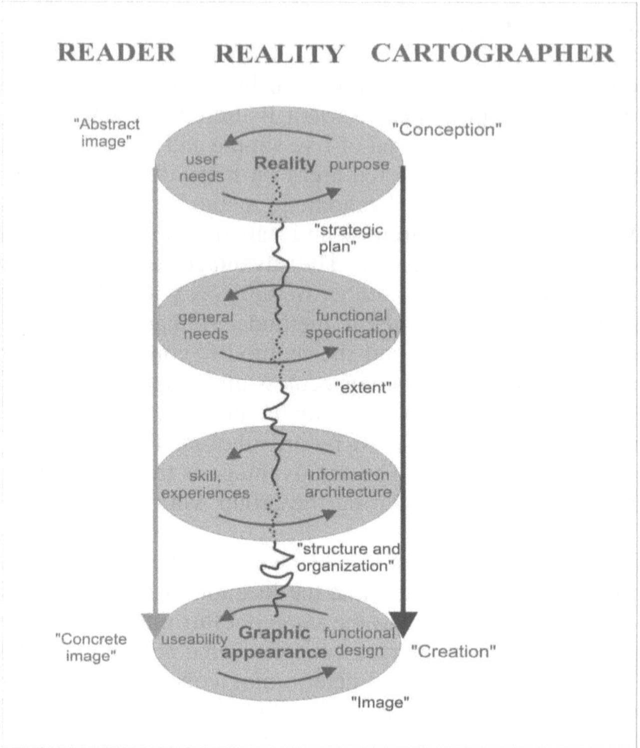

Fig. 3. Example of maps showing task 3 of test variant B (*top*) and task 4 of test variant C (*bottom*)

Communication between participants during the test was observed and recorded. Some additional interviews with individual participants were also conducted. The participants were asked questions in certain cases, for example what happened when they failed, why they chose certain areas, how they perceived the task or what they found difficult and why.

6 Results

6.1 Examples of qualitative analysis

Qualitative methodology is often used in pilot studies to provide deeper understanding of issues and to choose variables to be subsequently monitored by nomothetic methods. Qualitative researchers aim to gain an in-depth understanding of human behaviour and the reasons that govern such behaviour with survey questions asking 'How?' and 'Why?'. Our goal was to use both methods concurrently.

We used the qualitative method as a supplement to the quantitative analysis in order to better interpret our data. We also used qualitative methods for meta-analysis of the research process itself. The aim was to evaluate the possibilities and restrictions associated with our research design for future development of automated testing software. We obtained additional data using idiographic methods.

Our qualitative observations allowed us to determine that some participants counted the single points to find the correct area while others only tried to estimate the density, as evident in the following conversation:

Participant 1: 'Don't lose time with the counting!'

Participant 2: 'I have to count it.'

Participant 1: 'That's useless, you should just guess.'

The participants had no legend at their disposal in task 1. We hypothesised that some participants would mark the brightest area in the choropleth map (variant A) as the area with the highest quantity of the given phenomenon. We also expected that most of the participants would automatically choose the darkest area. The following conversation gives an example of the former hypothesis:

Participant 3: 'What are you waiting for? Mark this area!'

Participant 4: 'But it could be the white one, too.'

6.2 Statistical analysis results

Objective evaluation of perception of different cartographic methods is needed. The first data we collected was the time necessary to accomplish the task. Variant A provided organised information; therefore we expected the quickest performance with this type of map. In general, time to complete a task increases with the complexity of the task. The shortest times were found for variant A (choropleth map), except during task 1. Because there was no legend in task 1, participants had to decide whether darkness or brightness signified intensity. Variants B (dot map) and C (simple bars) showed times that were longer but in most cases comparable to those from variant A. Exceptions were task 1, with significant differences in standard deviation, and task 3, with significant differences in average time to complete the task (Fig. 4).

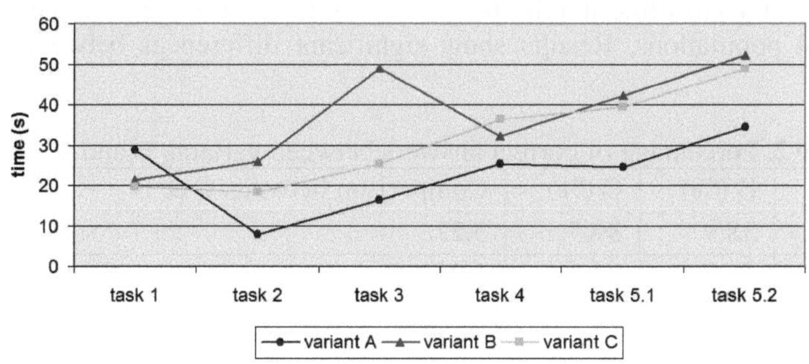

Fig. 4. Average time needed to accomplish the tasks

Figure 5 shows the percentage of wrong answers in each task. Task 2 was not included because of ambiguous determination of correct answers.

We tested for the homogeneity of two binomial populations (see Table 1). Values below 1.96 ($p < 0.5$) indicate no significant difference in achievement between the three tested map representations.

Task 1 was very simple, so there was no significant difference between variants. High differences between variants were found in tasks 3, 4 and 5. Variant A had the lowest error rate in all cases. The time needed for accomplishing the task was not involved in this evaluation.

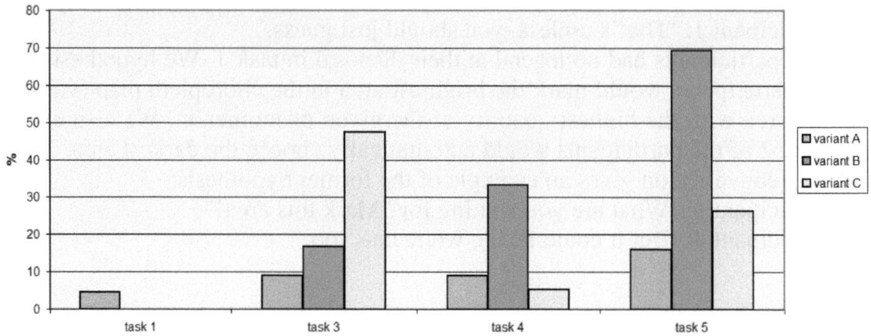

Fig. 5. Percentage of wrong answers

Table 1. Comparison of error rate between variants of the test

	Comparison of A & B	Comparison of A & C	Comparison of C & B
Task 1	0.91	0.94	0
Task 3	0.72	**6.07**	**1.99**
Task 4	**4.77**	**1.99**	**2.18**
Task 5	**8.92**	1.49	**2.18**

The next step was the comparability analysis. Table 2 shows the difference in error rate between variants B and C using the percentages of correct answers for tasks 1 and 4–absolute maximum. We again tested for the homogeneity of two binomial populations. Results show significant differences between variants in task 1.

Table 2. Percentage of correct answers between variants B and C

	B (%)	C (%)	Comparison between C & B
Task 1	38.9	89.5	**3.22**
Task 4	16.7	33.3	1.00

7 Conclusions and further research

The results from Chap. 6 validate the expectation of high levels of achievement using choropleth maps (variant A), except when a legend is absent (task 1). On the contrary, using this method can misrepresent the real distribution of a phenomenon. Comparison between variants B and C showed differences in the ability of participants to find absolute numbers, with variant C being advantageous. The results also indicate that the concrete applied map method can very strongly influence human perception and mental representation of a situation. Those findings correspond to the cognitive continuum theory mentioned in Chap. 4. Note, however, that the small number and heterogeneity of test participants may lead to erroneous conclusions and our results should be further verified.

We believe that the combination of idiographic and nomothetic methods can eliminate the deficiencies associated with each method. Strong statistical data can be obtained from nomothetic methods, while idiographic methods can be used to simultaneously provide deeper understanding. Additional variables that would be lost by statistical analysis alone can also be explored. As an example, consider a test response where two districts were marked together instead of one district. If we had used only a nomothetic method, we would have evaluated it as a mistake by the test subject. Our additional qualitative analysis, allowed us to discover that the border between the districts was covered with dots and that the 'mistake' was actually a consequence of the cartographic method.

There is a need for a special tool to use with this research design. Such a tool should be able to analyse achievements of participants immediately on completion of the test and then to choose the suitable individuals. These participants could subsequently be explored as representatives of different categories. Physiological correlates of the participants in the PC-based map tasks could also be measured, enabling observation of stress levels.

The findings of this research are very important for software currently being developed that will be used for testing various map methods and their usability. Primarily, the results have shown that statistically significant disparities in performance are associated even with very simple types of tasks (e.g., simple dot or linear markings) and thus testing software could be based on these types of tasks. Analysing the behaviour of test subjects during the experiment will allow us to optimise the user interface and presentation of stimuli in future testing programmes.

Acknowledgements

This research has been supported by funding from Project No. MSM0021622418, Dynamic Geovisualization in Risk Management.

References

Bertin J (1967) *Sémiologie Graphique. Les diagrammes, les cartes,* Paris Gauthier-Villars, Translation 1983, *Semiology of Graphics* by William J. Berg.

Bourne LE, Yaroush R (2003) Stress and cognition: a cognitive psychological perspective. Retrieved December 20, 2008 from http://humansystems.arc.nasa.gov/eas/download/non_EAS/

Dhami MK, Hertwig R, Hoffrage U (2004) The role of representative design in an ecological approach to cognition. Psychol Bull 130(6):959–988.

Dror IE, Busemeyer JR, Basola B (1999) Decision making under time pressure: an independent test of sequential sampling models. Mem Cognit 27(4):713–725.

Figueredo AJ, Hammond KR, McKiernan EC (2006) A Brunswikian evolutionary developmental theory of preparedness and plasticity. Intelligence 34:211–227.

Gohm CL, Baumann MR, Sniezek JA (2001) Personality in extreme situations: thinking (or Not) under acute stress. J Res Pers 35:388–399.

Hammond KR (2000) Judgments Under Stress. Oxford University Press, New York.

Koláčný A (1968) Cartographic information – a fundamental concept and term in modern cartography. Cartogr J6(1):47–49.

Kostroň L (1997) Psychologie vytváření úsudku. Masarykova univerzita, Brno. (In Czech)

Morita T (2004) Ubiquitous Mapping in Tokyo. International joint workshop on ubiquitous, pervasive and internet mapping (UPIMap2004), Tokyo, Japan.

Pravda J (1990) Základy koncepce mapového jazyka. Bratislava: Geografický ústav Slovenskej akademie vied. (In Slovak)

Slobounov et al. (2000) Neurophysiological and behavioral indices of time pressure effects on visuomotor task performance. Cogn Brain Res 9(3):287–298.

Slocum TA (2005) Thematic Cartography and Geographic Visualization. United States of America, 2nd ed., 518s.

Sternberg RJ, Zhang LF (2001) Perspectives on Thinking, Learning and Cognitive Styles. Lawrence Erlbaum Associates, Mahwah, New Jersey.

Vedhara et al. (2000) Acute stress, memory, attention and cortisol. Psychoneuro-endocrinology 25:535–549.

Zeelenberg R, Wagenmakers EJ, Rotteveel M (2006) The impact of emotion on perception: bias or enhanced processing? Psychol Sci 17(4):287–291.

Research of Students' Cartographical Knowledge in Early Warning and Crisis Management

Temenoujka Bandrova, Milan Konecny and Monika Rusnakova

Abstract

European Union and other developed countries are discussing ways to reinforce their disaster response capacity. The wide activities portfolio is intended to improve the European civil protection response capacity. A European Disaster Response Training Network has been developed with high standards of preparedness, self-sufficiency and interoperability in order to increase the European level of training and scientific knowledge dealing with natural hazards and disasters issues (COM 2008). We suppose that the fundamental way to sustainably improve the training process should involve education and preparation tailored to different levels of schools. In order to accomplish this, we must first know the level of knowledge, understanding, skills, ability and preparation of students for crisis situations. This chapter describes the results of an investigation of the possible response of students for early warning and crisis situations as regards their cartographical knowledge. The study was conducted in four European countries: Austria, Bulgaria, the Czech Republic and Slovakia, among three student groups: 12–13-year-old students in secondary schools, 17–18-year-old students in high schools and university students. We present the study and analyses of the results based on questionnaire investigations. Using these, we draw conclusions about the students' knowledge and preparedness for crisis situations. The study clearly shows that mapping and cartography can greatly help these youths. To achieve this, cartographers should use the students' ideas in map design.

1 Introduction

Children need to understand and know how to work with cartographic materials in crisis management situations. At the same time, cartographers need to know if students know about early warning and crisis management (EW and CM), if they know certain definitions. They must also know whether students know how to

M. Konecny et al. (eds.), *Geographic Information and Cartography for Risk and Crisis Management*, Lecture Notes in Geoinformation and Cartography,
DOI 10.1007/978-3-642-03442-8_25, © Springer-Verlag Berlin Heidelberg 2010

behave, use spatial geographic information, orient themselves, work with maps in crisis situations and read and understand special symbols. Furthermore, cartographers must learn how students learn and understand each of these concepts. We recommend using students' ideas in the future development of symbol systems for map compiling in the case of early warning and crisis management. Cartographers must make an effort to be closer to their users and recognise real needs. We attempt to do this using questionnaire-based experimental work in three user groups in four countries: Austria, Bulgaria, the Czech Republic and Slovakia, with a total of 210 students (see Table 1).

The results of the questionnaire experiments will help teachers and professors of geography, cartography and GIS to prepare **their students for crisis situations. Negative responses or misunderstandings can** point out the need for lessons with information on certain topics.

Only a few researchers have investigated the role of cartography for children in the context of early warning and crisis management. More children-related research can be found in environmental sustainability and education areas. For example, Taylor et al. (2003) investigated the views of primary and secondary teachers about sustainability education in New South Wales. Other studies are directed at children's understanding of object presentations on maps (Bandrova and Deleva 1998), gender comparisons for understanding and extracting information from atlases (Bandrova and Nikolova 2005), and students' knowledge of maps (Bandrova and Nikolova 2000). All of these studies aim to connect cartographers and children and to find a way for these studies to help cartographers to make maps more attractive and understandable to children. We conduct similar research on a detailed subject connected with EW and CM: how cartographers can help children and students in disaster situations to make their lives safer and minimise damage using maps created for these purposes.

The cooperation between two ICA (International Cartographic Association) commission/working groups on cartography and children and on cartography for EW and CM has led to several important steps: maps for natural disasters in school atlases were published, standards for the colours of disaster representations on maps were proposed and a classification of natural hazards and disasters was completed (Konecny and Bandrova 2006). The research presented in this chapter is a continuation of these studies.

2 Background

Disasters are happening in all four countries within this study; they are sometimes similar, sometimes specific to the region. The flood protection practices in Bulgaria comprise a range of activities, more or less intensive in different regions, depending on the level of flood hazard. Other natural hazards in Bulgaria are earthquakes, landslides, forest fires and avalanches. It is difficult to say which disaster is most frequent, floods or earthquakes. The total number of disaster events in Bulgaria is shown in Fig. 1.

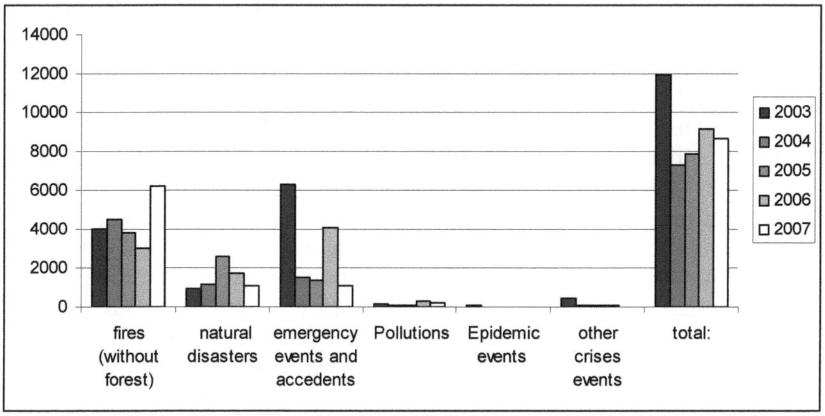

Fig. 1. Number of crisis events in Bulgaria in 2003, 2004, 2005, 2006 and 2007

Reading many studies in the field of cartography, it is clear that non-cartographers, students and children should be educated in different cartographic subjects related to thematic mapping and special cartographic topics related to extracting information from maps quickly and easily. This can be seen in information brochures distributed by the UN (Anderson and Innes 2008, Anderson and Vasconceilos 1995) and in many other scientific studies. The conclusions confirm that people are not sufficiently educated in how to use maps, what to do and how to behave when a disaster happens. The systems for EW and CM have just started to be developed; the responsible organisation is the Ministry of Emergency Situations. The emergency telephone number 112 started to work in the country during the last year.

In Austria, the most frequent disasters are avalanches and floods. However, Staudinger and Lehner report in MeteoRisk (*www.meteorisk.info*) that storms, lightning, hail, flooding snowstorms, avalanches, arctic cold, icy streets, extreme heat with chronic water shortages and forest fires comprise additional dangers. This is a very good example where the EU regional programme covers the Alpine countries Austria, Germany, Italy, Slovenia and Switzerland. In the provided website we can see online maps that are designed in EU standard colors for EW and CM (Bandrova and Konecny 2006). Staudinger and Lehner in their published brochure in *www.meteorisk.info* show symbols and texts describing the current weather hazards (Staudinger and Lehner). The warning levels are precisely calibrated for the different climatic regions. For example, the emergency level red is reached after 20 cm of fresh snowfall in cities, but only after 60 cm in the mountains. The provided information is clear and readable for users using international standards and associated symbols. This could be a good example for other regions in the EU for preparing necessary public information in EW and CM. No similar brochures or maps exist in Bulgaria, the Czech Republic or Slovakia.

In the Czech Republic and Slovakia, the most frequent disaster is flooding. During August 7–16, 2002, Prague was plagued by a disastrous flood, during which the level of culmination and maximum flow rate and seriousness were the biggest in the history of Czech floods. The flood originated from two waves of extremely intensive rainfall on the 7th and 11th of August. Sixteen people died, 225 000 of citizens were evacuated and the overall flood damages reached 3.5 billion €. Other dangerous meteorological phenomena include exceptionally strong extratropical storms connected with extremely strong rainfalls and drainage. The worst of these in the Czech Republic were the storms Kyrill in January 2007 and Emma in February 2008. The maximum wind speed during Kyrill was, e.g., 60 m/s (220 km/h) in Sněžka and 47.4 m/s (170 km/h) in Dukovany. Kyrill was the worst natural disaster in Europe since 1999. The death toll in Europe was 47, out of which 4 were in the Czech Republic. The direct damages in Europe were 4 billion EUR, 80 million in the Czech Republic. Emma caused less damage: 14 Europeans died, out of which 2 were in the Czech Republic. Damages in the energy industry in the Czech Republic were approximately 50 mil. EUR, which is more than during Kyrill, but damages to forests and public property were only one-third those of Kyrill.

Floods in Slovakia are very common, especially in the eastern part. The biggest flood took place in August 1998 due to extreme rainfall that raised the water level of all the rivers. The water flowed with an estimated return period of 1000 years and involved the entire eastern part, especially the small village of Jarovnice, where 2 hours cloudburst killed 50 people. Flood damages there reached 2.5 million. €. On September 19, 2004, one of the biggest wind disasters in the High Tatras, which destroyed 3 million m^3 of wood over an area of 12,000–14,000 ha, took place. The reason for this was the wind 'Poljak', which reached a speed of 90–170 km/h and uprooted trees 2 – 7 m high. It will take about 100 years for the High Tatras to return to its pre-disaster conditions.

This detailed information shows that the responsible organisations gather all details of disasters after the fact, but we cannot find similar information before the disaster hits. One of the tasks of cartography is to gather, analyse and visualise the information that shows the existing disaster risk and to teach and prepare people for coming events. This could be one way to save human lives and reduce material loses.

In 2008, 236, 000 people died worldwide as a result of natural disasters. The International Strategy for Disaster Reduction (ISDR) U.N. organisation consultant and hazards expert Ben Wisner said there were an estimated 34 million children living in the 20 countries that registered the most deadly earthquakes during the 20th century. Over 150,000 people died in the tsunamis, and at least one-third of the dead are children, according to the United Nation's Children Fund (UNICEF). Because of so many disasters in all of the countries mentioned above and many children casualties, we decided to investigate the preparedness of children and students in crisis situations. The aim of the following experimental work is to understand more about students' knowledge in the field of EW and CM and on this basis to make recommendations to teachers, cartographers and other specialists. The study can be considered a first step towards more fundamental research.

Research experiments and practices from all over the world will help authors develop some plans for working with children in EW and CM. A good step in this direction can be seen in (Wortley, D. 2009), where serious games, virtual worlds and geo-spatial technologies can help educate children about early warning and crisis management. Many studies in the field of 3D can increasingly improve the presentation of information in 3D forms and their users needs (see e.g., Zlatanova, 2008). These two last approaches are the most attractive in children's education in EW and CM. National attempts in these directions should develop new approaches using international achievements and best practices.

3 Research experiment – questionnaire

The research experiment was based on a questionnaire and analyses of the answers. The goal was to understand more about students' knowledge, early warning, crisis management, mapping and thematic maps related to specific subjects. Negative answers will direct cartographers understand how to make maps for early warning and crisis management, specifically how to visualise information on the maps.

The most important task in the questionnaire is the last one, connected to students drawing maps. The resulting maps will show cartographers whether students understand how the necessary information should be represented on the maps. The cartographers will learn which cartographical symbols and methods or cartographical representations and visualisation are most closely linked to students' thoughts.

The questionnaire consists of the following twelve questions:

1. Do you know what 'early warning' is? Please explain.
2. Do you know what 'crisis management' is? Please explain.
3. Have you ever heard of a system for early warning and crisis management in your country? What do you know about it?
4. Do you know who is responsible for crisis situations in your country?
5. Do you know what to do if you are in a place where a forest fire starts?
6. Have you ever seen maps for early warning? Where? Please describe them.
7. Do you think a map could be useful for you in crisis management? How?
8. How would you represent the following phenomena on a map: fire, landslide, flood, earthquake, and volcano? Please draw the symbol.
9. Do you have lessons in school about early warning and crisis management?
10. Do you use maps on the Internet or on your mobile phone? If yes, for what purposes?
11. What meaning do you associate with the symbols below (which object or phenomenon do they represent)?
12. How do you imagine a map that could be helpful to people? Please draw it using your own symbols. Explain them in a legend. You can use the Google map of Vidin city in Bulgaria as a background for your map. The city is situated on the Danube River (Fig.2). (The map is given in Fig. 10.)

Fig. 2. Map symbols (wwww.sokwanele.com/map/all_breaches)

These questions are constructed with several intentions. The first two questions aim to clarify the understanding of early warning and crisis management; researchers need to know if students understand the terminology. The next three questions are directed at knowledge about administration and correct behaviours in emergency situations. From all five questions, we can extract general knowledge related to using and understanding maps and cartography in these specific cases. Questions six and seven are chosen to clarify whether students need maps in EW and CM. The last four questions are directed at map making, map using and cartography. The responses here should show the level at which students and cartographers are ready to work together. Some questions specify some cartographical elements and can be used in proposals for future work: for example, questions eight and eleven may help cartographers in legend construction and symbol design.

Students that participated in the experiment were distributed into groups as shown in Table 1. The youngest students were from primary schools. Only the Slovak students are from a region in which floods often happen. The students from high schools study in gymnasiums without special preparation for this study. The university students study geodesy in Bulgaria, geography in Austria and different subjects in Slovakia and Czech Republic. This means that the research was done with students who do not have preliminary preparation in drawing or working with materials for EW and CM (the UN directs more attention to preparing students living in the regions of the world with the most disasters). In this way, the results can be used as a general conclusion about the knowledge and preparation of an average student.

Table 1. Students' distribution by country and age

Country/ Questions	**Austria, Vienna**	**Bulgaria, Sofia**			**Czech Rep., Brno**			**Slovakia, Kosice, Cana, Preso**		
Age groups	***C***	A	B	C	A	B	C	A	B	C
Number of participants	10	22	24	14	20	20	20	40	20	20
Total number	**10**	**60**			**60**			**80**		

A – 12–13 years old; B – 17–18 years old; C - University students

Students had between 20 min (for the youngest ones) and 30–40 min for the high school and university students to complete the questionnaire. This time is common for similar tests (Alexander 2004). When a disaster occurs and starts to develop, the time for disaster managers as well as all people living in the region to react is limited. Because of this, the time limitation will also influence the students' reactions in disaster situations and questionnaire answers. Their responses are given in written form, anonymously.

4 Research experiment – responses

The **first question** will show whether students understand the topic of this research. We use the ISDR definition: 'The provision of timely and effective information, through identified institutions, that allows individuals exposed to a hazard to take action to ...' (ISDR 2004) and the definition used by the UN Staff College's program on 'Early Warning and Preventive Measures: 'The process of collecting and analyzing information for the purpose of identifying and recommending strategic options for preventive measures.' (Fabrizio Bilucaglia, personal communication), http://www.allacademic.com/meta/p_mla_apa_research_citation. The percentages of correct responses in different countries and groups are shown in Fig. 3.

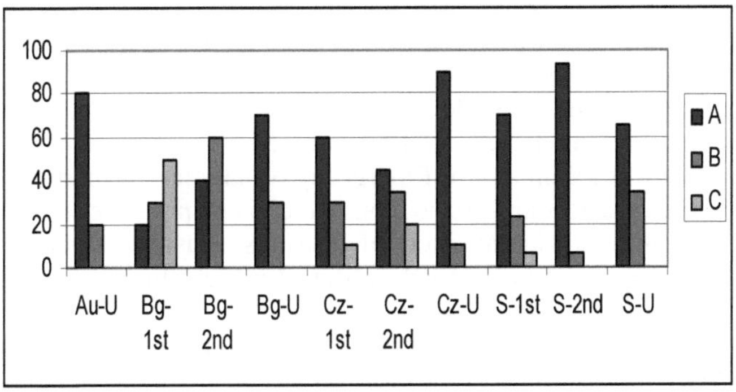

Fig. 3. Responses to the first question (where A, B and C obtain values as explained in Table 2)

The **second question** is: 'Do you know what 'crisis management' is? Please explain'. If we again use the Web definition: 'Crisis management is the systematic attempt to avoid organizational crises or to manage those crises events that do occur' (Pearson and Clair 1998) or 'CM is the process of measuring or assessing risk and developing strategies how to manage it' (Reznik 2008), we can say that students in high schools and universities give correct answers (correct contexts and meanings). More difficulties appear with students at secondary schools, for example some of them connect the term with financial crises, most likely influenced by the media frequently speaking about the world finance crisis (end of 2008). This means that they are not educated enough about this topic and do not have theoretical knowledge.

These two questions and students' answers show that the participants have clear idea about the topic. One unexpected fact is that all of the youngest students in Slovakia and the Czech Republic tried to write correct answers, in contrast to the older students. This is because the students who participated in the research live in an area where floods happen often. They are very active in this subject, having their own experience and more lessons given by teachers and parents.

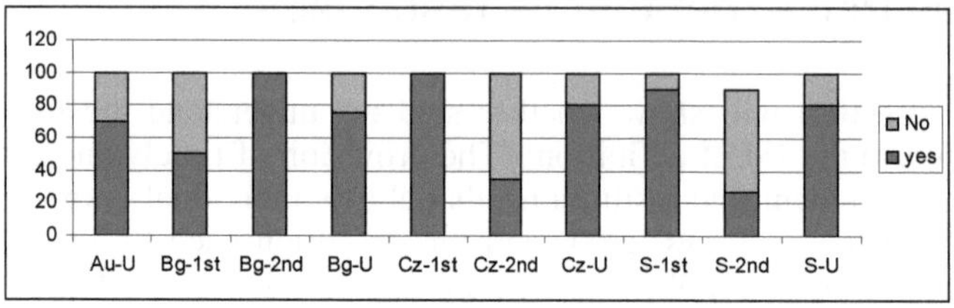

Fig. 4. Results' presentation of fourth question responses

The **third question** is directed to the administration in the participant countries. Almost all participants from Bulgaria report that they do not know about such a system and if it exists, it does not work effectively. It is interesting that in the youngest group from Bulgaria, 50% declare that they know about such a system, which may be influenced by the media (for some, TV and radio announcements attempt to describe the local development of an EW system, for example in the Rousse city region), and they do not understand what exactly it is. In Bulgaria, there still is no working system for EW and CM, except for some local attempts. In Bulgaria, there is a Ministry of Emergency Situations (formerly the Ministry of State Policy for Disasters and Accidents), which is responsible for developing EW and CM systems for different kinds of disasters. Austrian students explain this question with some examples of EW system for the Alps, floods, local systems, etc. The students in the Czech Republic and Slovakia think that is a system of co-operation of police with the fire brigade and rescue teams. They criticize the lack of information – they want to know more about this system. This means again that the education in EW and CM is not sufficiently developed.

'Do you know who is responsible for crisis situations in your country?' is the **fourth question**. Many students – more than expected – do not know the answer, as is visible from Fig. 4. The smallest percentage of correct answers comes from high school students from the Czech Republic and Slovakia and university students from all countries. This result shows that there is no information given to students about organisations responsible for different crisis situations. Students from the Czech Republic and Slovakia gave very different answers; some of them think that depends on how large the involved area is.

The **fifth question** is *'Do you know what to do if you are in a place where a forest fire starts?'* 70% of Austrian students gave the answer that they would call the emergency institutions that are responsible in such situations of that country. One of the interesting results is that of the Bulgarian participants, 80% of the youngest participants, 70% of high school participants and 60% of university students gave the correct answer. This means that the youngest group is better educated than the others. All of the students in the Czech Republic and Slovakia know what to do. The youngest students in particular learn each year how to react correctly in this situation.

The **sixth question** is *'Have you ever seen maps for early warning? Where? Can you describe them?'* About 50% of students declare that they know about such maps. Many Bulgarian participants connect these maps mainly with plans for evacuations, which are situated in buildings. Most of the students from the Czech Republic and Slovakia have never seen these maps. Some university students gave examples like maps of places with the frequency of fire in the USA or South America, and the youngest mentioned student evacuation plans. It is clear that such maps are not used in the teaching of geography and students are not familiar with them or have never been educated using maps for EW and CM. This means that the maps for early warning as one of the newest cartographic products are not popular and people do not know about them. These maps are mainly represented by animation cartography and need special software and computer equipment. Cartographers should find ways in which these maps can be used by a wider range of users, including non-professionals.

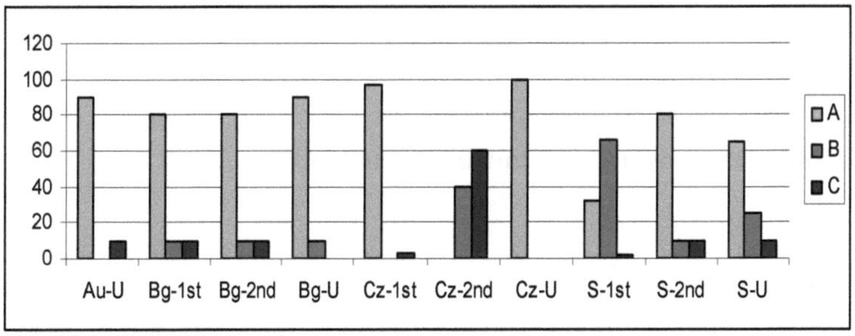

Fig. 5. Responses to the seventh question (where A, B, C are given values as explained in Table 2)

The **seventh question** and its positive answers are evidence that everyone needs maps in EW and CM and it is the duty of the cartographers to make the necessary maps. Students from the Czech Republic and Slovakia wrote for example that maps are very useful in case of emergency when it is important to find a way to get away from a dangerous place and for dispatching of trace. University students wrote that flood maps can reduce the effects and size of the potentially affected area or in case of forest fire maps can notify of tree species. The positive results are shown with column 'A' – 'Call emergency institutions', **'B'** shows correct behaviour in the place, **'C'** – wrong response (see Fig. 5 and Table 2).

Fig. 6. Symbols drawn by a 13-year-old Bulgarian student

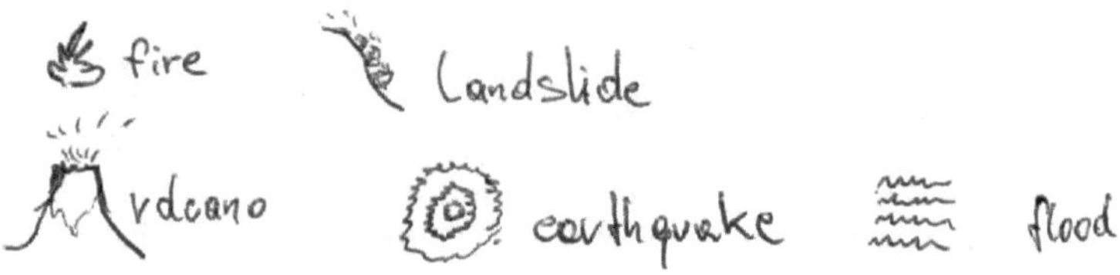

Fig. 7. Symbols drawn by an Austrian university student

Fig. 8. Symbols drawn by a Czech university student

Fig. 9. Symbols drawn by a 17–year-old Slovak student

The **eighth question** is *'How would you represent the following phenomena on a map: fire, landslide, flood, earthquake, and volcano? Please draw the symbol'* and the proposed symbols help cartographers in the visualisation process for map making. Most of students prefer pictorial associative symbols. The form of symbols depends on the age group of the students. The youngest students drew very detailed and colored pictures, while older students drew simple and pertinent symbols. Some examples are given in Fig. 6, 7, 8 and 9.

The **ninth question** *'Do you have lessons in school about early warning and crisis management?'* shows that many students do not recognise lectures concerning EW and CM. Some of them declare that they have evacuation training every year. Bulgarian and Slovak students remember their lessons from primary school. Czech and Slovak students learned about EW and CM in school during one lesson of geography and Slovak students have evacuation training several times per year.

It is clear that the simple lessons given in primary school are not enough. Lessons and training in primary schools are good first steps, but the education stops and is not developed in secondary and high school. Students do not remember everything from their lessons in primary school and are not given any more information about EW and CM. We need to develop special lessons for secondary school and high school and lectures for university students.

The **tenth question** is *'Do you use Internet or mobile phone maps? If yes, for what purposes?'* The responses show that students use Internet and mobile phone maps mainly for orientation, to identify certain locations, do homework and find addresses. This is one of the mapping areas where cartographers could direct attention to EW and CM mapping. Cartographers continue to require children's input to produce proper maps and help children in EW and CM. Cartographers also have started to develop technology for EW and CM based on mobile phones, SMS, with instructions for children in disaster situations and navigation maps (Bandrova and Vasilev, 2008). This new cartographical direction could continue to develop, as students and children use Internet and mobile phone maps more in the future.

The **eleventh question** is related to symbol recognition. *'What meaning do you associate with the symbols below (which object or phenomenon do they represent)?'* The symbols are taken from 'Mapping the election conditions in Zimbabwe' (Sokwanele Article: March 11th, 2008) and their correct meanings are: 1. *Looting, 2. Voter Registration, 3. Abduction, 4. Political Cleansing and 5. Unlawful Detention.* Moreover, they have very long definitions given in the legend (Table), given below:

1.  Looting and Destruction of Property: Incidents where personal belongings are stolen or 'confiscated', or where property is destroyed.
2.  Voter registration: Events related voter registration; for example, ineffectual efforts to register voters, refusing to allow people to register, fraudulently registering voters, and a lack of information in relation to the voter registration.
3.  Abduction: Forcing people, often violently, to go somewhere against their will.
4.  Political Cleansing: This term and icon covers attempts to clear or 'cleanse' an area of political opposition. It covers a wide range of atrocities. In some instances people are literally driven from their homes: in others intimidation and threats are used to achieve the same 'cleansing' objective. The ruling party has used the words 'reorientation' and 're-education' as part of its political strategy before.
5.  Unlawful detention: Activists and political and civic leaders are regularly arrested for no reason at all in Zimbabwe.

Students' answers to the above symbols are

1. **fire**;
2. **information point**/note, school, map, editor note, crisis centre, laboratory, voting station plan, penalty ticket, matchsticks, office, school, information point;
3. **police**, arrest, prison/crime zone, theft, criminality, criminal offence;
4. **emergency exit**, evacuation, kindergarten, disaster, refuge;
5. **prison**, telephone box, no exit, info point, crime.

The bold, first words in each line above is the most common in students' answers. (Students did not know the original purpose of the symbols, so nobody gave the original meaning of the symbols in the map).

These symbols are also used in emergencies. The children interpreted the symbols very well. They were put in the context of emergencies and they did so. If we told them that these were elections, they could have represented the phenomena and objects in a different way. The results clarify that these symbols have popular meanings regarding phenomena in EW and CM and that cartographical symbols should have associated meanings. The same symbol should not be used for multiple purposes, objects and phenomena.

It is clear that the cartographers need to work with their users to produce understandable and quickly readable maps, especially in EW and CM. The best practice is when cartographers check the maps' understandings before publishing them.

The last, **twelfth task** *is connected with drawing a map*. The background is the Google map of Vidin city in Bulgaria. The city is situated on the Danube River. The results are visible in Table 1. Half of the students have no training in drawing maps. The other half understands the situation and proposed a map that could help people in a crisis situation. Students from the Czech Republic and Slovakia did not draw the expected maps from a cartographical point of view. This could be because they do not use enough maps in their education or because teacher explanations have not been clear enough about the expected results. Only some of them gave a full explanation and a map with a legend. Most of them know that people on the image need an early warning system, but they cannot imagine or draw this map.

Usually when cartographers are not sure if users will accept information given by maps for various reasons (too young users, blind people or too specific a topic as EW), they start with experiments about users drawing symbol systems, maps, plans, etc. This task was the most difficult for the students but the most useful for cartographers. It is clear that if cartographers use symbols used by students in their maps and cartographers represent the data and information in the way shown by students, then the cartographical products will be clear and easily understandable. This is the way to make the gap between the cartographer and the user shorter. By examining many drawings, we can say that different nations and age groups need different approaches to maps: more pictorial associative symbols for young students and more topographic objects for older ones; more schematic presentations for students living in regions with frequent disasters.

Some examples are given in Fig. 10.

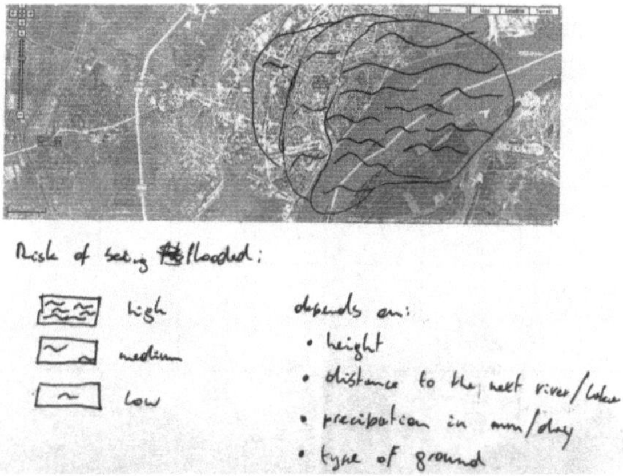

(a)

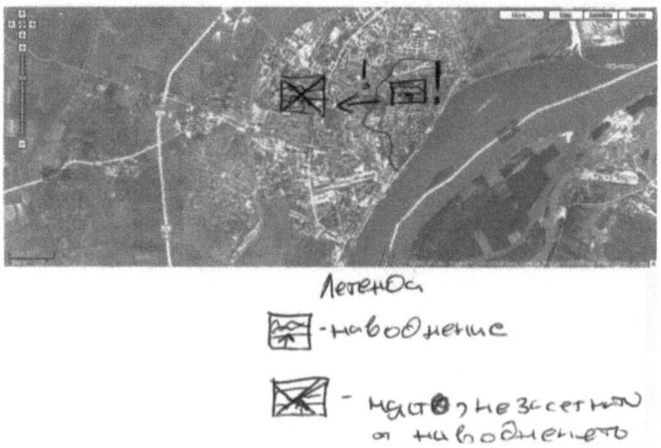

(b)

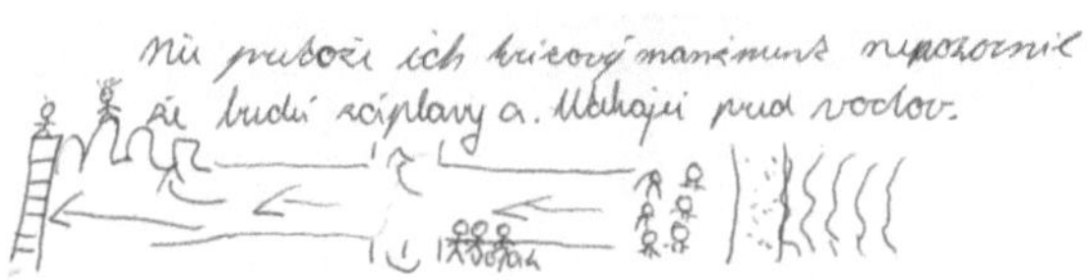

(c)

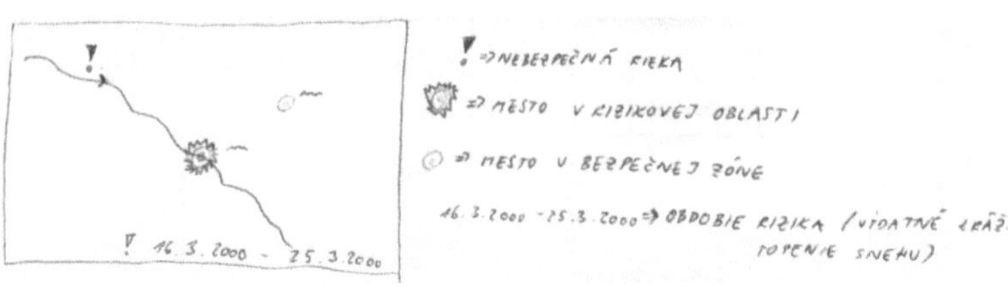

(d)

Fig. 10. Maps – plans for evacuations in a flood situation drawn by (**a**) Austrian University; (**b**) Bulgarian University; (**c**) Czech- secondary and (**d**) Slovak- high school students

Table 2. Distribution by group and country of students who participated in the questionnaire research and their responses

Countries/ Questions	Austria, Vienna	Bulgaria, Sofia			Czech Rep., Brno			Slovakia, Kosice, Cana, Presov		
Age groups	*University Students*	*12-13 years old*	*17-18 years old*	*University Students*	*12-13 years old*	*17-18 years old*	*University Students*	*12-13 years old*	*17-18 years old*	*University Students*
1	80% - A 20% - B	20% - A 30% - B 50% - C	40% - A 60% - B	70% - A 30% - B	60% - A 30% - B 10% - C	45% - A 35% - B 20% - C	90% - A 10% - B	70% - A 23% - B 7% - C	93% - A 7% - B	65% - A 35% - B
2	80% - A 20% - B	20% - A 30% - B 50% - C	40% - A 60% - B	70% - A 30% - B	62% - A 33% - B 5% - C	50% - A 10% - B 40% - C	80% - A 10% - B 10% - C	64% - A 31% - B 5% - C	50% - A 40% - B 10% - C	75% - A 20% - B 5% - C
3	100%-Yes	50% - Yes 50% - No	10% - Yes 90% - No	100% - No	88% - Yes 12% - No	15% - Yes 85% - No	80% - Yes 20% - No	92% - Yes 8% - No	50% - Yes 50% - No	50% - Yes 50% - No
4	70% - Yes 30% - No	50% - Yes 50% - No	100% - Yes	75% - Yes 25% - No	100% - Yes	35% - Yes 65% - No	80% - Yes 20% - No	90% - Yes 10% - No	37% - Yes 63% - No	80% - Yes 20% - No
5	70% - A 30% - B	80% - A 10% - B 10% - C	70% - A 20% - B 10% - C	60% - A 20% - B 20% - C	100% - A	100% - A	100% - A	100% - A	100% - A	100% - A
6	50% - A 20% - B 30% - C	50% - Evacuation 50% - C	50% - Evacuation 50% - C	20% - A 30% - Evacuation 50% - C	90% - Evacuation 10% - C	20% - Evacuation 80% - C	20% - A 20% - B 60% - C	97% - A 3% - C	27% - B 73% - C	15% - A 85% - C

7	90% - A 10% - C	80% - A 10% - B 10% - C	80% - A 10% - B 10% - C	90% - A 10% - B	97% - A 3% - C	40% - B 60% - C	100% - A	32% - A 66% - B 2% - C	80% - A 10% - B 10% - C	65% - A 25% - B 10% - C
8	90% - A 10% - B	60% - A 40% - C	80% - A 20% - C	80% - A 20% - C	55% - A 30% - B 15% - C	80% - A 20% - C	85% - A 15% - C	50% - A 40%- B 10% - C	75% - A 25% - C	95% - A 5% - C
9	30% - Yes 70% - No	80% - Yes 20% - No	50% - Yes 50% - No	70% - Yes 30% - No	77% - Yes 23% - No	25% - Yes 75% - No	100% - No	100% - Yes	47% - Yes 53% - No	100% - No
10	80% - A 20% - C	50% - A 20% - B 30% - C	30% - A 20% - B 50% - C	80% - A 10% - B 10% - C	92% - A 8% - C	70% - A 15% - B 10% - C	90% - A 10% - C	63% - A 13% - B 24% - C	80% - A 20% - C	90% - A 5% - B 5% - C
11	100% - C	100% - C	100% - C	100% - C	100% - C	100% - C	100% - C	100% - C	100% - C	100% - C
12	80% - A 10% - B 10% - C	10% - A 20% - B 70% - C	40% - A 20% - B 40% - C	40% - A --- 60% - C	25% - A 35% - B 40% - C	5% - A 5% - B 90% - C	10% - A 40% - B 50% - C	35% - A 62% - B 3% - C	20% - A 30% - B 50% - C	35% - A 40% - B 25% - C

For questions 1 and 2: Answer quality is **A** if you find a good complete definition; **B** – good definition; **C** – poor or missing definition.

For questions 3, 4 and 9: Answer can be '**Yes**' or '**No.**'

For question 5: Answer is **A** if you find 'Call emergency institutions,' **B** – appropriate behaviour, **C** – wrong response.

For question 6: Answer is **A** if you find 'Yes with description of maps', **B** – 'Yes', **C** – 'No', **Evacuation** maps = maps for CM.

For questions 7 and 10: Answer quality is **A** if you find answer 'Yes and reason / details', **B**– 'Yes', **C** – 'No'.

For question 8: Answer is **A** if you find 'Pictorial associative symbols', **B** – 'Animation presentation', **C** – 'No proposals'.

For question 11: Answer quality is **A** if you find the correct response; **C**- incorrect response.

For question 12: Answer quality is **A** if you find a full explanation and map with legend, **B** – incomplete explanation and not very good map, **C** – bad response.

5 Conclusions

Our investigations show that students in the mentioned countries have some knowledge about EW and CM without education, but this is not enough for them to be well prepared when a disaster occurs. This conclusion is documented by all of the responses of our questionnaire. The goal of all international and national organisations working on this topic is to educate a large number of students and to follow the good practice and examples of the educated Australian students from special schools 'where three-quarters of children reported participating in a hazard education program. These programs were generally carried put in school by Civil defense personnel or a teacher' (Finnis et al. 2004).

To propose national or international education programs, more similar studies should be carried out and more good practices should be investigated. Institutions such as FEMA and EMA are well developed for American and Australian conditions. Taking everything that could be adopted for Europe as well as all results from such studies should help to investigate ways to develop good education programs for students in EW and CM.

To improve the present situation in countries where the research was done, several directions of work with children were planned:

1. New appropriate lessons with tests and exercises about fire escape planning, smoke alarms and general home fire safety, etc. Ideas for such lessons can be found in (Bandrova and Konecny 2008);
2. Publishing books, brochures and information materials in an attractive way for different age groups;

Using the best examples published by UN children commissions and organisations (Bandrova and Konecny 2006) and gaps in children education visible from the studies, textbooks and pamphlets should be prepared. In Bulgaria, some books for teachers are published; this work should continue in writing books for students (http://zadeca.mes.bg). The published books will help teachers because they are not educated a priori in these topics.

3. Developing web pages helping students in disaster management.

This task was developed in its first step in Bulgaria by the Ministry of Emergency Situations (http://zadeca.mes.bg). This first step could be further developed and cover the needs of students of different ages. The web page should give detailed answers to all questions connected with EW and CM that were poorly answered by students. This research shows some weaknesses in students' knowledge. Others may be found in future studies. Some web exercises could be developed as serious games and the best ideas could be taken by www.ema.ohio/gov/Kids_Page/index.htm and www.fema.gov/kids/, where 'Each game teaches your child safety precautions to take in the event of a natural disaster. The more children know about how to protect themselves from natural disasters, the better prepared they are...and being prepared is their best defense.'

Designing and distributing lessons, books and web pages are steps that could be developed by a common program for each of the countries that participated in the experiment. Special national specificity should be taken into consideration, but the UN and all international directives (including EU ones) should be followed as main directions in program development that will help students' education in EW and CM.

References

Anderson JM, Vasconceilos R (1995) Maps for and by children: possible contributions by cartographers. 17th International Cartographic Conference. Proceedings II, Barcelona, Spain, ICA.

Anderson PS, Innes LM (2008) Globe Maps: Novelty or Cartographic Education Necessity? In http://lazarus.elte.hu/cet/publications/311.pdf

Alexander DE (2004) Cognitive mapping as an emergency management training exercise. J Contingencies Crisis Manage 12(4), December 2004, Blackwell Publishing Ltd., Oxford, OX4 2DQ UK &Malden, MA 02148, USA.

Bandrova T, Deleva A (1998) Contemporary Cartography for Children in Bulgaria. The Joint Seminar 'Maps for Special Users'. Wroclaw, Poland, pp. 59–76.

Bandrova T, Nikolova V (2000) Knowledge of Maps and Information Extraction from Them in the Bulgarian Schools', Conference on Teaching Maps for Children: Theories, Experiences and Perspectives Beginning the 3rd Millennium, Budapest, Hungary, pp. 19–28.

Bandrova T, Nikolova V (2005) Gender Comparison of Understanding and Information Extraction from School Atlases in Bulgarian Schools, XXII International Cartographic Conference, CD ISBN 0-958-46093-0, A Coruna, Spain.

Bandrova T, Konecny M (2006) Mapping of Nature Risks and Disasters for Educational Purposes, ISSN 1333-896X, Zagreb, Br. 6, pp. 4–12.

Bandrova T, Konecny M (2008) 'Cartography and GIS' – Training School for Teachers in Geography. 5th International Conference on Geographic Information Systems. Istanbul, Turkey, Vol. 2, pp. 721–727.

Bandrova T, Vasilev S . (2008) Creating and Updating Navigation Maps for Children Purposes (in Crises Situations). Proceedings AutoCarto 2008. The 17th International Research Symposium on Computer-based Cartography. Shepherdstown, West Virginia, USA, Sept. 8–11, CD-ROM.

COM (2008) 130, Communication From The Commission To The European Parliament And The Council on Reinforcing the Union's Disaster Response Capacity, Brussels, 5.3.2008

Finnis KS, Standring D, Johnston, Kevin R (2004) Children's understanding of natural hazards in Christchurch, New Zealand. In: The Australian Journal of Emergency Management, 19(2), 11–20.

ISDR (2004) 'Living with Risk–A Global Review of Disaster Reduction Initiatives,' United Nation, 2004 Version – Vol. 1.

Konecny M, Bandrova T (2006) Proposal for a Standard in Cartographic Visualisation of Natural Risks and Disasters. SMF/UPI Map 2006, Proceedings of the Joint Symposium of Seoul Metropolitan Fora & Second International Workshop on Ubiquitous, Pervasive and Internet Mapping. pp. 157–164, Seoul, South Korea, October 23–25, 2006I.

Reznik T (2008) Web Mapping Technologies for Crisis Management. Second International Conference on Cartography & GIS, Borovets, Bulgaria, Proceedings 2, UACG, ISBN: 978-954-724-038-4, pp. 105–113.

Staudinger M, Lehner G, MeteoRisk, 14 European partners assess Alpine weather risks, www.meteorisk.info

Taylor N, Nathan S, Coll RK (2003) Education for Sustainability in Regional New South Wales, Australia: An Exploratory Study of Some Teachers' Perceptions. International Research in Geographical and Environmental Education. Vol. 12: 4, IGU, Great Britain, Cambrian Printers Ltd., ISSN 1038-2046, pp. 291–311.

Wortley D (2009) Serious Games and Virtual Worlds for Emergency Management, Joint Symposium of ICA Working Group on CEW & CM and JBGIS Gi4DM 'Cartography and Geoinformatics for Early Warning and Emergency Management', January 19–22, 2009, CD Proceedings, ISBN: 978-80-210-4796-9, pp. 658–662.

Zlatanova S (2008) SII for Emergency response: the 3D Challenges. The International Archives of Photogrammetry, Remote Sensing and Spatial Information Sciences. Vol. XXXVII. Part B4, Beijing, China, pp. 1631–1737.

www.unisdr.org/eng/library/lib-terminology-eng%20home.htm, Early Warning WEB Definition

www.womenwarpeace.org/node/12, Early Warning UN definition

www.wikipedia.org/wiki/Crisis management, Pearson & Clair, 1998, Crisis Management WEB Definition

www.fema.gov/kids/

www.ema.ohio/gov/Kids_Page/index.htm

www.usfa.dhs.gov/kids/flash.shtm

http://kids.earth.nasa.gov/

http://www.usfa.dhs.gov/kids/flash.shtm

http://kids.earth.nasa.gov/

http://www.allacademic.com/meta/p_mla_apa_research_citation

http://www.sokwanele.com/map/all_breaches

Geovisual Analytics Tools for Communicating Emergency and Early Warning

Mikael Jern, Monica Brezzi and Patrik Lundblad

Abstract

The large and ever-increasing amounts of multi-dimensional, multi-source, time-varying and geospatial digital information represent a major challenge for the analyst. The need to analyse and make decisions based on these information streams, often in time-critical situations, demands efficient, integrated and interactive tools that aid the user to explore, present and communicate visually large information spaces. This approach has been encapsulated in the idea of Geovisual Analytics, an emerging interdisciplinary field based on the principles from Visual Analytics that facilitates analytical reasoning and decision making through integrated and highly interactive visual interfaces and creative visualization of complex and dynamic data. Geovisual analytics supports geo-information for emergency and early warning systems through a science that augments analyst and decision-maker capabilities to assimilate complex situations and reach informed decisions. Geovisual analytics originates from geovisualization and information visualization but also growing particularly on a high degree of synergy from scientific visualization. In this context, we introduce a web-enabled toolkit GeoAnalytics Visualization (GAV) and associate demonstrators developed in close collaboration with SMHI and OECD, composed of GAV components facilitating a broad collection of dynamic visualization methods integrated with the Adobe© Flash© and Flex© development platform. We also seek to support collaborative knowledge sharing.

M. Konecny et al. (eds.), *Geographic Information and Cartography for Risk and Crisis Management*, Lecture Notes in Geoinformation and Cartography,
DOI 10.1007/978-3-642-03442-8_26, © Springer-Verlag Berlin Heidelberg 2010

1 Introduction

Geovisual Analytics tools are designed to support interactive explorative spatial-temporal data analysis (Andrienko and Andrienko 2005, Guo et al. 2005, Chen et al. 2006, Guo et al. 2006, OECD 2009, Roberts 2004) of large geospatial data. They enable analysts to look at geospatial data from multiple perspectives and explore interesting and important relationships that may exist over a wide array of spatial, time and multivariate scales. Geovisual analytics research focuses in particular on integrating cartographic approaches with visual representations and interactive methods from the information visualization community and on exploring data and enabling collaboration and sharing of gained insights over the Internet.

The major tenets of Web 2.0 are collaboration and sharing, be it of content or technology. The term 'Web 2.0' has become undisputedly linked with developments such as blogs, wikis, social networking and collaborative software development. Web 2.0 can make dramatic impact on developing interactive and collaborative Geovisual Analytics tools for the Internet. A challenge here is to increase the awareness and requirement for more advanced geovisual analytics methods, i.e. tools beyond simple static graphs. Tools are needed that advances humans ability to exchange gained knowledge and develop a shared understanding with other people (Thomas and Cook 2005). Stimulating brainstorming and problem-solving through creative and incremental discovery and developing a contextual collaborative understanding – commonly referred to as geospatial 'analytics reasoning' – are important tasks to solve.

While the benefits of geovisual analytics tools are many, it remains a challenge to adapt these tools to the Internet and reach a broad user community through sharable knowledge. In this context, we introduce a web-enabled toolkit GeoAnalytics Visualization (GAV) (Jern et al, 2007) and associate demonstrators composed of GAV components facilitating a broad collection of dynamic visualization methods integrated with the Adobe© Flash© and Flex© development platform (Fig. 1). GAV is based on the principles behind the Visual Analytics research program (Thomas and Cook 2005). providing explorative, communicative and collaborative visualization. We demonstrate the GAV Flash toolkit in the course of three demonstrators *OECD eXplorer* (Fig. 2), *Ship and Weather Information Monitoring* (Figs. 3,4) and *Swedish Road Warning Prediction* (Figs. 5,6) related to visualization of emergency and early warning scenarios. These demonstrators are developed in close collaboration with domain experts from OECD (OECD eXplorer. 2009) and the Swedish meteorological and hydrological institute (SMHI) and focus on the analytics reasoning aspects enabling analysts to explore spatial, temporal and multivariate data from multiple perspectives simultaneously using dynamically linked views, discover important relationships, share their incremental discoveries with colleagues and finally communicate selected relevant knowledge and publish early warnings if required. These discoveries often emerge through the collaboration between expert domains and operators with diverse backgrounds and are precious in a creative analytics reasoning process.

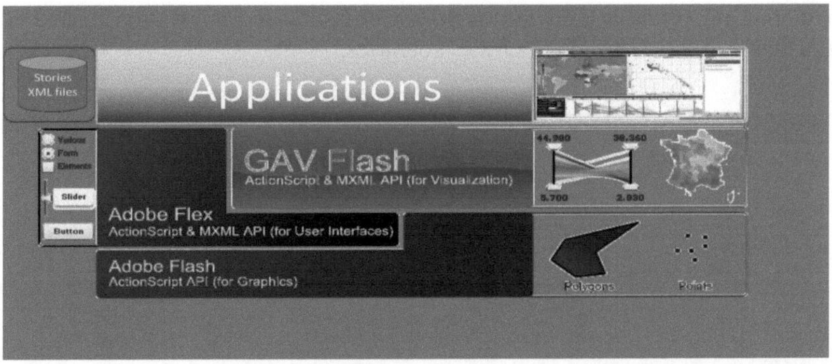

Fig. 1. GAV Flash visualization components are integrated and embedded with Adobe Flash and Flex and programmed in Flash's ActionScript

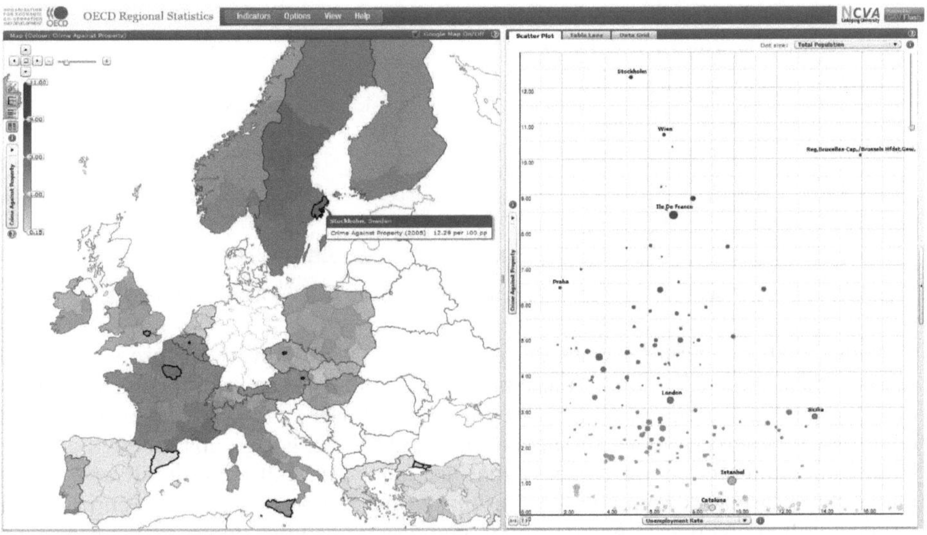

Fig. 2. A snapshot of OECD eXplorer (http://stats.oecd.org/OECDregionalstatistics/) displaying an indicator '*crime against property*' for small regions across Europe and Turkey. Data is simultaneously explored in three dynamically linked views Map, Scatter Plot and Parallel Coordinates (see Fig. 6). Several regions of interest, Stockholm (highest crime rate in Europe), Vienna, Brussels, Paris (Ile De France), London and Istanbul) are highlighted in all views. Most of the high crime rate against property (darker regions) appears in Sweden, France and the northern regions of Italy

European countries have experienced a growing interest in regional statistics in recent years (OECD 2009). Geovisual analytics of local regions in Europe could also become an important tool of geo-visualization for emergency and early warning scenarios related to, for example, epidemic or natural disaster events. The *OECD regional database* provides a 'socio-economic layer (indicators)' to the geographical one and gives the opportunity to integrate different information to increase the knowledge. It contains yearly time-series for around 50 indicators on demography, economy, health and labour market opportunities, environment, social issues and innovation related activities for more than 1,700 regions of the OECD countries.

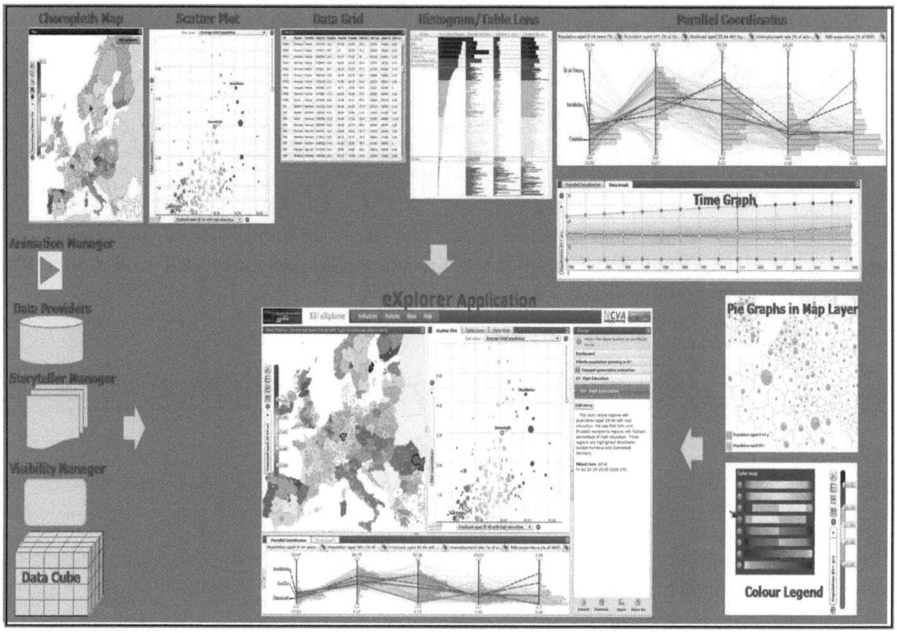

Fig. 3. A geovisual analytics application is assembled from GAV Flash components

Many analyst operators want to create content and express themselves through 'user-created knowledge' and a more pro-active, collaborative role in content creation, distribution and shared use to citizens. More active users and user-centred innovation could have increasing social impact and importance. Target groups for such a knowledge-generating collaborative geo-visual web analytics tool are quite diverse. A potential target group is regional decision-makers in charge of regional security development policy, who can make use of geovisual analytics tools in their decision process. Citizens and the media would also be able to get informed and at the same time participate in increasing the knowledge on an emerging epidemic disaster scenario from region to region. Because of the different expertise and needs of the target-groups, the tool should be flexible and adaptable to different audiences.

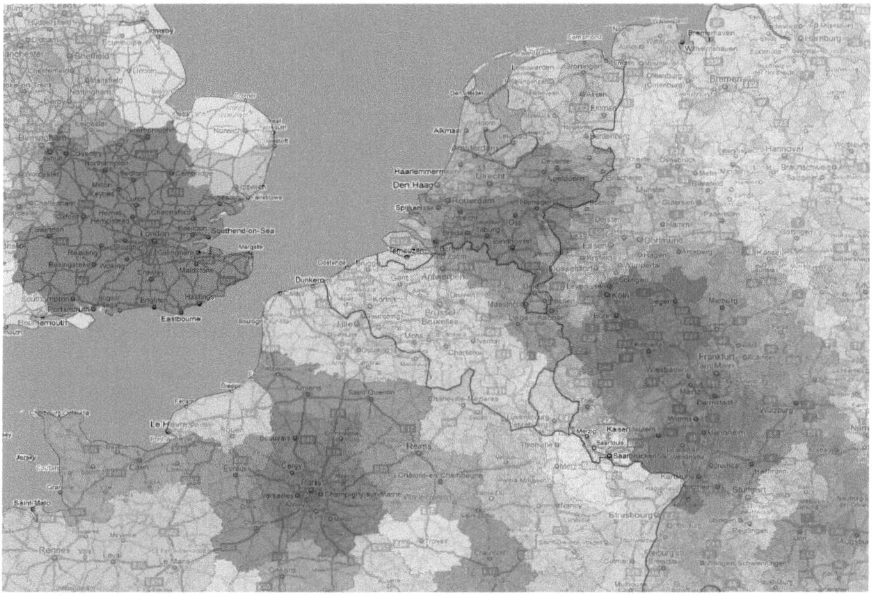

Fig. 4. The GAV Flash Choropleth map component applied to TL3 (small) European regions showing a simulated epidemic disaster scenario in west Europe. We see regions in darker colour around larger cities such as London Paris developing an epidemic disaster. The GAV map component is here overlaid on top of a Google map using the GAV map layer approach

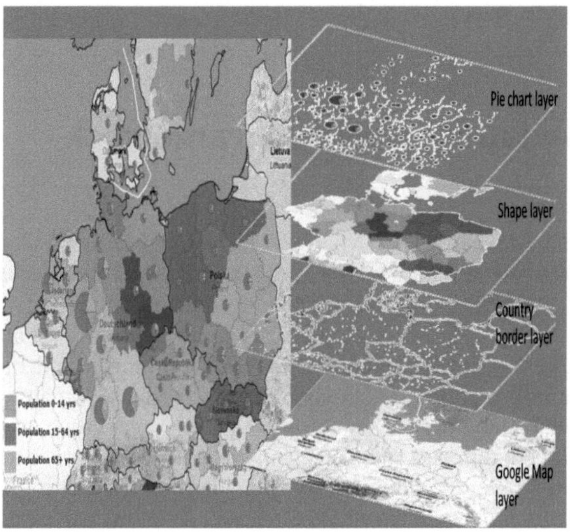

Fig. 5. The GAV Flash map component facilitates a map layer architecture that can integrate with, for example, a Google map. The location of a colour-shaded region can now more easily be identified from the Google map layer, while the regions are coloured by GAV (Figs. 2 and 4). The same Mercator projection is applied for both layers to guarantee correct overlapping. The opacity level is specified by the user. Additional glyph layers such as ships (Figs. 10,11) and wind symbols (Fig. 12) can be added on top of the map layers

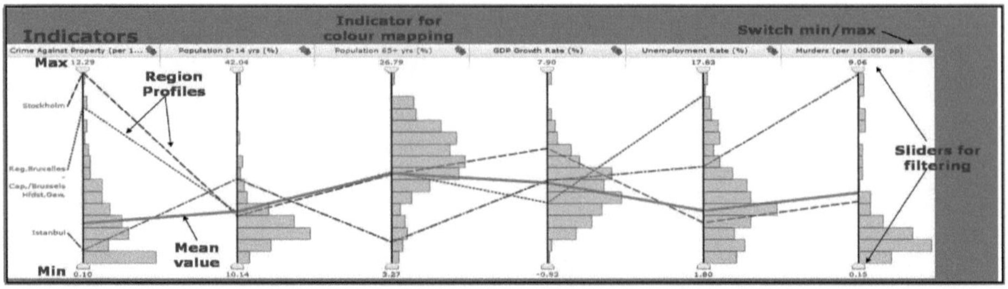

Fig. 6. This PCP is linked the two views in figure 2 ('*Crime against property*') and shows three highlighted regions Stockholm (*top profile*), Brussels (*middle*) and Istanbul (*lower*)

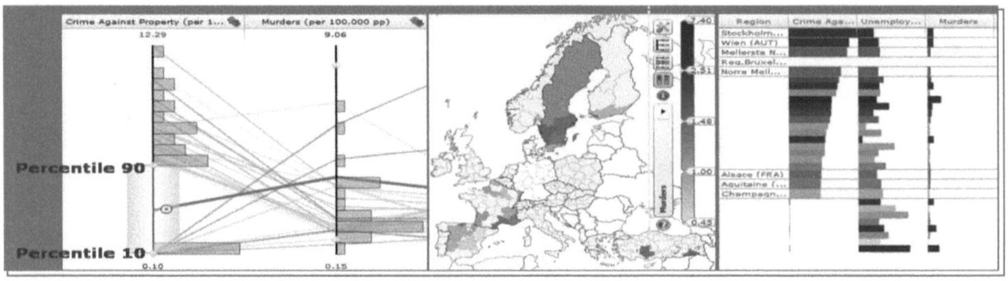

Fig. 7. Regions (outliers) with a low and high population share for '*Crime against property*' are discovered using the statistical method percentiles. Regions above the 10% percentile and below the 90% percentile are removed (filtered out). The Table lens (*right view*) displays the list of regions from highest values (*top*) to lowest values

2 The Geo-visual analytics framework

GAV was originally developed for Microsoft's .Net and DirectX (Jern et al. 2007, Jern and Franzén 2007). Visualisation of *Emergency and early warning tasks*, however, requires a Web 2.0 integrated solution facilitating dynamic geovisual analytics. The new GAV Flash uses Adobe©'s Flash basic graphics and Flex 3 for user interface layout, programmed in Adobe's object-oriented language ActionScript and facilitating common methods from the information and geovisualization research domain. Interactive features that support a spatial analytical reasoning process are exposed such as pan, zoom, tooltips, brushing, highlight, conditioned filter mechanisms that can discover outliers and dynamically linked multiple views adhering to Shneiderman's mantra (Shneiderman 1996) '*Overview and zoom, filter and details-on-demand*'. The key implementations of the toolkit are described in more detail below.

2.1 Layered component architecture

The GAV Flash framework enables deployment of customized and user-centric applications by assembling visual and manager GAV Flash components (Fig. 8). Each component according to its nature in the context of object-oriented ActionScript programming performs a small specific task in the overall geovisual analytics process. The functional components in Fig. 8 is the level that is constituted by the combination of one or more atomic GAV and Flex components and typically implements the functionalities in a dynamically linked views environment (Fig. 2). This component way of thinking could enable a shorter development time, scalability, extensibility, reusability and a key contribution for architecture supporting storytelling.

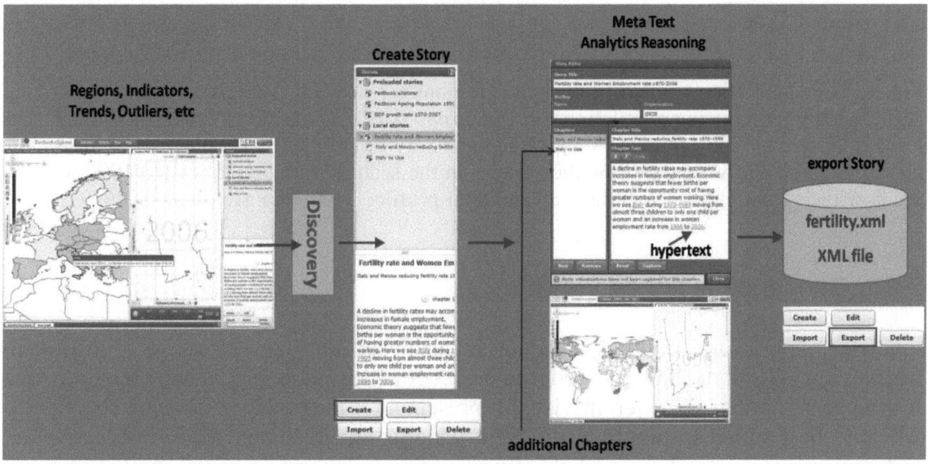

Fig. 8. During an explorative session, the operator first selects data to be analysed. Then a search for trends, outliers, discovers important observations, highlights regions to be compared etc. - a discovery is made! Secondly, open the Story Panel, use button ***Create*** a Story, a Story Editor panel comes up, fill in the required information and associate reasoning text and finally press ***Capture***, the entire current exploration scenario (all views and attributes) are saved together with selected data. The user can now start a second Chapter (***New***) and create a new scenario and repeat the process or ***Close*** and then use the button ***Save as***, give the Story a name 'my story nr 2'.xml. The Story is now saved locally on your computer and can be reused ***Open*** or sent to a colleague for review

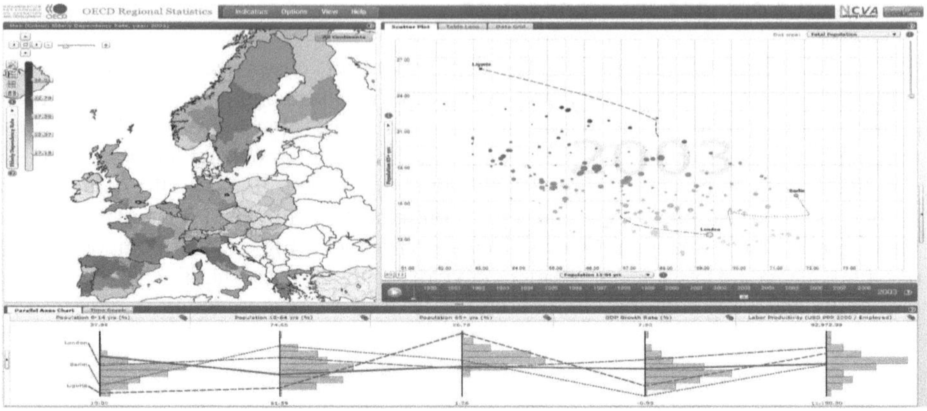

Fig. 9. Time animation applied to European regions for growing '*elderly population 65+*' and at the same time reducing population for '*age group 0-14*'. Three views (choropleth map, scatter animation plot and optional PCP) are dynamically time-linked and animated. Four regions are highlighted and we can follow their traces in the scatter plot from time step 1990 to 2006, see how their colour changes in the map and their profiles in the PCP simultaneously

2.2 Dynamically linked views

Spatial, temporal and multivariate data are analysed through the use of dynamically linked views (Roberts 2004). In order to detect complex patterns it is convenient to view it through a number of different visual representation methods simultaneously, each of which is best suited to highlight different patterns and features. These alternative and different views on the data can help stimulate the collaborative visual thinking process that is so characteristic for geovisual analytics reasoning. The views are separated by interactive splitters allowing the user to scale the individual views and allocate more space to visual representation that is most important. The views are coordinated using a data linking method where the visualization components use the same cube data model, transformation and colouring scheme, and where any dynamic filtering or picking made in one of the linked visualization components propagates to all the others. Visual highlighting is another feature exemplified through visual representations that use pre-attentive highlighting styles like colour, shape and transparency. Figure 2 shows highlighting implementation in the choropleth map and scatter plot views.

2.3 Dynamic choropleth map using map layers and dynamic colour legend

The choropleth map (Figs. 2, 9, 10, 3, 4 and 5) provides the overall context such as data-dense summary of a selected indicator for thousands of regions. Moving the mouse over the map, tooltips display actual indicator values for comparison. Each region is coloured according to the current assigned colour map indicator. Colour coding is a fundamental important technique for mapping data to visual representations and is widely used in a large variety of geographical visualizations. GAV Flash supports a dynamic colour legend based on percentile distribution that splits the data into four intervals based on the 25th, 50th and the 75th percentiles. This provides a better automatic distribution of colours related to the data based on an accurate statistical method and the visualization is improved in a way that differences between values can be more easily detected. Colour scale values can also be adjusted dynamically by moving the handlers. The percentile and corresponding data values are shown to guide the user when moving any of the handlers attached to the colour scale legend.

Colour coding is a fundamental technique for mapping data to visual representations. We use perception-based colour schemes, which have been evaluated by domain experts and proven to be appropriate and facilitate better insight into the data. By considering data characteristics we combine our colour scale legends with embedded fundamental statistics methods. Usually, colour mapping functions perform a linear interpolation based on the minimum and maximum of values of a chosen indicator, a method that leads to problems if the values are not uniformly distributed, e.g. if outliers are plotted. In this case a wide range of the colour scale represents a small number of values and the majority of values have to cope with only a narrow range on the colour scale. Our (default) colour scheme is divided into four ranges, separated at the values of the 25th, 50th and the 75th percentiles (Fig. 2). This provides a better automatic distribution of colours related to the data based on an accurate statistical method and the visualization is improved in a way that differences between values can be more easily detected. Alternatively, the 10th, 50th and 90th percentiles are used to discover outliers (Fig. 11). Colour scales can also be adjusted interactively and dynamically (e.g. all linked views are updated immediately). The percentile value and corresponding data value is shown to guide the user when moving one of the three range sliders attached to the colour scale legend.

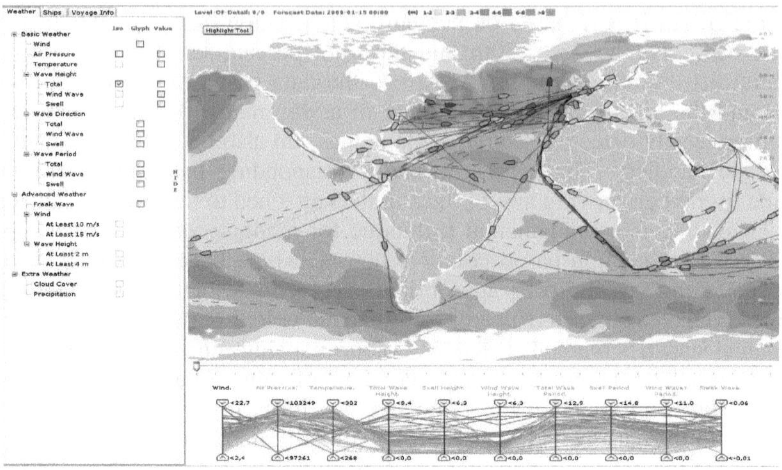

Fig. 10. Tool for Ship and Weather Information Monitoring visualizing weather data combined with data from ship voyages developed by NCVA and SMHI. The GAV choropleth world map ships are visualized using glyphs and their routes are plotted as lines. Significant wave height is displayed using an iso-surface layer (GAV map layer technique) where warmer colour indicates higher waves. The time step can be changed using the time slider positioned underneath the world map. Using the weather attribute panel, the operator can select which attribute to visualize and which visual representation to use. At the bottom weather attributes are plotted in the PCP showing all voyages with selected time step. Each voyage is represented as a line where each axis corresponds to a specific weather parameter. The operator gets a forecast of the weather for all the voyages at a specific time step. Advancing the time slider will update all views simultaneously with new weather information for each voyage at selected time step. Weather forecasts are thus generated for each voyage and early warnings and alternative route recommendations can be communicated to the ship by the operator

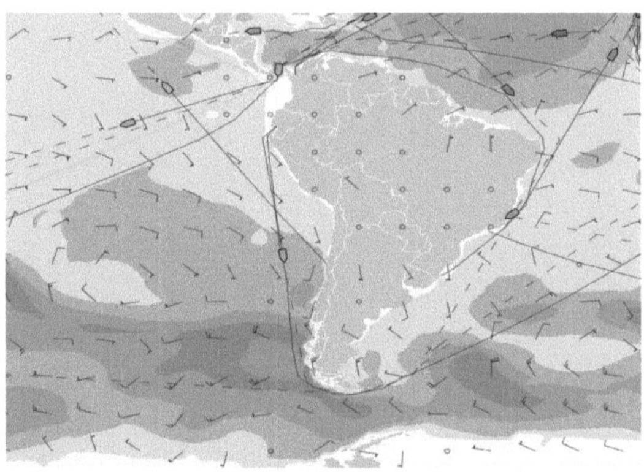

Fig. 11. The world map is the main view where all weather attributes originating from corresponding GRIB files (data format used in meteorology to store weather data) can be presented. Options are provided in the panel on how to visualize an attributes, e.g. iso-line, iso-surface, glyph or numerical values. The ships are represented by ship glyphs which are positioned and rotated according to an interpolation based on their pre-planned waypoints, reported positions and present time. Weather data for a certain date and time provide ten days forecast with a time interval of 6 hours the first 5 days and an interval of 12 hours the later 5 days. This enables the temporal aspect to be taken into account and it is represented by a forecast time slider

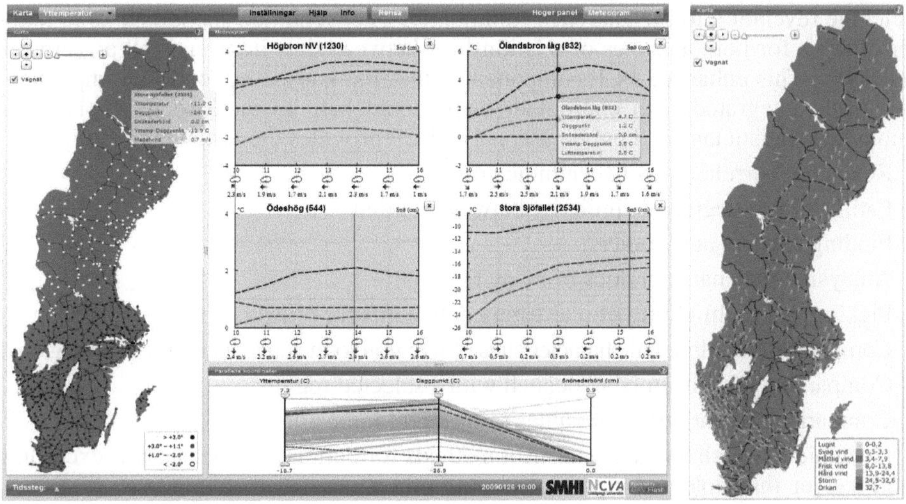

Fig. 12. This extended (Jern and Lundblad 2008) Flash-based version of Swedish Road Warning Prediction System (SMHI 2009) now also includes a meteogram plot with 24h forecast calculation with dynamic time animation. During the winter months, the Swedish roads are dangerous for driving. Ice, rain and snow make the roads slippery and hundreds of people are killed or injured. The Swedish road administration based on measurements from 700 road stations prevents accidents by preparing the roads based on real-time weather forecasting. Today, this is done by accessing static pictures over Internet and direct contacts with meteorologists. The Road Warning Prediction application was developed in collaboration with the SMHI and assembled from GAV Flash components (map with layer support, glyph layer, PCP, line diagram) demonstrates that advanced geovisual analytics tools are suitable for developing highly interactive and tailor-made applications applied to early warning management. The GUI is developed with Flex components. The map shows stations coloured according to temperature or ice slipperiness. The left map displays the wind direction and speed. Each station is represented by a line in the PCP where the colour is dependent by the attribute *Surface temp*. Three selected stations are highlighted and provide each individual station with a profile allowing the analyst to compare *Surface temp.*, *Dew point* and *Snow condition* next to each other, and can now explore how the selected profiles show the potential for condensation occurring or frost depending on the surface temperature. The time slider at the bottom of the window controls a 24 hourly forecast. This application will be published on the Web this fall by the Swedish Road Administrator: http://produkter.smhi.se/trafik/land/roadwizard/Test/RoadFlash.html

2.4 Parallel coordinates

The parallel coordinates plot (PCP) (Inselberg 1985) technique has a long history within the information visualization research community and has demonstrated to be useful in many scientific environments (Andrienko and Andrienko 2005, Andrienko et al. 2006, Guo et al. 2006). The PCP enables visual representation of spatial multivariate attribute data and hence an important explorative mechanism in a Geovisual Analytics application. The relationships (or patterns) formed in the PCP have, over the years, been extensively evaluated and are today well known by users in the research community – whereas it has, through this joint project, proven a new and welcome revelation to the analytics community. The technique supports a large number of tasks for both analyses of relationships between data items as well as between indicators. This enhanced PCP component is the first known ActionScript implementation and integrated with choropleth maps could help to increase its exploitation among the statisticians and public users. The PCP provides the following features:

- Revealing correlation between indicators;
- Estimation of degree of similarity between regions;
- Finding clusters and outliers;
- Analysing the characteristics of many regions;
- Picking and highlighting of interesting data items for profile and comparison;
- Comparison of individual characteristics of a region to the characteristics of all regions;
- Comparison of indicators associated with a selected region;
- Comparison of variations of values of different indicators;
- Dynamic range sliders and statistical methods for defining events such as exceeding of a given threshold and identification of outliers;
- Dynamic visual inquiries, filter operations using familiar statistical methods;

Each region is represented by a string (polyline) that intersects the axes at the values of the attributes for that region. Each axis represents a single indicator (e.g. Crime against property, Unemployment rate and GDP per capita etc.) in the statistical data set. Each indicator axis corresponds to a column in the spreadsheet. The scaling of the individual axes typically ranges from the indicators minimum values at the bottom to their maximum values at the top. The string forms a visual representation of the characteristics of one regional area – see the highlighted thick black line representing Paris in the figure below. Differences between selected regions can be found by visually comparing the strings representing them and plotted with different line styles (Figs. 2, 7, and 5). This new PCP Flash version has been further extended with histograms, thresholds and filter operations using range sliders and percentile statistics.

2.5 Visual Inquiries and filter operations

Range sliders are attached to each PCP axis (Fig. 7) and the user can interactively select or combine filter methods thus altering constraints on indicator values. The range of an indicator can be specified by moving the handles on the top and bottom of the corresponding range slider (Fig. 7). Regions with values for selected indicator that fall outside of the specified range are filtered out. A combination of range slider movements can be used to dynamically formulate complex visual inquiries. These visual conditions and constraints will immediately reflect the visual contents in all views.

3 Collaborative geovisual analytics through storytelling

Operators with diverse background and expertise participate in creative discovery processes that transform information into knowledge. Recognized scenarios, however, that cannot be captured remain within the mind of a single user and are not easily accessible to external analysis (Takatsuka and Gahegan 2002, Wohlfart and Hauser

2007). Tools that integrate the geovisual analytics process with collaborative means are necessary in many emergency operations to streamline a knowledge exchange process of developing a shared understanding with other experts and for early warning to the public. The GAV Flash Framework integrates tools for both collaborative interactive visualization and sense-making. A story indicates a successful suggestion and subsequently fosters additional suggestions based on similar considerations. This learning mechanism allows our storytelling system to improve the accuracy of its suggestions as well as to dynamically adapt to particular users, tasks and circumstances. Colleagues can review a story arrangement and respond with suggestions and comments and subsequently foster additional suggestions based on similar considerations.

Our demonstrators facilitate the GAV tools and architecture to support means of capture, add descriptive metatext, save, package and share the discovery and results of a geovisual analytics process in a series of snapshots 'Story'. When the button 'Capture' in the *Story Editor* is pressed, the state of each GAV view is saved together with user-defined metatext. Before closing the application, the operator exports the story into an XML formatted file. Team members can through metatext combined with interactive visualization follow the analyst's way of logical reasoning by loading selected stories. At any time a team member can access stories and apply them in the current demonstrator or any other GAV application assembled from the same component. A comprehensive story in the context of a remote collaborative sense-making activity can thus be created by the operator through a set of linked snapshots (chapters). Operators can thus discuss relevant issues through storytelling based on solid evidence, thus raising awareness and increasing the common knowledge on a certain phenomenon.

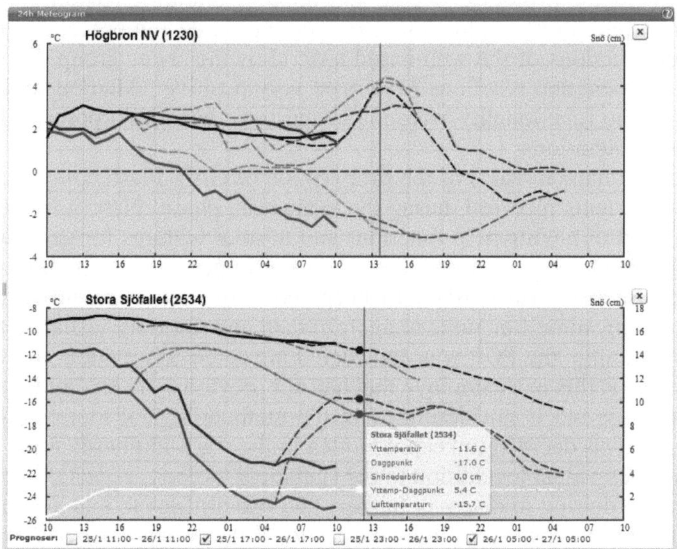

Fig.13. Meteogram plot for two selected road stations. The *left* vertical axis represents temperature, the right snow condition and the horizontal axis shows time. The *upper* line shows the air temperature, *middle* line road surface temperature and the *lower* line dew point. The various line types represent two 24h forecast scenarios. A tooltip in *lower* figure guides the operator with real numbers

4 Simultaneous spatial-temporal and multivariate animation

GAV Flash employs a data model optimized for handling spatial-temporal and multivariate data in a geovisual analytics context. This conceptual data model can be seen as a data cube with three dimensions: space, time and attributes. The spatial dimension is here (Fig. 10) represented by OECD TL2 regions and the indicators are various demographics measurements (GDP growth, elderly dependency rate, etc). Time is the data acquisition period. The bubble plot is demonstrated to be an effective Flash-based time animation method, but integrated and linked with a choropleth map and a multivariate indicator frequency histogram and dynamic filter operations embedded in a PCP, makes it even more useful. The GAV open architecture, handling of large data sets and integrated snapshot mechanism are other important extensions to this emerging animation technique. The spatial cognition for the time-linked views was evaluated by OECD and SMHI and found to be both intuitive and innovative. The operators discovered important trends in all views.

5 Evaluations

Our application development follows a user-centric design approach (Vos et al. 2000). A combination of domain experts from OECD and SMHI have been involved in the various stages of the prototype design and implementation, providing user feedback about usability and utility evaluation. The user-centric design process involved public access to beta versions of the web-based tool, allowing focus group discussions. The overall involvement and reactions have been very positive. Many useful suggestions for improving the functionality were made and have been incorporated in successive implementation iterations.

A number of characteristics of the first version of GAV Flash components were derived from comments received during the evaluation phase. First, it became clear that there was a need of having help functions and tutorial features for geovisual analytics tools targeted at audiences whose expertise is not in geo- or information visualization technologies. Second, users asked to keep the entire structure sufficiently simple, while maintaining some functions to analyze data and not only visualize them. In this context, for example, the PCP was considered not to be self-evident to traditional users of statistics, as this is a technique that has not previously been employed in the statistics community and is not described in the methodological literature on statistics, and therefore it was decided to keep it hidden in the start-up phase; at the same time it was regarded as a valuable addition to the statistical toolbox, especially the possibility of dynamically filtering to discover outliers and use profiles to make comparisons between highlighted regions. Finally, the dynamic links between views (context and focus maps scatter plot, PCP and table grid) were evaluated as very important.

5 Conclusions and future development

The authors expect that the three demonstrators introduced in this chapter and now in full operation will enhance the use and understanding of spatial-temporal and multivariate data, thus adding to sound, evidence-based policy decisions. At the same time, the availability of a comprehensive web-enabled GAV Flash toolkit will encourage the practical use of advanced geovisual analytics science technologies in emergency and early warning systems because of its easy accessibility and dynamic visualization on the Internet. It will enable the analyst and operator to take a more active role in the discovery process of exploring. The tool will increase the interest in and knowledge of structures and development patterns among specialist as well as non-specialist users. Feedback from domain experts and operators who have started using the tool shows that a sense of analytical reasoning and speed-of-thought interaction is achieved. Major achievements include

- Introduction of a novel geovisual analytics framework and toolkit developed in the object-oriented language ActionScript with 100% deployment to Internet;
- Three proof-of-concept demonstrators developed in close collaboration with domain experts from OECD and SMHI facilitating case studies from emergency and early warning scenarios based on real data and to be deployed;
- A Flash implementation of an extended PCP with embedded fundamental statistics based on dynamic percentile inquiry and filtering with virtually instantaneous response time and also attached histograms;
- An architecture facilitating the operators to explore data and simultaneously save important snapshots of discoveries or create a continuous story of snapshots (storytelling) to be communicated and shared with team or public;
- Dynamically time-linked views for simultaneous animation of multivariate data;

Acknowledgements

These applied research case studies were carried out by NCVA (http://ncva.itn.liu.se) in close collaboration with OECD and SMHI who supplied data and comprehensive evaluation of the application. The research is in part supported by funding from the Swedish Knowledge Foundation. The authors thank colleagues Tobias Åström, Markus Johnsson and Gunnar Krig at the NCVA at Linkoping University.

References

Andrienko G, Andrienko N (2005) Visual Exploration of the Spatial Distribution of Temporal Behaviors. In: Proceedings of the International Conference on Information Visualisation IEEE Computer Society, pp. 799–806.

Andrienko G, Andrienko N, Fischer R, Mues V, Schuck A (2006) The parallel coordinate plot in action: design and use for geographic visualization. Int J Geogr Inform Sci 20(10):1149–1171.

Chen J, MacEachren AM, Guo D (2006) Visual Inquiry Toolkit – An Integrated Approach for Exploring and Interpreting Space-Time, Multivariate Patterns. In: Proceedings AutoCarto, Vancouver.

Guo D, Gahegan M, MacEachren AM, Zhou B (2005) Multivariate analysis and geovisualization with an integrated geographic knowledge discovery approach. CGIS 32(2):113–132.

Guo D, Chen J, MacEachren, AM, Liao K (2006) A visualization system for space-time and multivariate patterns (VIS-STAMP). IEEE Visual Comput Graph 12(6).

Inselberg A (1985) The plane with parallel coordinates. The Visual Comput 1(2):69–92.

Jern M, Johansson S, Johansson J, Franzén J (2007) The GAV Toolkit for Multiple Linked Views. Reviewed Proceedings, CMV07, Zurich, Published by IEEE Computer Society.

Jern M, Franzén J (2007) Integrating GeoVis and InfoVis components. Reviewed proceedings, IV07, Zurich, Published by IEEE Computer Society.

Jern M, Johansson S (2007) GeoAnalytics Visual Inquiry and Filtering Tools in Parallel Coordinates Plot. In: Seattle, Nov 2007, ACM GIS2007.

Jern M, Lundblad P (2008) Geovisual Analytics Applied to a Swedish Road Warning Prediction System. Reviewed Paper in Proceedings, GI4DM Conference on Geo-Information Systems for Crisis Management – Harbin, China Aug 2008.

OECD eXplorer (2009) www.oecd.org/GOV/regionaldevelopment.

OECD (2009) Regions at a Glance: 2009 ed.

Roberts J (2004) Exploratory Visualization with Multiple Linked Views. In: A. MacEachren M Kraak J, Dykes J (eds.), Exploring Geovisualization. Elseviers, Amsterdam.

Shneiderman B (1996) The eyes have it: a task by data type taxonomy for information visualizations. In: Symposium on Visual Languages Boulder, CO, USA, pp. 336–343.

Takatsuka M, Gahegan M (2002) Sharing exploratory geospatial analysis and decision making using GEOVISTA studio from a desktop to the web. J Geogr Inform Dec Anal (JGIDA), 5(2): 129–139

Thomas J, Cook K (2005) Illuminating the Path: The Research and Development Agenda for Visual Analytics. http://nvac.pnl.gov/agenda.stm

Voss H, Andrienko N, Andrienko G (2000) Common GIS – common access to geographically referenced data. ERCIM News, 41: 44–46.

http://www.oecd.org/pages/0,3417,en_40033426_40033828_1_1_1_1_1,00.html

Wohlfart M, Hauser H (2007) Story Telling for Presentation. In: Volume Visualization EuroVis2007.

SMHI http://www.smhi.se

SMHI (2009) Wind measurement

http://www.smhi.se/cmp/jsp/polopoly.jsp?d=10059&a=29936&l=sv, 2009-05-27

http://weather.unisys.com/surface/meteogram/

Kernel Density Estimations for Visual Analysis of Emergency Response Data

Jukka M. Krisp and Olga Špatenková

Abstract

The purpose of this chapter is to investigate the calculation and representation of geocoded fire & rescue service missions. The study of relationships between the incident distribution and the identification of high (or low) incident density areas supports the general emergency preparedness planning and resource allocation. Point density information can be included into broad risk analysis procedures, which consider the spatial distribution of the phenomena and its relevance to other geographical and socio-economical data (e.g., age distribution, workspace distribution). The service mission reports include individual points representing the x/y coordinates of the incident locations. These points can be represented as a continuous function to result in an effective and accurate impression of the incident distribution. The continuity is recognized by kernel density calculations, which replaces each point with a three-dimensional moving function. This method allows to control the degree of density smoothing by the search radius (also referred to as bandwidth) of the kernels. The choice of the kernel bandwidth strongly influences the resulting density surface. If the bandwidth is too large the estimated densities will be similar everywhere and close to the average point density of the entire study area. When the bandwidth is too small, the surface pattern will be focused on the individual point records. Experimentation is necessary to derive the optimal bandwidth setting to acquire a satisfactory case-specific density surface. The kernel density tools provided in standard GIS (like ArcGIS) software suggest a default calculation of the search radius based on the linear units based on the projection of the output spatial reference, which seems to be inadequate to display incident density. Within this chapter we provide a flexible approach to explore the point patterns by displaying the changes in density representation with changing search radius. We will investigate how the parameters can be optimized for

M. Konecny et al. (eds.), *Geographic Information and Cartography for Risk and Crisis Management*, Lecture Notes in Geoinformation and Cartography,
DOI 10.1007/978-3-642-03442-8_27, © Springer-Verlag Berlin Heidelberg 2010

displaying incident distribution in relation to the service areas of Fire and Rescue stations.

1 Introduction

It is of growing importance to investigate and enhance risk models for the fire and rescue services. The preparedness in the fire brigades must be based on the municipal risk analysis. Such a risk assessment should create a basis for setting the target level of the preparedness for emergencies in each municipality (Lonka 1999). We consider the visual analysis on point densities, which are introduced further on, as an important assistance to the mitigation process for the fire & rescue services. Therefore it is essential to provide visualization techniques as a means of exploring spatial data in order to detect features of interest contained in them as suggested by Hearnshaw (Hearnshaw and Unwin 1994). Previous research by Krisp (Krisp et al. 2005) investigated the possibility to visually review the significance of a variable (e.g., population density) in the context of risk analysis.

Geographic information systems utilizing geo-visualization methods provide a wide set of tools and processes to create all kinds of maps, and this research indicates that making use of these tools helps the interaction between decision maker, general public and domain experts. Interaction with domain experts essentially assists the development of models and the creation of task-specific maps. Visualization is a crucial point in getting the message across. Visualizing spatial-functional relationships is a problem of growing importance for several research fields. The choice of the visualization, the orientation of the geographic space in relation to the viewer, as well as the selection of the spatial area portrayed all similarly affect the overall message in these cases. These points also concern appropriate risk assessment and planning the emergency preparedness, which requires political backing and communication of experts from various domains.

Within this paper we investigate kernel density estimation methods which are included in a variety of applications: point data smoothing; creation of continuous surfaces from point data in order to combine these with other data sets that are continuous/in raster form; probability distribution estimation; interpolation or hot spot detection. Kernel density estimation methods can also be used in visualizing and analyzing temporal patterns, for example crime events at different times of day and/or over different periods, with the objective of understanding and potentially predicting event patterns (Smith et al. 2008). Bandwidth selection, as explained in the following, is often more of an art than a science, but it may be subject to formal analysis and estimation, for example by applying kernel density estimation procedures to sets of data where actual densities are known. An alternative to fixed bandwidth selection is adaptive selection, whereby the user specifies the selection criterion, for example defining the number of event points to include within a circle centred on each event point, and taking the radius of this circle as the bandwidth around that point kernel.

The definition of 'high' and 'low' point densities needs to be considered because those values differ greatly between different countries. Typically known values are average density calculations for entire cities or counties. It is difficult to compare point density values between the counties because of the size of the area may differ greatly. This influences the results of the calculations. Point densities can be displayed as classic choropleth map within a thematic map (Slocum 2005; Slocum et al. 2001). Generally as areas are bigger the level of generalization increases (Baxtor 1976). There can be a tendency either for increasing the size of the units to provide the lower measures for the point densities or to decrease the size of the base areas, so that both the peak values and the variance of the point density increases. Furthermore the location boundaries have in most cases no logical relationship to the density properties that are being mapped, as they take administrative boundaries into account (Langford and Unwin 1994). Previous research includes work done by Langford and Unwin (1994). They compare the point densities represented in a conventional choropleth map, in which the point density surface is shown as a continuous (or field data) representation. Unlike in this research the data used in their maps is based on an interpretation of a 30 m by 30 m pixel gird, which provides a density estimate at the zone centroid. Levine (2004) provides an excellent discussion of the various techniques and options, including alternative methods for bandwidth selection, together with examples from crime analysis, health research, urban development and ecology (Levine 2004; Smith et al. 2008).

Kernel density maps serve different purposes in the emergency preparedness planning process. Maps of small scales covering bigger region are needed for a high-level planning, On the other hand, maps of bigger scale unveiling more details in the spatial distribution of incidents are necessary for planning the actual preparedness of particular fire brigades. Such detailed maps are also used in further analysis of relations between the incidents and underlying environment. As a need of rescue missions in urban environment is often related to human activities, various socio-economical aspects are subject of such analysis. In addition, the incidents' occurrence forms a dynamic system. Changes in the time domain are therefore of the interest, while the most significant variations are expected for different hours of a day corresponding to the strongest patterns in the characteristic behaviour of inhabitants.

A relevant question for the fire and rescue services is 'Where and when the peaks in the amount of missions occur?' Within this chapter, we aim to provide an answer by investigating methods to model and to visualize the service mission records, which include x/y coordinates of the mission location as a point. We study the visualizations techniques as a means of exploring the spatial data on point densities in order to assist the Fire and Rescue services in their planning and preparation procedures. We suggest that viewing mission points as a continuous surface, by using kernel density estimations, and visualizing it as a 'landscape' (using the third dimension), support the identification of the areas, which are of particular importance to the fire and rescue services. Our aim is to investigate different parameters within the kernel density calculations to determine which settings would fit the needs and provide an appropriate visualization.

2 Methods and data - determining point density

2.1 Kernel density estimation – function and bandwidth

Point density is a continuous function and in order to present an effective and accurate impression of its distribution, a scheme that recognizes this continuity is needed (Langford and Unwin 1994). In this paper we investigate the calculation and visualization of point density using kernel function (Bowman and Azzalini 1997), which allows controlling the degree of smoothing by the search radius (bandwidth) of the kernels.

Given the point data of Helsinki including the geocoded fire and rescue service missions, the probability distribution of these variables at every location within the study area can be estimated by replacing each point with a kernel, giving them a 'spatial meaning'. The kernel is defined as a three-dimensional function, which is determined by the shape of the kernel, its bandwidth and the output size for the resulting raster cells. The density λ at each observation point s is estimated by

$$\lambda(s) = \left\{ \sum_{i=1}^{n} K_h(s - s_i)x_i \right\}, s \in U$$

where K is the kernel and h the bandwidth (Silverman 1986). According to this formula, the contribution of incidents to the density estimation corresponds to their distance from the kernel centre – the closer to the kernel centre, the bigger the influence.

A number of different density estimation methods have been compared in previous research (Bowman 1985; Cao et al. 1994; Hwang et al. 1994). According to Smith et al. (2008), the kernel function does not tend to have a major impact on the set of density values. Of much greater impact is the choice of the spread parameter, or bandwidth (Abramson 1982; Salgado-Ugarte and Pérez-Hernández 2003; Smith et al. 2008). In practice only the bandwidth, which represents the search radius, is adjusted. The choice of the kernel bandwidth strongly influences the resulting estimated density surface. If the bandwidth is too large the estimated densities will be similar everywhere and close to the average point density of the entire study area. When the bandwidth is too small, the surface pattern will be focused on the individual point records. Experimentation is required to derive the optimal bandwidth setting to acquire a satisfactory density surface (O'Sullivan and Unwin, 2003). Some investigations on the methods for optimal bandwidth selection have been conducted to some extent, although user experimentation is always needed (Park and Marron, 1990; Sheather 1992; Sheather and Jones 1991).

Smith (2008) argues that, when the density estimation is adjusted using normalized Gaussian kernel function, which often represents probability density, the resulting surface is described as a probability density surface, rather than a density surface. The density can be then interpreted as the probability that the point lies within any defined area, which is equal to the volume of the surface over that area. The volume for the entire study region is exactly 1, reflecting the fact, that the point is certain to lie somewhere (Smith et al. 2008). In this context, a suitable bandwidth would be a smallest value, which for different data collections yields a stable density map.

The use of kernel functions as a form of point weighting enables the creation of locally weighted means and variances. The locally weighted (kernel) statistics could be also supported within the geographically weighted regression (GWR) (Fotheringham et al. 2002).

2.2 Software packages for kernel density calculations

GIS packages support a variety of kernel functions and procedures. Within this study, we use ArcGIS Spatial Analyst, which provides kernel density estimation for point and line objects, but only supports one kernel function, which it describes as a quadratic kernel (a bounded kernel), but which is often described as an Epanechnikov kernel. MapInfo's grid analysis add-on package, Vertical Mapper, includes kernel density mapping. Additionally the spatial statistics package SPLANCS and the crime analysis add-on for MapInfo include a hotspot detective which are based on quartic kernels. TransCAD/Maptitude supports what it describes as density grid creation with the option of count (simple), quartic, triangular or uniform kernels. Crimestat supports four alternative kernels to the normal, all of which have finite extent (i.e., typically are defined to have a value of 0 beyond a specified distance). These are known as the quartic, exponential, triangular and uniform kernel (Smith et al. 2008). Density calculations using Gaussian kernel functions are also supported within spatstat, and R package for analyzing spatial point patterns (Baddeley and Turner 2008).

Now how does ESRI's ArcGIS Spatial Analyst suggest finding a default value for the bandwidth and output grid size in its spatial analyst extension? The default settings are found based on the input features (points or lines) to calculate the density for. The search radius units are based on the linear unit of the projection of the output spatial reference. For example, if the units are in meters, to include all features within a 1 mile neighbourhood, set the search radius equal to 1609.344 (1 mile = 1609.344 m). The default is the shortest of the width or height of the extent of point features in the output spatial reference, divided by 30. The cell output size defines the output raster that will be created. This is the value in the environment if specifically set. If the environment is not set, then cell size is the shortest of the width or height of the extent of point features in the output spatial reference, divided by 250 (ESRI 2008).

2.3 Classification for the kernel density output raster

Kernel functions applied to point data results in a continuous density surface. Due to limited ability of human eye to discriminate shades, it is desirable to classify the density values into several categories for visualization, while a suitable number of classes should not exceed seven (Gilmartin and Shelton 1989). Classification of the data then leads to more perceptual image and also saves a processing time

(Slocum 2005). Naturally, the aim of classification is to approximate the original surface as closely as possible by preserving characteristic patterns of the phenomenon. As methods used to classify the data can introduce an 'error', which in some cases substantially alters the final appearance of the map, definition of the class limits considering the data distribution is a challenging task (Jenks 1977; Unwin 1981). A method based on Jenks algorithm (Jenks 1967), also called natural breaks or goodness of variance fit (GVF) classification, offers a suitable solution. The method uses a measure of classification error (sum of absolute deviations about class means) to keep similar data values in the same class. In this way the classification gives in the most accurate and objective overview of the original data. The method is also commonly implemented in the GIS software.

2.4 Data – background and Fire and Rescue Service mission point data

This study covers the Helsinki Metropolitan Area, which is as the most densely populated region in Finland of particular interest to fire and rescue Services. Finland is one of the countries where data collection, digital databases and information society in general is well developed and organized, but this is no guarantee that information is available and interoperable. Organizational boundaries are clearly visible also in the domain of emergency planning (Jolma et al. 2004). Generally the unequal quality of spatial data requires individual models and scales for a risk analysis in each community. Additionally municipalities in Finland are obliged to gather register data on their point, buildings and land use plans. Due to the uniform character of the region it is important for the planners and decision makers to have reliable register data on the whole area, irrespective of these municipal boundaries. For this reason, Helsinki Metropolitan Area Council (YTV) has been working since 1997 on the production of a data package, SeutuCD, which covers the whole metropolitan area. It is a data package gathered from the municipalities' registers and other sources. It is updated once a year and includes register data on buildings and land use plans as well as the enterprises and agencies located within the metropolitan area (YTV 1999). In addition to register data, SeutuCD includes maps for different scales and metadata software.

Finnish fire and rescue Services document all their rescue missions in 'Pronto', an accident database maintained by the Ministry of the Interior. The provided data form a point data set, where each of the missions is used as a spatial object, which has a position (X, Y coordinate tuple) and a set of attributes, such as time of announcement, address, incident type and response time. This enables to ignore the municipal boundaries of metropolitan area in the analysis. We use the data from years 2005–2007. Attributes we are initially interested in are the location, mission types and time stamp.

Records of the past incidents provide a valuable source of information to be analysed and used to improve the effectiveness of rescue services. In Finland, the incident data are recorded into the register via an electronic report filled by the mission commander after each mission. The textual attributes are complemented by coordinates of the incident location given by clicking a map before the report is finished. This process should ensure the highest accuracy of the registered data, as it guarantees the completeness of the data and eliminates frequent errors caused by misprints. The main responsibility of rescue brigades, however, is in protecting the public safety and we may doubt about the precision of inserted incident coordinates, which may not correspond to the associated addresses. The issue of positional data errors and their identification was tackled by Krisp et al. (Krisp et al. 2007) proposing removal of erroneous points from the analysis. In the following, the full set of fire and rescues service mission points has been used for the density calculations. Generally we are dealing with probability densities; therefore in future data, e.g., about the next period (2008–2010), the points are likely to appear at different places, but the interpolated density map should look quite similar to the one of 2005–2007. Using the data for each individual year 2005, 2006 and 2007 the interpolated densities do show a comparable distribution of hotspots at similar locations. It can be suggested that a good bandwidth would be the smallest value that satisfies this criterion. It yields a meaningful density map, one that provides a 'signal' as detailed as possible without too much statistical 'noise'.

3 Results

We try to restrain the large number of possible analysis usages in the paper to the determination of an optimal bandwidth for displaying fire and rescue service incidents to the rescue services. One question occurring is 'Where are the locations of incidents hotspots?' Using kernel density maps we can compute different maps with one, two, etc. hotspots displayed in the map. We apply the kernel density estimation to the fire and rescue mission data points using ESRI's ArcGIS 9.2 and analyze the suitability of the resulting map in relation to changing calculation parameters. As the search radius controls the strength of contribution of the incident points to the resulting density, it has the most significant influence on the smoothness of the surface. Figure 1 shows the kernel density estimation with the settings suggested by the ArcGIS Spatial Analyst extension based on the default values (see above).

Fig. 1. Kernel Density map for Helsinki fire and rescue service missions with a scale 1:100.000 applying ESRI's ArcGIS default parameters for bandwidth 703.6 m and output grids size 84.4 m

As described above the software calculates default settings based on the extent of the mission data points. The software suggests settings for the bandwidth with 703.6 m and output grids size with 84.4 m. The output raster is classified using a natural break (sometimes referred to as Jenks) classification with six classes. They are visualized with a colour range from red, orange to yellow and transparent for zero values. Generally a 30% transparency is applied to show the underlying background map.

The parameters of the search radius can be altered. Theoretically we can acquire an infinite number of different density maps. Figure 2 illustrates four examples of a map series with 18 maps that have been produced to investigate an 'optimal' bandwidth for this data set for the scale of 1:100.000.

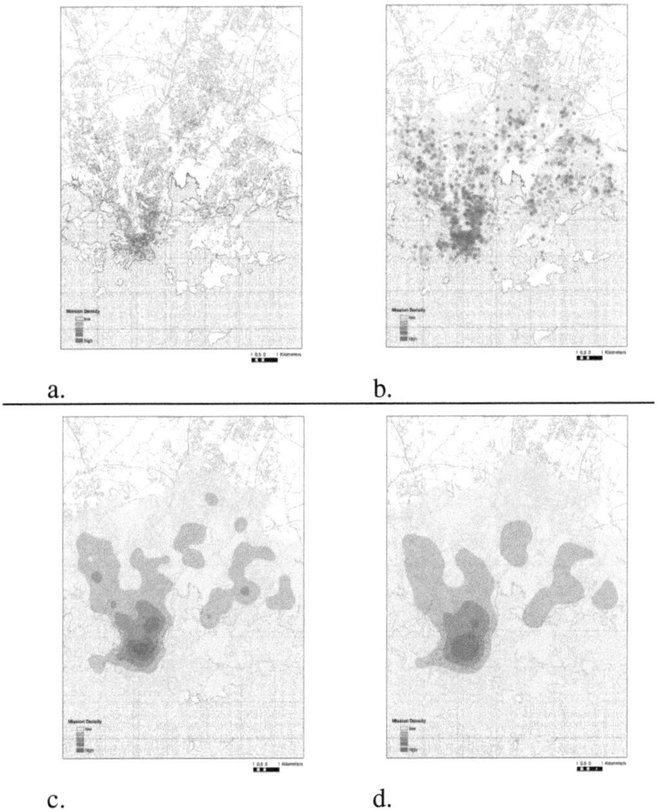

Fig. 2. Mission points density maps with a scale of 1:100.000, setting different bandwidth; (**a.**) bandwidth 50 m, (**b.**) bandwidth 200 m, (**c.**) bandwidth 1000 m, (**d.**) bandwidth 1500 m

The maps actually show the number of incidents that happen within the selected bandwidth from any point in Helsinki. It is important to notice that larger values of the radius parameter produce a smoother, more generalized density raster. Smaller values produce a raster that shows more detail. In this way we can control the amount of information contained in the resulting maps. The same test has been done for a map with a scale of 1:25.000 shown in Fig. 3.

The meaning of the data is difficult to interpret and requires close collaboration with the responsible fire and rescue services. Generally the services use classic statistical methods to analyze the mission data. The case for the Helsinki Fire and Rescue services indicates that the desirable density maps should not be too focused on details on one hand, but on the other hand they should also not be too general. The required level of detail also depends on the map scale and purpose of its use. Having a large scale for example 1:10.000 we might want to set a smaller search radius to have a less generalized density surface. For a map with a scale smaller than 1:100.000 it might be useful to select a bigger search radius to get a more generalized density surface. For this particular mission data set and specified area we suggest a search radius between 1000 and 1500 m for a density map of scale 1:100.000. For the larger scale 1:25.000, we suggest to select a bandwidth between 700 and 1000 m.

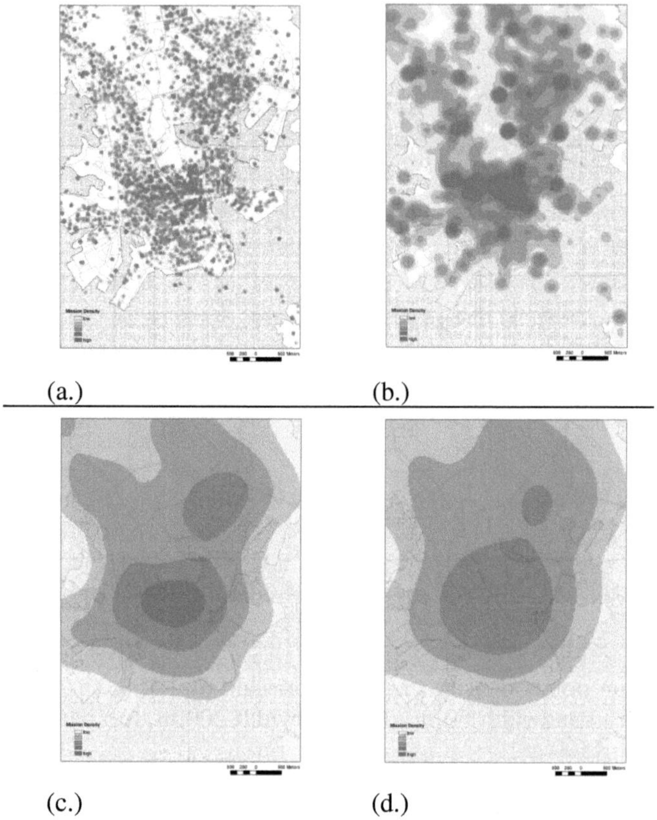

Fig. 3. Mission points density maps with a scale of 1:25.000; different bandwidth; (**a.**) bandwidth 50 m, (**b.**) bandwidth 200 m, (**c.**) bandwidth 1000 m, (**d.**) bandwidth 1500 m

4 Conclusions

Generally the calculation of a density surfaces gives the users a better overview of the point data compared to a dot map, where we have to deal with overprinting. Besides visualization purposes of point data sets, density maps are often subject to further spatial analysis, for which a suitable density map and understanding its meaning is crucial. Using the kernel functions for density estimation allows controlling information presented in the resulting map by changing the kernel search radius parameter. Suitable visualization of the density maps then supports communication of such information, which leads to finding adequate parameter values for the given purpose.

This study illustrates that the choice of parameter settings for density calculations is case specific. The primary question is what message the map should provide. The answer should be established according to the user requirements. It is important to consider the scale of the output map when calculating the point density, as maps with different scales may require to present different level of detail to the user. The user should be also aware of the physical meaning of the resulting maps. An additional finding of this investigation is the importance of scale, the research area and the distribution of the incident points. It seems to be difficult to make universal statements about the computation of an incident kernel density map. The optimal map needs to incorporate the common planning scale(s) used by the rescue services, the unique distribution of incident locations and the characteristics of the study area (e.g., clustered by water areas, rivers, mountains). Therefore each incident density map may have unique parameters depending on these features.

The visualization classified from red to yellow proves to be useful in visually identifying hot spots; it also gives useful results in a black/white printout. When combining the output map visually with a raster grid, it is important to consider this size when setting the output cell size. The output density map and the overlaid grid correspond visually better when the density cell size is the same (or a derivate) with the raster size. For example a density map with a 62.5 m cell size is overlaid with a 250 m raster matches well in a visual way. The search radius is depended on the data set, the scale of the map and the study area.

It is difficult to give a general suggestion on parameter settings, as they are dependent on user requirements. However, information represented by the resulting density surface depends on the choice of the kernel bandwidth and the output grid size. It is therefore necessary to experiment with these parameters to acquire map suitable for the user needs. Presentation of the results using expressive visualization techniques supports understanding the differences and thus enables the user to make the decision. The existing software support seems to be insufficient for these purposes.

In our test case related to the fire and rescue missions, we find a bandwidth 1000 m adequate to display the density of incidents on a scale 1:100.000. In this case, the grid cell size is set to 25 m. To display the incident density on a large scale 1:25.000, a bandwidth 700 m is more appropriate to encompass more details.

5 Further research

Density analysis as presented in this study is one step in often very complex risk assessment process. We may also think of less frequent incidents, which will affect rather an area than a point, e.g., forest fires, floods. In this case these incidents are not suitable to be visualized in mission density maps and they should be dealt with separately. These results are open for further interpretation and can be used by the fire and rescue services in a risk model.

Past incident records are a valuable source of information in the risk analysis. Investigation of changes in the incident density patterns over time can provide an insight into the dynamics of the phenomena, either by visualizing the development of incidents distribution over years and decades, or showing daytime variations to support the actual preparedness of fire brigades. It is also important to investigate possible trends considering especially the point density. These can be studied to find more information for future situations, e.g., how the point mission patterns are likely to change within the Helsinki metropolitan area in the coming years. Additionally, further research has to consider individual records in the data, the time of each record and the time of the point density calculations. The integration of a time variable (daytimes, nighttimes etc.) seems to be essential when relating the point density to incident densities. The density patterns can be also compared to the distribution of other geographical or socio-economical attributes to understand the relations existing behind the incidents.

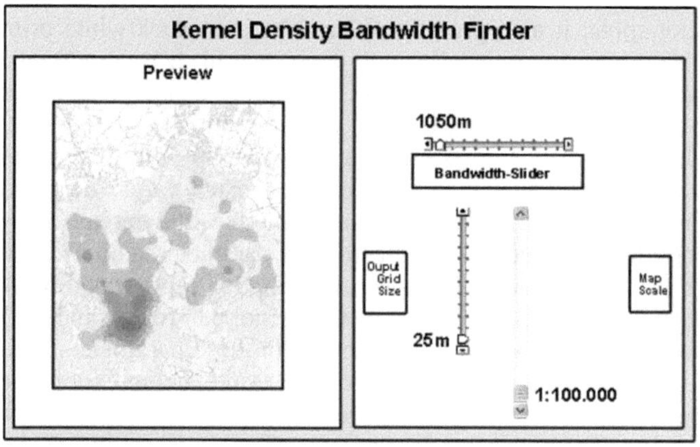

Fig. 4. Illustration kernel density slider tool mock-up

A proper density estimation using suitable parameters for the particular purpose is crucial for all of these applications. As experimentation is required to select a suitable kernel bandwidth, we aim to develop a 'slider-tool', which allows user to visually explore the data and find appropriate parameter settings. This needs to include selection tools for the bandwidth, the output grid size and possibly the scale of the output map. The parameter sliders are interconnected with a map preview window, where the selected options are visualized using a simple classification method. A proposed tool layout is shown in Fig. 4.

Future work will investigate the possibilities for this slider tool and the technical implementation into existing geographic information systems. Fire and rescue Service mission records are stored for a reasonably long time, within the fire and rescue service München, Germany for about 5 years. Eventually these records will be deleted. We may have to investigate how these databases can be stored for a longer (perhaps infinite) time. These records provide a valuable 'window into the history'. These missions have been geocoded now starting around the beginning of the century with increasing location accuracy. Eventually these records can be used in 20, 30, 100 years to investigate the 'historical' hotspots of fire and rescue service missions. Therefore we suggest strongly investigating sustainable methods to store this data.

References

Abramson IS (1982) On bandwidth variation in kernel estimates–a square root law. Ann Stat 10(4):1217–1223.

Baddeley A, Turner R (2008) Spatial Point Pattern analysis, Model-Fitting, Simulation, Tests, On-Line Manual, Available from: http://www.spatstat.org (accessed 17 December 2008).

Baxtor RS (1976) Some methodological issues in computer drawn maps. Cartogr J 13: 145–155.

Bowman AW (1985) A comparative study of some kernel-based nonparametric density estimators. J Stat Comput Simulat 21:313–327.

Bowman AW, Azzalini A (1997) Applied Smoothing Techniques for Data Analysis: The Kernel Approach with S-Plus Illustrations. Oxford University Press, New York.

Cao R, Cuevas A, Gonzalez Manteiga W (1994) A comparative study of several smoothing methods in density estimation. Comput Stat Data Anal 17(2):153–176.

ESRI (2008) ArcGIS 9.3 Online Manual. Environmental Systems Research Institute Inc., http://webhelp.esri.com/arcgisdesktop/9.3/index.cfm?TopicName=Kernel_Density (18.12.2008).

Fotheringham AS, Charlton M, Brunsdon C (2002) Geographically Weighted Regression: The Analysis of Spatially Varying Relationships. Leicester (UK), Wiley, 282 pp.

Gilmartin P, Shelton E (1989) Choropleth maps on high resolution CRTs: the effects of number of classes and hue on communication. Cartographica 26(2):40–52.

Hearnshaw HM, Unwin DJ (1994) Visualization in geographic information systems. Chichester (UK), Wiley, 268 pp.

Hwang J, Lay S, Lippman A. (1994) Nonparametric multivariate density estimation: A comparative study. IEEE Trans Signal Process 42(10):2795–2810.

Jenks GF (1967) The Data Model Concept in Statistical Mapping, International Yearbook of Cartography Vol, 27, pp. 186–190.

Jenks GF (1977) Optimal Data Classification for Choropleth Maps. Department of Geography, University of Kansas.

Jolma A, Krisp JM, Virrantaus K (2004) Review of Geoinformatical Methods for a City Safety Information System, ISESS, 9th International Symposium on Environmental Software Systems, Harrisonburg.

Krisp JM, Ahola T, SpatenkovaO (2007) Visual error detection in geocoded point data, ICA commission on Visualization, workshop 'From Geovisualization toward geovisual analytic', August 2nd & 3rd, Helsinki.

Krisp JM, Jolma A, Virrantaus K (2005) Using explorative spatial analysis to improve fire and rescue services in Helsinki, Finland. In: P. Oosterom, S. Zlatanova and E. Fendel (eds.), Geo-Information for Disaster Management. Springer, Delft, The Netherlands, pp. 1282–1296.

Langford M, Unwin DJ (1994) Generating and mapping population density surfaces within a geographical information system. Cartogr J 31: 21–26.

Levine N (2004) CrimeStat: A Spatial Statistics Program for the Analysis of Crime Incident Locations (Ver. 3.0). Ned Levine and Associates, National Institute of Justice, Washington, DC. http://www.icpsr.umich.edu/NACJD/crimestat.html/, Houston.

Lonka H (1999) Risk Assessment Procedures Used in the Field of Civil Protection and Rescue Services in Different European Union Countries and Norway. SYKE, Helsinki.

O'Sullivan D, Unwin DJ (2003) Geographic Information Analysis. Wiley, New Jersey.

Park BU, Marron JS (1990) Comparison of data-driven bandwidth selectors. J Am Stat Assoc 85: 66–72.

Salgado-Ugarte IH, Pérez-Hernández MA(2003) Exploring the use of variable bandwidth kernel density estimators. Stata Journal 3(2):133–147.

Sheather SJ (1992) The performance of six popular bandwidth selection methods on some real data sets. Comput Stat 7:225–250.

Sheather SJ, Jones MC (1991). A reliable data-based bandwidth selection method for kernel density estimation. J Royal Stat Soc, B 53: 683–690.

Silverman BW (1986) Density Estimation for Statistics and Data Analysis. Chapman and Hall, London.

Slocum TA (2005) Thematic Cartography and Geographic Visualization. Pearson/Prentice Hall, Upper Saddle River, NJ, 575pp.

Slocum TA et al. (2001) Cognitive and usability issues in geovisualization. Cartogr Geogr Inf Sci 28:61–75.

Smith M J, Goodchild MF, Longley PA (2008) Geospatial Analysis – A Comprehensive Guide. Matador (an imprint of Troubador Publishing Ltd), www.spatialanalysisonline.com

Unwin DJ (1981) Introductory Spatial Analysis.Methuen, London.

YTV (1999) Establishments in the Metropolitan Area 1999. Helsinki Metropolitan Area Council (YTV), Helsinki.

Monitoring and Comparing: A Cartographic Web Application for Real-time Visualization of Hydrological Data

Christophe Lienert, Melanie Kunz, Rolf Weingartner and Lorenz Hurni

Abstract

As decision support in flood warning, experts within crisis management groups need readily available real-time data, and real-time visualization derived from them. Up until now, work steps including acquisition, harmonization, storage, processing, visualization and archiving of data have been accomplished manually or semi-automated. In a cartographic real-time application, these worksteps have to be achieved online and error-free. The presented web application and the generated real-time visualization products are based on a viable data model which is extendable by additional measurement or model ouput data. By means of a graphical user interface, users may overview the most current hydrological situation in the form of automatically processed real-time maps. In addition, these maps may be interactively compiled, depending on users' needs, on different levels of detail, and in various thematic combinations. In order to classify the most current hydrological situation in the historical context, i.e., to eventually learn from the past, the application also allows to easily create visualizations of past flood events. Data from a long-term, high-resolution archive undergo the same cartographic rules and abstraction as real-time data for the purpose of direct comparisons.

M. Konecny et al. (eds.), *Geographic Information and Cartography for Risk and Crisis Management*, Lecture Notes in Geoinformation and Cartography,
DOI 10.1007/978-3-642-03442-8_28,

1 Introduction

1.1 Motivation

Climate models suggest that extreme precipitation events will become more prevalent and future changes of such events, in response to global warming, may even be under predicted (Allan and Soden 2008). As a ramification of an anthropogenic warmed climate and increased rainfall events, recurrence intervals of natural hazards – such as flooding – are decreasing. Depending on the time period under investigation, flood events, such as the devastating one of 2005 in Switzerland, must rather be termed rare than extraordinary. Post-event analyses of this and other extreme events in Switzerland clearly indicate that it is imperative to prepare and adapt to more intense natural hazard events (Bezzola and Hegg 2007).

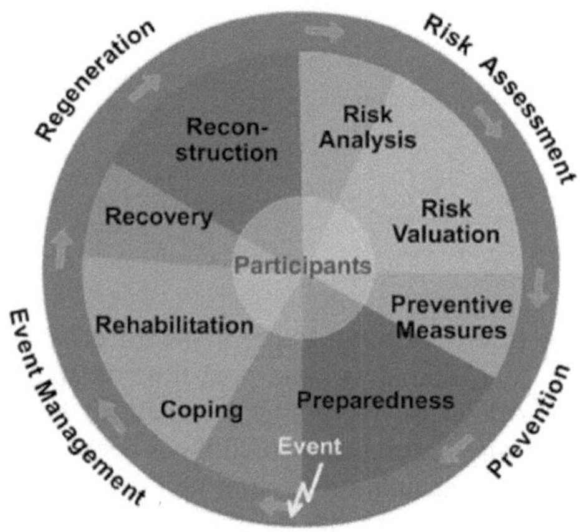

Fig. 1. Integral risk management circle (Kienholz et al. 2004)

In order to be better prepared for future floods, methods aimed at detecting flood events at an early stage must inevitably be improved and refined. Flood monitoring can significantly contribute to the reduction of flood-induced damages as any time saving is crucial when preparing counter-measures in flood-prone areas. This primarily includes the provision of a place-independent access to real-time data. Flood specialists need such real-time data, ideally visualized in the form of interpreted and well-readable visualizations, in order to be capable of optimally advising crisis committees in emergency situations.

1.2 Early warning context and application specifications

The ultimate goal of any monitoring infrastructure is to aid in detecting at an early stage a potential flooding. From the viewpoint of integral risk management (Fig. 1), a monitoring system is positioned in the domain of preparedness, in which availability and accessibility of time-critical information is eminently important to revaluate the most current hydrological situation. In consequence of including most modern hydro-meteorological forecasting to the integral risk management, the domain of coping shifts much closer to the domain of preparedness, or even ahead of the event (Bezzola and Hegg 2007).

Currently, a lot of efforts are made in real-time data assimilation for hydrometeorological *nowcasting* and *forecasting* purposes (the former denotes shortest-term forecasting). Forecast models are run as ensembles and the combined output is used to feed hydrological models (i.e., Bartholmes and Todini 2005; Jaun et al. 2008). There are applications dealing with the use of water resources that facilitate flood risk and emergency management (i.e., Foraci et al. 2005; Jha and Das Gupta 2003), but regardless of whether we deal with real-time data or with real-time forecast data, methods to visualize real-time data are essentially the same. Still, they often fail to meet today's user needs and have to be improved. In order to observe cartographic rules when handling real-time data, work steps such as data acquisition, storage, processing, visualization and archiving must be achieved on-line, error-free and entirely automated. A robust system is needed, capable of displaying real-time data on different spatially and temporally related levels of detail. Time-sensitive data have to be represented dynamically and the system must be able to classify data for which the accuracy, range and absolute values are not known in advance. In addition to these requirements, maps are to be delivered to any computer and displayed in different web browsers. In order that the generated cartographic products look as intended in the map users' preferred browsers, they necessarily have to comply with current web standards.

Compared to other existing cartographic solutions, supplementary value is added by presenting data interactively as well as by providing efficient analysis tools in a graphical user interface (GUI). The GUI should permit to retrieve processed and adjusted data of past flood events in a user-friendly manner. This way, users (i.e., operational hydrologists) may instantly draw on the experiences and knowledge from past floods and may better assess the current situation. Finding comparable data is time-consuming and disadvantageous in emergency situations. Making available historical data archives in the application itself is certainly more efficient.

1.3 Monitoring and comparing – similar approaches

This chapter takes up this idea of comparing real-time hydrological data with historical data. The presented results are based on the ongoing project 'Real-Time Cartography in Operational Hydrology', supported by the Swiss National Science Foundation. Besides flood hydrology this project apparently crosses various other research fields such as web cartography, real-time applications in geosciences or decision support systems (Lienert et al. 2007).

Merely few similar approaches exist that allow for simultaneous visualization of real-time and historical data. The Swiss Federal Office for the Environment, for example, publishes raster-based maps on its website on which the most actual river discharges at gauging locations are related to their extreme value statistics. Gauging locations are symbolized by a coloured dot representing a predefined recurrence interval. This idea has been included in the application described here (Fig. 3).

The new project Swissrivers (http://www.swissrivers.ch) may serve as an additional cartographic example with which real-time data from dozens of Swiss catchments are shown. Historical data merely cover the past couple of hours back from the actual time. The application presents graphical forecast information at numerous river gauges, located on an interactive Google map mash-up. The conveyed information is of great importance to experts. The cartographic implementation of the application is yet expandable as to legibility and discriminability of the map symbolization and in terms of interaction between the user and certain map elements and forecast graphs.

Decision support systems employed as operational and strategic components in decision making are widespread to address a variety of hydrological questions. Yet, some of these systems are built upon historical, averaged data and do not contain real-time data at all (e.g., Berlekamp et al. 2007; de Kort and Booij 2007). Other systems handle real-time and historical data for the purpose of real-time data assimilation, calibration and validation of forecast models (e.g., Todini 1999; Vrugt et al. 2006). A rare example of a database that contains a chronology of hydrological extreme events is discussed in Black and Law (2004). The system stores hydrological information that dates from times when no measurements were made. The database, however, solely contains textual information about water levels and extents of past flooding. As a result, queries are neither being georeferenced nor cartographically displayed.

2 Basic data and spatial extent

2.1 Qualified and used data

The overall project was designed to visualize observed real-time data as well as manually collected archive data and records. Both types of data are stored in a database and managed by a database management system. The difference between the two types is their accuracy. Real-time data are automatically collected from the data supplier's data servers using routines that are run at certain intervals. Before insertion into the database takes place, data are rudimentarily checked for their integrity (i.e., duplicate measurement times, data type such as text or number, data range or sign). If data fails to meet such conditions, either no data or predefined substitute values standing for unavailable data are inserted. These integrity checks, however, do not entirely free real-time data from inherent uncertainties. Expert users should therefore consider varying degrees of confidence of real-time data when using them for comparing with archived data.

Table 1. Overview of the archive data collection

Parameter	Archive from-to	Spatial extent	Temporal Resolution
River discharges, lake levels	01.01.1974–31.12.2007	All of Switzerland	10 minutes
Precipitation, air temperature, air pressure, air humidity	01.01.1981–31.12.2007	All of Switzerland, plus additional gauges in and around *Thur* basin	10 minutes
Precipitation radar images	01.01.1991–31.12.2007	All of Switzerland	1 hour
Groundwater levels	01.01.1996–01.05.2007	*Thur* basin	5 minutes

Archived data, in turn, are delivered annually by network operators after having passed quality controls. Such control activities may involve the homogenization of data, the adjustment of faulty outliers or the adaptation of the rating curve (on which the real-time river discharges used in this project are based). Archived data have been inserted into the project data base manually. To draw comparisons, data have to be available at any gauging location in form of long, high-resolution time series as well as in real-time. Apart from being accessible through a network, the criteria for real-time data were that they had to have a temporal resolution of ≤1 hour and a delivery interval of ≤2 hours. All data are measured and provided by the Swiss Federal Offices and State Offices. Spatially, they cover all of Switzerland with some densification in the north-eastern part of the country, where the basin of the *Thur* river is located. In Table 1, an overview of the archive data collection is shown. Real-time data, in contrast, feature coarser temporal resolutions but are identical to archive data as to the number of parameters and spatial extent.

Data are assigned to the metric ratio scale, with the exception of air temperature data (which have metric interval scale properties). Apart from radar images, all data stem from individual point gauging stations. Different real-time point symbolizations have been developed for these quantitative data. Additional automated processing and visualization functions were developed for more sophisticated depiction of precipitation radar, point temperature and point river discharge (Lienert et al. 2008; Lienert et al. 2009b).

In addition to the thematic data mentioned above, different sets of static topographic and land use data have been collected and processed, including a multi-resolution relief, hypsometric distribution, forest, and settlement areas. These data are used for the base map and are pre-loaded in the application by default. Static hydrological data in the application covers lakes, groundwater areas, the complete 1:200000 and 1:25000 river network, and hydrological balance area geometry. Since the amount of data is continuously increasing an object-relational database management system has been installed on a separate server from the project start. The database stores numerical and vector data. Raster data, in turn, are not stored in the database but in web-enabled formats on the server's file managment system.

2.2 Geographical focus

As shown in Table 1, data are collected from all parts of Switzerland, though the project's geographical focus is on the Thur basin in the north-eastern part of the country (Fig. 2). The basin has an area of 1700 km^2 and the altitude ranges from 336 to 2501 m above sea level. With regard to early warning and monitoring activities, the basin's size is adequate for testing real-time cartographic concepts as lead times for warnings still allow for counter measures. From a hydrological point of view, the basin's upper part is dominated by snowmelt in spring time and thus attributed *nivo-pluvial* (Weingartner and Aschwanden 1992). Climate change studies on the Thur basin's hydrology suggest – even when examining past decades – a substantial reduction of the extent and duration of the snow cover. Therefore, the seasonal runoff pattern is reckoned to shift, with earlier peaks due to snowmelt (Jasper et al. 2004).

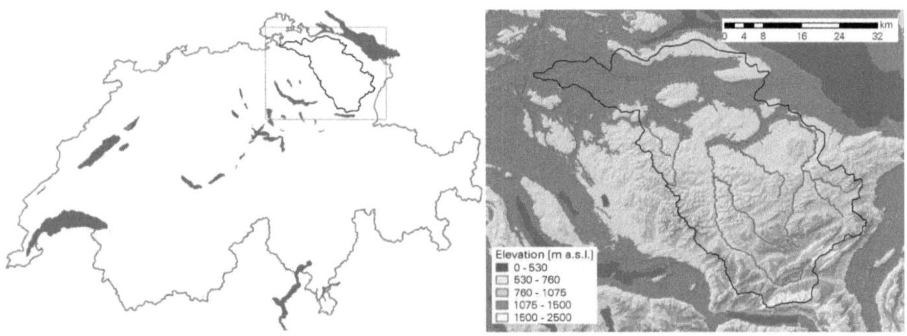

Fig. 2. Location of the Thur basin in north-eastern Switzerland (*left*) and enlarged view (*right*)

3 Design and methodology

3.1 Conceptual considerations

The visualization system consists of eight main, modular components (Fig. 3): the prototypical, web-based GUI (#1) that accesses data based on three different visualization approaches (see below). Real-time data are treated as a self-contained component (#2) and are either delivered to the project server infrastructure or picked up on the data supplier's server. Data are delivered in various formats and inserted into the real-time partition (#3) of the overall database by individual program routines. Elementary quality checks, as noted above, are performed at this stage. The most actual and pre-processed data is stored in the database, ready to be used for further processing or immediate visualization. Additional database partitions store static topographic and landuse vector data (#4) as well as archive measurement data (#5) which have previously been quality-checked by the network operator and delivered once a year (see above). The combination of data in this partition and data in the real-time partition form the basis for comparing these two types of data. A fourth distinct database partition contains the metadata (#6) which store information about the overall data collection. Two interfaces are connected to the database: the first (#7) aims at interconnecting additional real-time data processing steps prior to visualization (e.g., point-to-area interpolations, inclusion of statistical or hydrological models). The second interface serves the purpose of importing new measurement data (#8) in order to annually update the archive with the network operator's quality-checked data. This interface may also be used to store qualitative data in the database such as textual data (e.g., media products and written eye-witness accounts) of past flood events.

The overall concept of the project and the data visualization in the GUI is also based on results obtained from discussion with potential expert users working as advisors in relevant federal or state administration offices. It illustrates what visualizations and functionalities are needed in order that these specialists may

1. quickly and comprehensively get an idea of the current (flood) situation ('*monitoring*', Fig. 3)
2. retrace short-term developments of the current or any past situation ('*retracing*')
3. learn from past extreme events and relate them to the current situation ('*comparing*').

In the following, emphasis is put on methodological aspects of '*comparing*' real-time hydrological data with historical flood data.

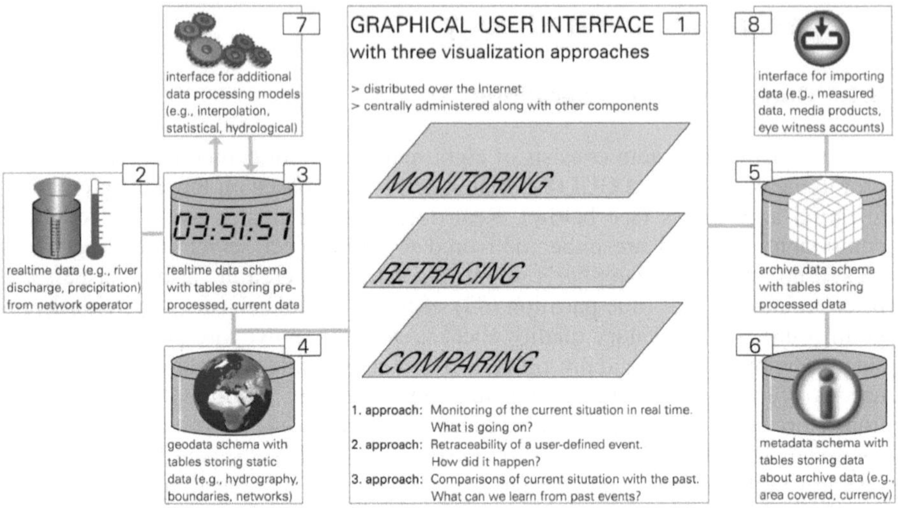

Fig. 3. Conceptual framework of the real-time visualization application

3.2 Comparing real-time data with historical data – how and what can we learn from the past?

To generate maps and graphical representations of historical floods, an arbitrary point in time (in form of a timestamp) is input to the application via the GUI. Subsequently, all user-preset data associated with this time stamp are retrieved. In doing so, each table of the database (Fig. 3, #5) is searched for the very record containing the timestamp that matches the input timestamp (see below). The result of this filtering operation is a thematic map showing the hydrological situation of the time that has previously been passed to the application. It is generated using identical cartographic abstractions and symbolizations that are applied to real-time data. In order to visually compare real-time and historical data, the respective maps are arranged in different tabs of the web browser. Users may then toggle between these two maps or they may decide to toggle between real-time and historical visualizations within the GUI itself. When exploring the data in the GUI, particularly on higher levels of detail, the configuration and the development patterns of past flood events are quickly conveyed (Figs. 4, 5 and 6). Comparing may include contrasting spatio-temporal relations of different parameters or interactive exploration of time series at various gauging locations (i.e., displaying of additional information when moving the mouse over such time series graphs). Such interactive methods facilitate the access to historical data and contribute to generation of knowledge about past events which, in turn, may again be incorporated in the current decision-making process. Two ways of making accessible and representing such knowledge are conceivable:

- quantitative (logical data level)
- qualitative (descriptive-textual, multi-media level)

Further interactive methods have to be developed in order to support straightforward navigation (i.e., shifting, toggling, resizing, contrasting, overlaying, etc.) of the two above mentioned types of data. Ultimately, the main questions are how to make relevant information intuitively available and how to present it in a clear and space-saving way.

In addition to these interactive methods incorporated in the GUI, comparing real-time data with similar, historical data call for algorithm-based search methods on the application's logical level, taking into account uncertainty ranges of the data parameters in question. Looking up similar historical situations via one particular measuring point must result in the formation of multi-dimensional search terms linking locations, time points and value ranges of data measured at several adjacent gauges.

3.3 Data model and visualization methods

Generating real-time maps means to apply automated working steps to an underlying, consistent data model. To achieve this, both measured data and automatically delivered real-time data are modelled in the database along with their related time of measurement. Measured values are therefore not only distinguishable by their value and location but also by their temporal occurence. In doing so, the time itself is treated as an attribute of the measurement data (e.g., Valpreda 2004) and, in terms of timestamps, defined as a unique identifier for every single measurement record. Joining of data records is accomplished using these unique timestamps and additional foreign keys. Beside the spatial database management system, all involved technologies for data processing and cartographic data representation are based on free open-source software (Lienert et al. 2009b).

Visualizations are produced at the end of a processing chain. Functions of an in-built webmap server are used when displaying raster data. Raster data are made available over the internet subject to map scale and actual map section requested by the GUI. Measurement and static, topographic base data are combined by scripts and then sent over the internet as *Scalable Vector Graphic (SVG)* code. *SVG* is the web standard for two-dimensional graphics and is suitable for most web-cartography projects (Neumann and Winter 2003). The index *SVG* file that gathers all the data to be viewed by the user is further supplemented by including a server-side software module, making the index file more dynamic. It is able to add newly created graphical elements or handle conditions when variables – such as timestamps related to past floods – are passed to the application. *JavaScript* is a well-known scripting language for web applications and is used to provide interactivity. By means of these scripts intrinsic user events such as clicks or mouse-overs can be handled in order to release additional work steps. Data is transferred asynchronously between the server and the browser so that merely the necessary parts of the index file have to be reloaded instead of the complete file. This mechanism proves to be advantageous as the application organizes thematic data in layers that can be turned on and off independently.

4 Exemplary results

The GUI with its default data is accessible by typing a uniform resource locator (URL) in the browser. The automated creation of web-based real-time maps using true real-time data was one major project specification. Another specification was to offer interactive and user-friendly data access so that users do not have to deal with underlying workflows. Further data for visualization, processes and functionalities are automatically loaded on the project server after the URL has been typed in.

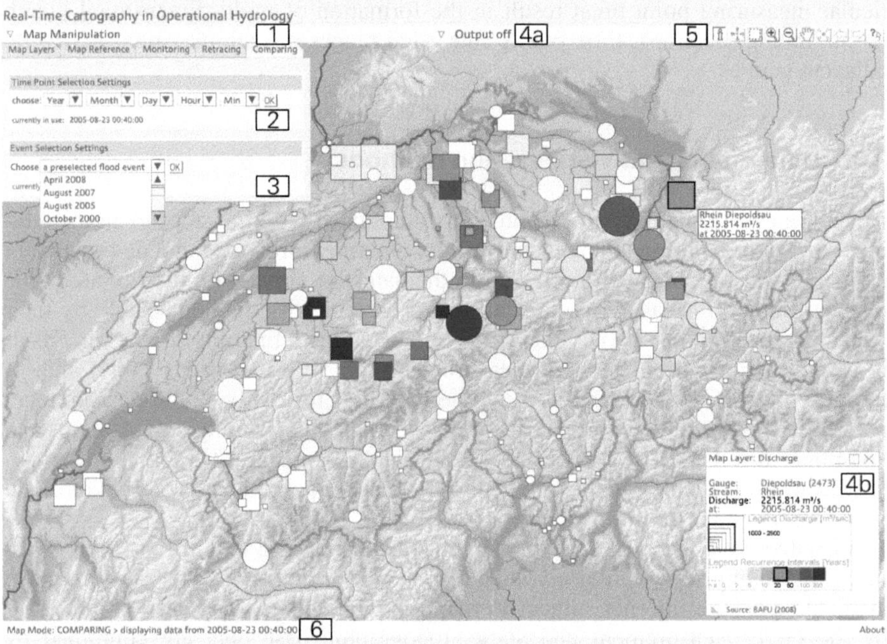

Fig. 4. Graphical user interface with comparing functionality and a map showing river discharge (*square point symbols*) and 24-hours precipitation sums (*circle point symbols*) of August 23, 2005 (see text for explanations)

By means of the *Map Manipulation* button (Fig. 4, top left), a window opens up containing four tabs. One of them is called *Comparing* (#1 in Fig. 4) and comprises two different settings to choose from. The *Time Point Selection Setting* (#2) allows for visualizing an arbitrarily chosen point in time, accurate to 10 minutes. To make comparing activities simpler and faster, the *Event Selection Setting* (#3) lets a user directly retrieve data of a pre-selected extreme event. Eleven preselected significant extreme flood events since 1977 that affected large parts of Switzerland are part of this list: April 2008, August 2007, August 2005, October 2000,

May 1999 (beginning), May 1999 (middle), September 1993, July 1987, August 1987, August 1978 and August 1977.

Information about a map object on a higher level of detail is obtained using an output window which can be opened, closed or shifted (#4a and b). It contains the map legend and the time series graphs of measurements. Such additional information is directly generated when moving or clicking over the symbolization (#4b). Additional map navigation controls allow for activities such as map zooming, panning or re-centering (#5). In order to know which point in time a map refers to, the time of the data along with the map mode is indicated in the status bar (#6). The map mode is either set to *monitoring* or *comparing*. The map presented in Figure 4 shows the hydro-meteorological situation on August 23, 2005 at 00:40:00. This is when a considerable number of both rain and discharge gauges measured their maximum during the devastating 2005 event in Switzerland. The square and circle sizes represent classified discharge quantities, and 24-hour precipitation sums respectively. The colours, for both parameters, denote the recurrence intervals, expressed in years. The darker the colour is, the smaller is the occurrence probability of the shown value. The occurrence probability is obtained from the long-term extreme value statistics, available for each gauge.

In Fig. 5, two maps are shown that cover the Thur basin and its surroundings. Lake levels and 10-min precipitation sums are represented as bars, 24-hour precipitation sums are represented as circles and river discharges are depicted as rectangles. The map on the bottom shows the real-time situation while the one on top represents the situation of August 08, 2007 at 23:00:00. In case the map is set to the *monitoring* mode and a map symbol is clicked, time series graphs with data of the past 24 hours are displayed. If the *comparing* mode is on, the previous and the following 12 hours of the input timestamp are displayed. Users may interactively increase or decrease this 24 hours time span via a dropdown list so that another time span is immediately displayed (Fig. 5, top, graph in the middle).

Another example of comparing real-time maps with historical maps is shown in Fig. 6. The layer that contains hypsometric data is switched off and the precipitation radar layer is on instead. To some original data – like these raster-based radar images or point temperature data – complex, automated processing functions are applied in real time. The radar image shown in Fig. 6 is re-coloured for better readability and underlain by a grid, specifically created for that reason. This grid contains indexed colours of each radar image pixel. Functions in the GUI let users explore the values and the new colour legend when the mouse is being moved over the image. In addition to the radar image, the map in Fig. 6 contains the 0°C isotherm which is a valuable indicator as to whether and to what extent precipitation falls in fluid or solid form. This line symbolization of temperature is based on the integration of automated interpolation of measured data from ground gauges and the extraction of the contour line for which 0°C was modelled. The workflows and functions that eventually result in improved precipitation radar and temperature visualizations are discussed in Lienert et al. (2009a).

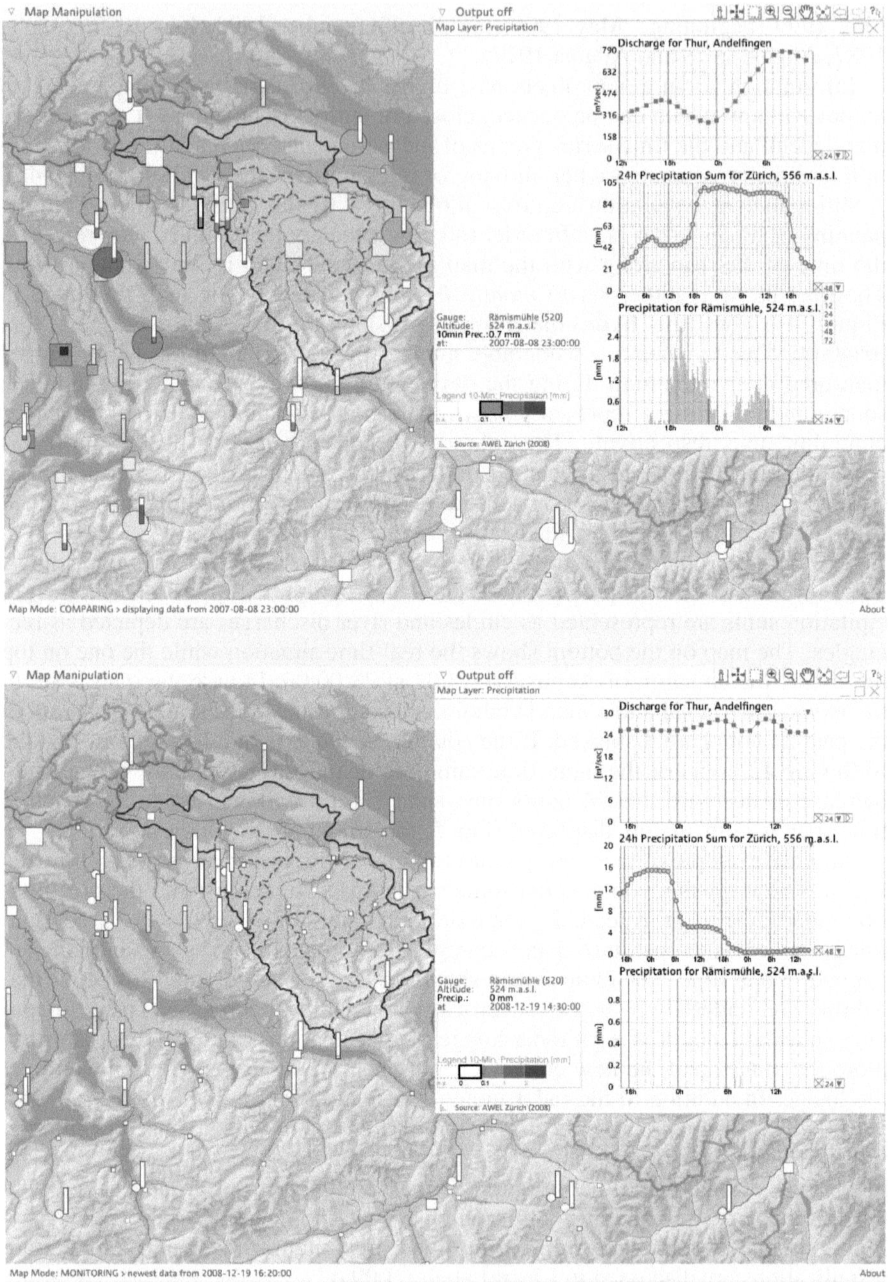

Fig. 5. Comparison of two maps showing the real-time situation (*bottom*) and the situation at 2007-08-08 23:00:00 (*top*) in and around the *Thur* basin (see text for explanations)

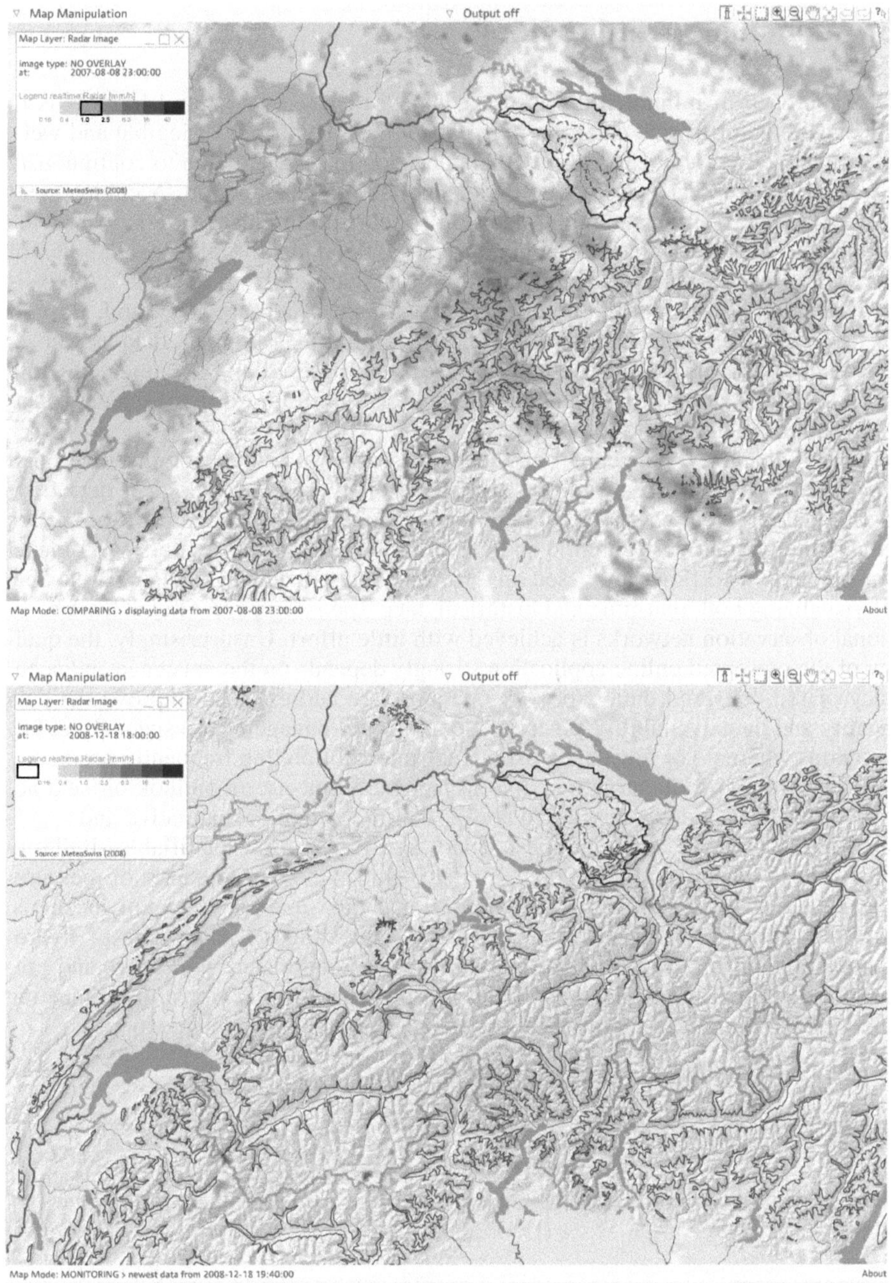

Fig. 6. Comparison of two maps showing the real-time situation (*bottom*) and the situation at 2007-08-08 23:00:00 (*top*) with the 0°C isotherm and the radar image

5 Conclusion and outlook

One approach to minimize devastating damages caused by natural hazards is to enhance monitoring and early warning capabilities. By means of legible and well-interpretable hydrological real-time data decision makers are able to continuously revaluate impending hydrological floods and take targeted action. A complementary approach to offer additional decision support in an emergency is to provide immediate visual comparisons with historical flood data. Quick availability of such visually processed archive data allow for 'learning from the past'. Analysis of past floods can contribute to a better assessment of an ongoing flood. By collecting and maintaining quantitative, high-resolution measurement data as well as qualitative, descriptive data past events are both captured and documented. Both types of information provide invaluable help in flood hazard management, planning and the realization of protective measures.

In this chapter, a conceptual and technical framework has been discussed that generates web-based, real-time visualizations of hydro-meteorological data from Swiss measurement networks. Experts within flood crisis management groups are enabled to interactively compile and explore relevant data parameters. Automated cartographic workflows are based on a flexible and extendable data model capable of visualizing both real-time and historical data. The integration of data from additional observation networks is achieved with little effort. Unsurprisingly, the quality of the presented online application strongly depends on the existence and accuracy of the delivered data. Network operators are addressed to improve the data quality and data availability aspect of flood hazard management, especially at key locations where no or few data exist but detrimental flooding frequently occurs.

Apart from consolidating a couple of technical aspects, an outlook on next activities could include, as discussed above, the integration of qualitative and multimedia data (i.e., text, photographs, video) and research on qantifiable similarity measures to detect historical floods. Latter includes the consideration of measurement data uncertainties and multi-dimensional data queries (adjacent locations, different parmeters, varying time and data value ranges). Also, data of hydro-meteorological forecasts may be integrated in the automated workflows and prepended to visualization functions. In all, the presented application aims at meeting the increased time-critical information needs of flood management experts.

Acknowledgements

This project is supported by the Swiss National Science Foundation (Grant No. 200020-122291).

References

Allan RP, Soden BJ (2008) Atmospheric warming and the amplification of precipitation extremes. J Sci 321(5895):1481–1484.

Bartholmes J, Todini E (2005) Coupling meteorological and hydrological models for flood forecasting. J Hydrol Earth Syst Sci 9(4):333–346.

Berlekamp J, Lautenbach S, Graf N, Reimer S, Matthies M (2007) Integration of MONERIS and GREAT-ER in the decision support system for the German Elbe river basin. J Environ Model Software 22(2):239–247.

Bezzola GR, Hegg C (2007) Fazit. In: Bezzola GR, Hegg C (eds.) Ereignisanalyse Hochwasser 2005 – Teil 1 Prozesse, Schäden und erste Einordnungen. Swiss Federal Office for the Environment and Swiss Research Center for Forest, Snow and Landscape, Berne and Birmensdorf, pp. 211–215.

Black AR, Law FM (2004) Development and utilization of a national web-based chronology of hydrological events. Hydrol Sci J 49(2): 237–246.

de Kort IAT, Booij MJ (2007) Decision making under uncertainty in a decision support system for the red river. J Environ Model Software 22(2):128–136.

Foraci R, Bider M, Cox JP, Lugli F, Nicolini R, Simoni G, Tirelli L, Mazzeo S, Marsigli M (2005) M3Flood: An Integrated System for Flood Forecasting and Emergency Management. In: Oosterom PJMv (ed.) Geo-Information for Disaster Management. Springer, Berlin, pp. 1193–1210.

Jasper K, Calanca P, Gyalistras D, Fuhrer J (2004) Differential impacts of climate change on the hydrology of two alpine river basins. J Climate Res 26(2):113–129.

Jaun S, Ahrens B, Walser A, Ewen T, Schar C (2008) A probabilistic view on the August 2005 floods in the upper Rhine catchment. J Nat Hazards Earth Syst Sci 8(2):281–291.

Jha MK, Das Gupta A (2003) Application of Mike basin for water management strategies in a watershed. J Water Int 28(1):27–35.

Kienholz H, Krummenacher B, Kipfer A, Perret S (2004) Aspects of integral risk management in practice – considerations with respect to mountain hazards in Switzerland. J Österreichische Wasser- und Abfallwirtschaft 56(3–4):43–50.

Lienert C, Kunz M, Weingartner R, Hurni L (2009a) Methods for Real-Time Visualization and Exploration Improvements of Radar and Temperature Data in Web-Cartography (in preparation). In: Proceedings of the 24th International Cartography Conference, Santiago, Chile.

Lienert C, Schnabel O, Hutzler E, Hurni L (2008) A Real-Time Flow Map of the Swiss 1:200’000 River Network. Considerations on Its Preparation and Its Dynamic Integration into a Web-Based Prototype. In: Proceedings of the 6th ICA Workshop on Mountain Cartography, Lenk, Switzerland.

Lienert C, Weingartner R, Hurni L (2007) Real-Time Cartography in Operational Hydrology. In: Proceedings of the 23rd International Cartography Conference, Moscow, Russia.

Lienert C, Weingartner R, Hurni L (2009b) Real-time visualization in operational hydrology through web-based cartography. J Cartograp Geograp Infor Sci 36(1):45–48.

Neumann A, Winter AM (2003) Webmapping with Scalable Vector Graphics (SVG): Delivering the Promise of High Quality and Interactive Web Maps. In: Peterson MP (ed.) Maps and the Internet. Elsevier, Amsterdam, pp. 197–220.
Todini E (1999) An operational decision support system for flood risk mapping, forecasting and management. J Urban Water 1(2):131–143.
Valpreda E (2004) GIS and Natural Hazards. In: Casale R, Margottini C (eds.) Natural Disasters and Sustainable Development. Springer, Berlin, pp. 373–386.
Vrugt JA, Gupta HV, Nualláin BO, Bouten W (2006) Real-time data assimilation for operational ensemble streamflow forecasting. J Hydrometeorol 7(3):548–565.
Weingartner R, Aschwanden H (1992) Discharge Regime – the Basis for the Estimation of Average Flows. In: BWG (ed.) Hydrological Atlas of Switzerland. Federal Office for Water and Geology, Berne, pp. Plate 5.2.

Designing Map Keys for Crisis Management on the Regional Operational and Informational Centre Level: Monitoring Transport of Dangerous Goods via Contextual Visualisation

Lucie Friedmannová

Abstract

The research presented in this chapter is part of the project called Dynamic Geovisualization in Crisis Management (GEOKRIMA – Masaryk University, Brno, Czech Republic, 2005–2011). The project is focused on the use of existing data, verification of the data's timeliness and integrity, analysis of the data's qualitative features, interpretation, presentation and implementation of its accessibility to the users who are crisis management personnel. The chapter revolves around contextual visualisation that is designed, above all, for the regional Operational and Informational Centre of Integrated Rescue System. The goal is to offer maps in a form and structure that leads to a higher efficiency and a clearer communication process between the dispatching and rescue squads. The system of map keys consists of topography in four versions and two thematic contexts (one for monitoring and one for the incident). The topography is designed to be used in all crisis situations that are defined by the National Security Council. The thematic content aims to monitor the transport of dangerous goods. The final proposal of the map key is a compact system where colours and shapes are the leading attributes that knit groups of symbols together. Individual symbols are stored in SVG format (compatible with SLD standard) to be used through WMS.

M. Konecny et al. (eds.), *Geographic Information and Cartography for Risk and Crisis Management*, Lecture Notes in Geoinformation and Cartography,
DOI 10.1007/978-3-642-03442-8_29, © Springer-Verlag Berlin Heidelberg 2010

1 Current situation in crisis management data visualisation and usage

American cartographer Ute J. Dymon (Dymon 2001, 2003) called the consistent set of map symbols *a missing ingredient for emergency mapping*. Nowadays, the set of symbols that is provided by the FGDC HSWG (United States Federal Geographic Data Committee – Homeland Security Working Group) is one of the most quoted in the context of crisis management and was also accepted as an ANSI Standard in 2006 (ANSI INCITS 415-2006). Some features of the US Homeland Security symbology library were also included into the Australasian All-Hazards Symbology Project (Australia and New Zealand activity that is currently in testing, http://www.anzlic.org.au/symbology.html).

In Europe, visualisation is included in projects that are devoted to data manipulation and technology improvement, such as OASIS (www.oasis-fp6.org) or ORCHESTRA (www.eu-orchestra.org). Visualisation also played a substantial role in projects such as ARMONIA (www.armoniaproject.net) or GeoWarn (www.geowarn.ethz.ch). Outside of the international projects, visualisation of crisis management in European countries is often restricted by national boundaries (Netherlands' CENVIS project). The reason is the complexity and sheer quantity of input information on one side and the request to satisfy traditional local visual approaches on the other side. Another aspect of the problem is the security-sensitive nature of the input data. A large part of the research that is in support of crisis management is naturally oriented to developing Geospatial Data Infrastructures (GDI), ontology, data mining and data processing (Borkulo et al. 2006, Halliday 2007, Ramesh et al. 2007, Ludík-Ráček 2008, etc.).

Part of the GEOKRIMA project (Dynamic Geovisualisation in Crisis Management – Masaryk University, Brno, Czech Republic, 2005–2011, http://geokrima.geogr.muni.cz) was to observe and interview the operators and GIS administrators of the Integrated Rescue System (about IRS – http://www.hzscr.cz). It was determined that, currently, there is no unified system to visualise relevant data in crisis management in the Czech Republic. Each region and each part of the IRS has its own GIS and is solely dependent on GIS administrators to maintain not only the data's quality but also its visualisation. Most of the progress has been made in the Fire Brigade, where visual presentations of data from the central database are nationally unified. The chief GIS administrator makes on-demand adjustments that are based on requests from the regional Operational and Informational Centers (OIC). Depending on regional particularities, the OIC are provided with additional data sets (for example, the South Moravian Center has additional information for the Dukovany Nuclear Power Station).

GIS administrators are specialists in information technology. Visualisation is a part of their work load but usually not a part of their expertise. Usually, new information is simply added to the existing map key by selecting the graphics that are available in the software that is used. There is only a limited amount of time that the GIS administrators can dedicate to cartographic issues, such as designing original symbols. The offered symbol sets do not always correspond to the operators' requests, their level of expertise and the rooted visual traditions. For GIS administrators, it is also difficult to coordinate all of the visualised features. For example, the topographical base ZABAGED that is used by Crisis Management includes more than 100 types of geographical objects. ZABAGED is the Fundamental Base of Geographic Data that is guaranteed by the Land Survey Service (http://www.cuzk.cz). Kohlhammer and Zeltzer pointed out that this *rich information environment tends to overwhelm the decision maker* (Kohlhammer-Zeltzer 2004) while Andrienko added that for "intelligent" visualisation, reduction of the information load is a necessary requirement (Andrienko-Andrienko 2006).

Operators basically use the GIS the same way as paper maps. They are certainly able to switch off and on layers with unnecessary or desired objects; however, this operation takes time and therefore, it is not done. Map key building in the frame of the project is task-oriented. The goal is to remove tens of map layers and to create "themes" that display only the information that is relevant within the defined context. The task consists of visualisation of topography and thematic or contextual information. Common topography is designed for all of the crisis situations that are defined by the National Security Council of the Czech Republic. The thematic part is a case study that focuses on the visualisation of monitoring the transport of dangerous goods and the subsequent response to a potential disaster.

2 Contextual visualisation in the communication process of crisis management

2.1 Communication process

The whole communication process during a crisis is complex and includes the participation of various organisations with different levels of importance (Šafr et al. 2008). Contextual visualisation would be most beneficial during the so-called *response phase*. After the crisis situation has arisen, the response phase immediately follows and deals with mobilisation of IRS units (Halliday 2007, Kubíček et al. 2009). A simplified schematic of the communication process on the regional OIC level during the response phase is presented in Fig. 1.

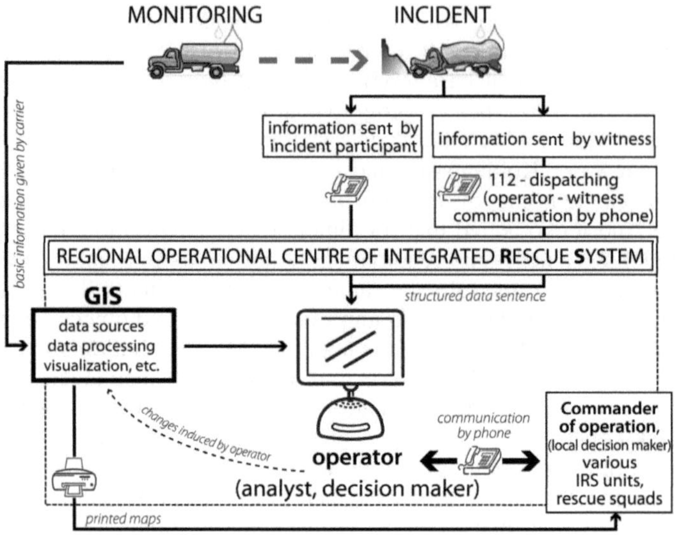

Fig. 1. Simplified communication process on the regional OIC level during a response phase

In this case, contextual visualisation will be directed mainly at the Operator and at the Commander of the rescue operation. The operator deals with two types of communication: receiving information about the crisis situation and alerting and monitoring the relevant rescue squads. Communication between the Operator and the Commander is done principally by phone. During the whole rescue operation, the Operator serves as the informational support of the Commander, who is the decision maker. The Operator also serves as a mediator between the IRS units (Fire Brigade, Police, Ambulance Service, etc.). Nowadays, operators often prefer to have a uniform view with a pre-set configuration of map features because of the quantity of displayed objects and the time that is necessary for configurational changes. Moreover, a pre-set map configuration is part of the operators' *situation awareness* (Kohlhammer-Zeltzer 2004). Nevertheless, there is a *common will for the development of an effective information support of commander* (Kubíček et al. 2009) that includes changes in the map's presentation.

2.2 Target group

Dynamic contextual visualisation requires a certain level of technical equipment. At least, there must be a computer furnished with the relevant software and databases and an Internet connection. The part of the project that is concerned with visualisation is therefore focused on the OIC operators and, in a wider scope, on the Commanders of the operations. A majority of the members of the target group have at least a secondary education. All of them are specially trained in communication with people in crisis situations and in work with the Crisis Management Communication System (including GIS). The commanders and part of the operators have practical experience with solving crisis situations. They also have relatively wide knowledge of technical and urban plans, traffic, etc. For Commanders, profound familiarity with the province of their usual competence is typical.

2.3 Definition of context in crisis management – dangerous goods transport case study

Contextual visualisation is task-oriented and adaptive to the user (Staněk et al. 2007; for the general idea of context in adaptive cartography see Fig. 2).

Fig. 2. Context in adaptive cartography (Konečný et al. 2006)

In Crisis Management, the term Context refers to a group of information and objects that are relevant to one particular task that solves one type of crisis situation. Understandably, crisis situations that are defined by the National Security Council of the Czech Republic had to be decomposed into scenarios that are related to particular tasks. Analysis of action plans for different types of crisis situations (fires, floods, etc., see Šafr et al. 2008) showed high repetition of tasks. A tasks' repetition also implies repetition of the required information and the relevant objects. Thanks to this fact, it was possible to define a group of common objects and to form a basic topographic map (compare to Dymon 2001 – Base map).

The second level was to select groups of objects that are relevant to the defined types of crisis situations and to specify their roles in reference to defined crisis solutions (on semantics, see Mulíčková and Stachoň 2009). For the case study that monitors the transport of dangerous goods, two contexts were specified: MONITOR and INCIDENT (Talhofer et al. 2007).

2.3.1 2.3.1 BASETOPO – Topography for crisis management

Topography should provide a means for orientation in the monitored space. Topography for Crisis Management should also provide specific information about the transit conditions for the rescue squads. From more than 100 object types included in ZABAGED, 24 were selected for the most-detailed view (BASETOTO MAX is the level of detail that is approximately equal to the scale 1:2000 – 1:10000). The object types represent detailed communication networks with potentially problematic objects (bridges, tunnels, railway crossings, etc.) and simplified land use (several types of built-up areas, forest and water bodies).

To enable the operators' individual preference, visualisation of BASETOPO was designed both in shades of grey and in a coloured version. There is also the possibility to use orthofoto images. For a comparison of ZABAGED, orthofoto and BASETOPO see Fig. 3.

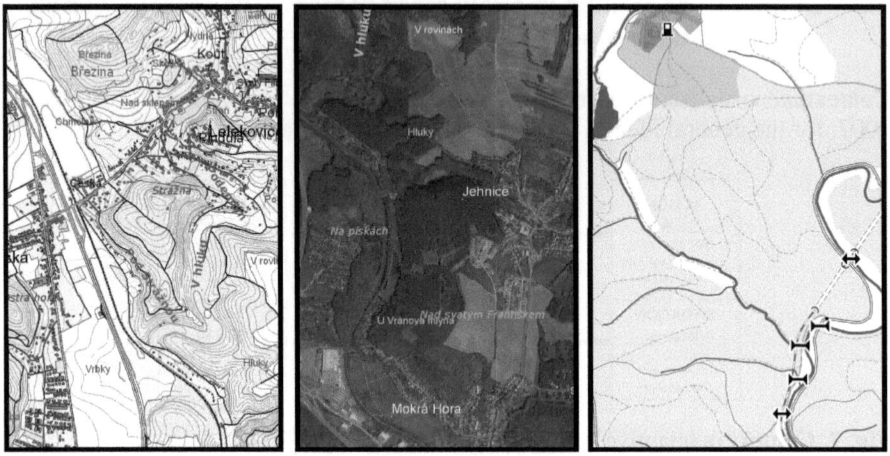

Fig. 3. A comparison of ZABAGED, orthofoto and BASETOPO, respectively (scaled down)

BASETOPO naturally includes objects with a potential to become context relevant, such as bodies of water. In these cases, the visualisation change (highlighting) is a part of the dynamic visualisation process.

2.3.2 Contexts relevant to the case study – MONITOR and INCIDENT

MONITOR

The case study that is devoted to Transporting Dangerous Goods treats its subject as a potential source of a crisis situation from the beginning of the transport. The monitoring itself comes under the pre-response phase of the crisis management process. The context MONITOR is primarily designed for Operators. The core of the context is tracking the vehicle's interaction with objects of Critical Infrastructure (CI) and additional traffic information (e.g., closures, places of frequent accidents).

The symbol that represents moving vehicles changes its visual attributes when interacting with CI objects (see Fig. 4). The term interaction refers to the passage of a vehicle through the CI object's safety zone. Objects of CI were divided into four thematic groups:

- SOCIAL (administration, schools, banks, medical facilities, etc.),
- TECHNO (technological objects, lines and safety zones),
- ENVIRO (preserved areas),
- WATER (water sources, water bodies, technological objects, etc.),

The map objects were specified with two levels of detail. The transition between them is made by object omission and point aggregation. The context MONITOR can also track the progress of the rescue squads. The direct connection of the underlying GIS to the police traffic information allows the rescue squad's route to be modified on demand.

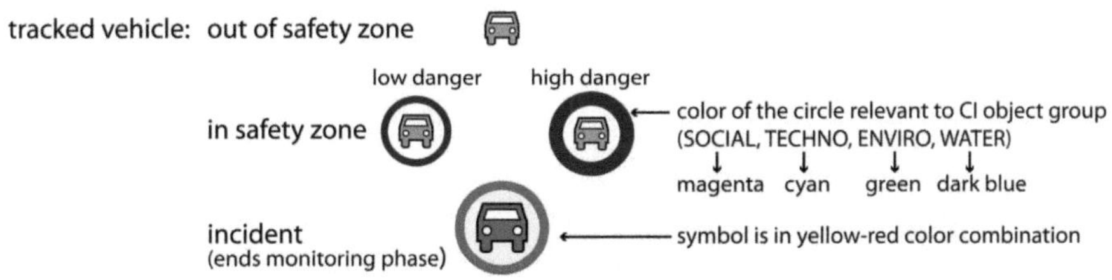

Fig. 4. Change of visual attributes that is activated by the interaction of the tracked vehicle with the CI objects

INCIDENT

The context INCIDENT is directly linked to the response phase of the Crisis Management process. In case of an accident, the tracking device that is located in the vehicle should directly send out a signal to the OIC. This signal should trigger a change of context from MONITOR to INCIDENT. Operators are of course also able to trigger the change themselves. INCIDENT has two sub-contexts:

- SECURITY is the sub-context that is primarily designed for OIC operators because the map provides the Operator with general knowledge about technical and medical sources in the vicinity of the accident (fire station, hospital). Depending on the type of cargo, a theoretical danger zone around the accident location is generated and the affected objects of the CI are evaluated as dangerous/endangered. The idea is to offer a structure of visual information that minimises the operator's need to manipulate the data.
- PERIMETER is the sub-context that is designed for the OIC operators and the Commanders of the operation alike. The content is a combination of MONITOR-DETAIL and INCIDENT-SECURITY. There is also a group of objects that are directly related to the Commander (operation lay-out). For an example of the visualisation see Fig. 5.

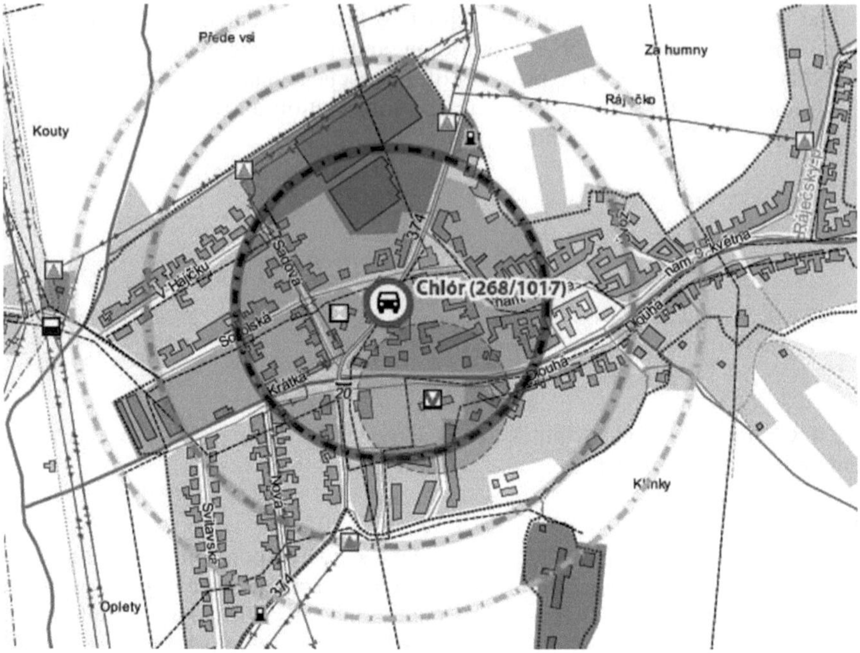

Fig. 5. INCIDENT–PERIMETR is the visualisation of the theoretical danger zone that is generated around the location of an accident and depends on the cargo type (cutout)

The system of proposed contexts can also work very well in other types of crisis situations. Groups of objects that represent CI and Operational lay-outs are universal in Crisis Management, as is the base topography map. Additional info is accessible with "On click", which contains detailed information about the transported cargo or objects of the CI that is presented in the pop-up windows. The operator instantly sees potential problems and is able to alert the commander about possible dangers even before the Commander reaches the location of the accident. Incorporation of the dynamic contextual visualisation into the process of crisis management communication is shown in Fig. 6.

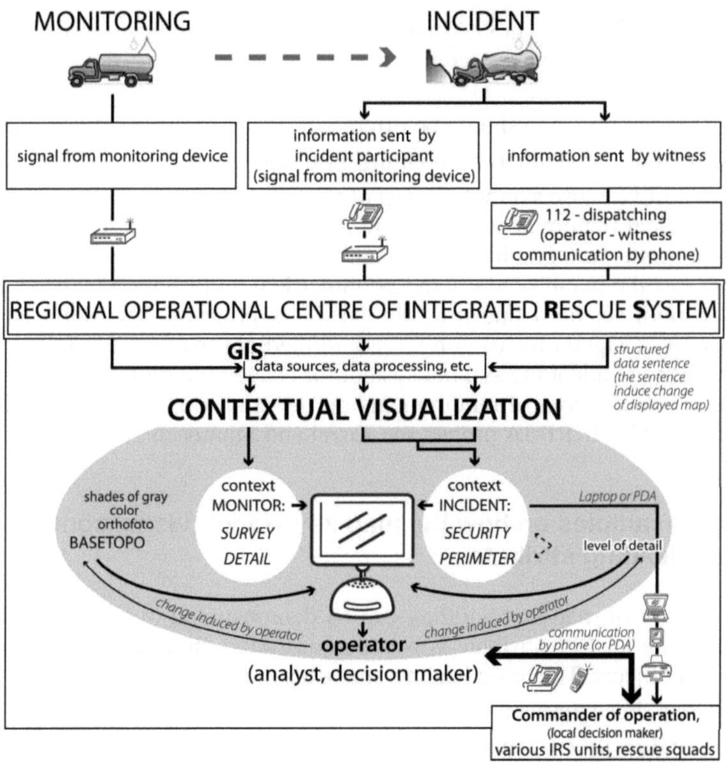

Fig. 6. Communication and visualisation processes in relevance to the situational change during the monitoring of the transport of dangerous goods

3 3. Building of the map key

3.1 Symbology background

The starting point in designing the map key was to create a list of objects with the definition of their requested geometry dimension in individual contexts and the definition of the roles they play in particular contexts. The second step was to name the requirements on the map key design, which were specified as

- The maximum use of symbols that are generally known to the target group (from a series of "Basic Maps of the Czech Republic", tourist maps, technical and urban plans, traffic signs, etc.)
- The maximum use of the leading attribute (the symbols in one group should have a common visual characteristic in order to allow aggregation)

- Visual connectivity (symbols that represent identical objects in different contexts or roles should have a common visual characteristic – for example shape or colour)
- A base topography map should be offered in shades of grey and in colour
- A universal use of symbols on both the base topography map versions
- The possibility to use symbols on top of the orthofoto map

Technological requirements (compatibility with SLD standard) restricted some of the most useful cartographic visualisation techniques. A particularly detrimental restriction was the inability to use the asymmetrical multi-lines and the inability to rotate a symbol set on a line towards the line's progress. The point and linear symbols were designed for use on large and medium map scales. There is a distinct possibility to adapt the symbol system for usage on mobile devices. For more information about technical requirements on the contextual map service that was created for the GEOKRIMA project see Kozel and Štampach 2009.

3.2 Two examples on how the map key works: visual connectivity and leading attribute

The philosophy of map key building can be demonstrated by two examples. The principle of the leading attribute is best demonstrated on a group of symbols that represent CI objects. The basic leading attributes in this case are

- Institutions' headquarters – pentagram
- Objects and facilities – square
- Lines (landlines, pipelines, etc.) – symbolised line
- Safety zones – dot-and-dash outline and semi-transparent fill (fill is optional).

Each of the CI thematic groups (see Sect. 2.3.2) is characterised by using a specific colour (see Fig. 4); thus, the institutions' headquarters, facilities, lines and safety zones, which comprise one thematic group, have different shapes but an identical colour. In the case of objects and facilities, the inner structure defines the type of object (school, medical facility, etc.). The inner structure, where possible, coincides with the symbols that are used for similar objects on technical maps or urban plans (compare to the first bullet in Sect. 3.1). An exception exists for gas stations. The symbol "gas station" that is used in BASETOPO is used also in contexts. The symbol is set off by a contrasting cyan edge (the colour that is assigned to the TECHNO thematic group of CI objects). Part of the map key for the CI object can be seen in Fig. 7.

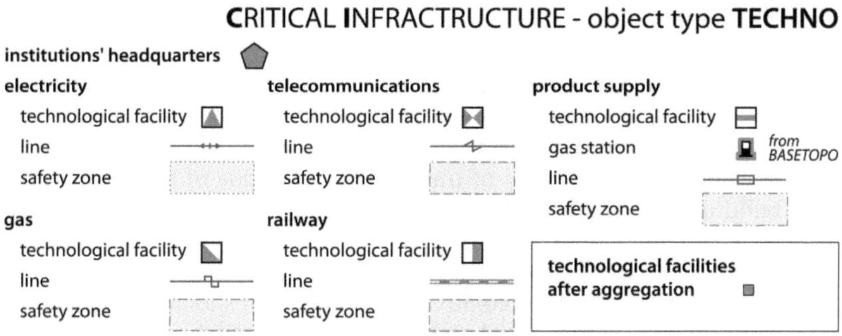

Fig. 7. Objects of Critical Infrastructure for the thematic group TECHNO (grey represents cyan)

A second example of the map key philosophy refers to visual connectivity. The term visual connectivity refers to similarity in graphical representation of objects with similar meaning across themes, scales, roles and map views. The principle of the leading attribute refers to groups of symbols, whereas visual connectivity refers to how one symbol changes through different contexts. The transfiguration of the Red Cross (Helvetian cross), a traditional representation for medical facilities, is a suitable example. All objects that are related to medical care have in their representation the cross, as can be seen in Fig. 8. The colour is what changes; for instance, an object of CI has a symbol that is a combination of black, white and magenta (magenta is the colour of the thematic group SOCIAL, see Sect. 2.3.2). When representing hospitals, the symbol has a traditional red colour on a white background. To better distinguish the medical area from the symbol for hospitals, the red and white colouring is inverted.

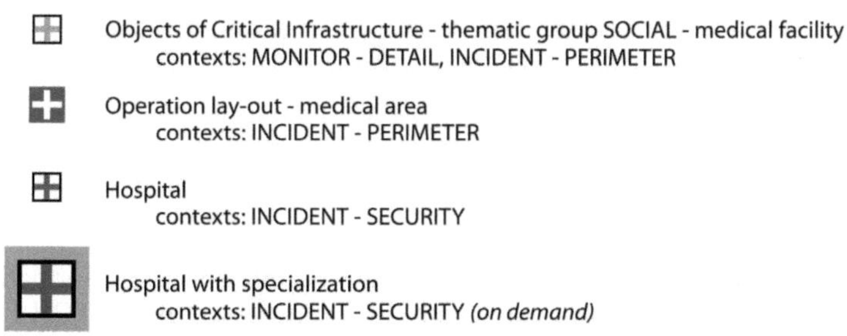

Fig. 8. Visualisation of objects that are related to medical care through the defined map contexts

4 Conclusion

The map key was built as a compact system with elements that are firmly tied to groups both by meaning and by visual representation. The main visual sources were symbol systems of the Basic maps of the Czech Republic, tourist maps, technical and urban plans, system of traffic signs, etc. One of the crucial goals of map key building was to join widely known symbols of various backgrounds into one system. The main approach that was used was contextual visualisation, which is task-oriented and adaptive to users, who will be the operators of the Integrated Rescue System. Newly designed contextual visualisation should improve communication between the operator and the Commander, which is mainly done by phone. Unlike the HSWG (FGD HSWG) approach, where the first phase was focused on designing the point symbols that cover all of the possible Crisis situations, in the GEOKRIMA project the process of designing visualisation for crisis Management support was task-oriented. The case, which focused on monitoring the transport of dangerous goods, comprised universal building blocks (base topography map, objects of Critical Infrastructure, operation lay out) and case-specific content (tracked vehicle).

The map key has undergone testing via experiments that were performed during the fall of 2008 and the spring of 2009. Some of the symbols and proposed visualisation will have to be adjusted. For example, the danger and critical zones were visually undervalued (danger zones on Fig. 5. have already been adjusted).

In the future, extension of the context system and consequently, the map key onto other crisis situations is expected.

Acknowledgement

The project (no.: MSM0021622418) is supported by the Ministry of Education, Youth and Sports of the Czech Republic.

References

Andrienko N, Andrienko G (2006) Intelligent Visualisation and Information Presentation for Civil Crisis Management. In: Proceedings of 9th AGILE Conference on Geographic Information Science, Visegrád, pp. 291–298.

ANSI – American National Standards Institute, http://www.ansi.org. Accessed 2 June 2009.

Borkulo E, Barbosa V, Dilo A, Zlatanova S, Scholten H (2006) Services for An Emergency Response System in the Netherlands. In: Second Symposium on Gi4DM, Goa, 6p.

Dymon UJ (2001) A Standardized Framework for Hazards Management Mapping. In: Proceedings 20th International Cartographic Conference, Beijing, Vol. 4, pp. 2760–2764.

Dymon UJ (2003) An analysis of emergency map symbology. Int J Emergency Management 1(3), 227–237.

Halliday Q (2007) Disaster Management, Ontology and the Semantic Web. UCLA Department of Information Studies.

FGDC HSWG (United States Federal Geographic Data Committee – Homeland Security Working Group) – symbology reference. http://www.fgdc.gov/HSWG/index.html. Accessed 2 June 2009.

Kohlhammer J, Zeltzer D (2004) Towards A Visualization Architecture for Time-Critical Applications. In: Proceedings IUI'04, Funchal, ACM Press, pp. 271–273.

Konečný M, Friedmannová L, Staněk K (2006) An Adaptive Cartographic Visualization for Support of the Crisis Management. In: CaGIS Publications – Autocarto 2006, Vancouver, pp. 100–105.

Kozel J, Štampach R (2009) Practical Experience with Contextual Map Service. In: Proceedings of Joint Symposium of ICA WG on CEWaCM and JBGIS Gi4DM – Cartography and Geoinformatics for Early Warning and Emergency Management: Towards Better Solutions. Masaryk University, Prague, pp. 304–312.

Kubíček P, Ludík T, Mulíčková E, Ráček J, Šafr G (2009) Emergency Management Process Support Using Geovisualization. In: Proceedings of Joint Symposium of ICA WG on CEWaCM and JBGIS Gi4DM – Cartography and Geoinformatics for Early Warning and Emergency Management: Towards Better Solutions. Masaryk University, Prague, pp. 338–348.

Ludík T, Ráček J (2008) Procesná analýza krízového riadenia a jej využitie pri tvorbe aplikačných ontológií, Riešenie krízových situácií v špecifickom prostredí. Žilinská univerzita, Žilina. in Slovak.

Mulíčková E, Stachoň Z (2009) Semantic Issues of Geodata Integration in Emergency Management. In: Proceedings of Joint Symposium of ICA WG on CEWaCM and JBGIS Gi4DM – Cartography and Geoinformatics for Early Warning and Emergency Management: Towards Better Solutions. Masaryk University, Prague, pp. 447–457.

Ramesh RR, Eisenberg J, Schmitt T, (Ed.) (2007) Improving Disaster Management: The Role of IT in Mitigation, Preparedness, Response and Recovery. National Academies Press, Washington.

Staněk K, Konečný M, Friedmannová L (2007) An Adaptive Cartographic Visualization for support of the Crisis Management. In: XXIII International Cartographic Conference – Cartography for Everyone and for You. Roskartografija, Moscow, 9p.

Šafr G, Karda L, Hon Z (2008) Struktura a legislativa Integrovaného záchranného systému (IZS), koordinace a návaznost činností složek IZS, mimořádné události a krizové situace. Jihočeská universita, České Budějovice. (In Czech).

Talhofer V, Kubicek P, Brazdilova J, Svatonova H (2007) Transport of Dangerous Chemical Substances and Its Cartographic Visualisation. In: M Wachowicz, L Bodum, (eds.) Proceedings of 10th AGILE International Conference on Geographic Information Science, AGILE 2007, Aalborg.

Index

M. Konecny et al. (eds.), *Geographic Information and Cartography for Risk and Crisis Management*, Lecture Notes in Geoinformation and Cartography,
DOI 10.1007/978-3-642-03442-8, © Springer-Verlag Berlin Heidelberg 2010

T

MIX
Papier aus verantwortungsvollen Quellen
Paper from responsible sources
FSC® C105338

If you have any concerns about our products,
you can contact us on
ProductSafety@springernature.com

In case Publisher is established outside the EU,
the EU authorized representative is:
Springer Nature Customer Service Center GmbH
Europaplatz 3, 69115 Heidelberg, Germany

Printed by Libri Plureos GmbH
in Hamburg, Germany